Earthquake resistant design

Earthquake resistant design

A manual for engineers and architects

D. J. Dowrick

Originally written for the Ove Arup Partnership

A Wiley–Interscience Publication

JOHN WILEY & SONS

Chichester · New York · Brisbane · Toronto

Copyright © 1977, by John Wiley & Sons, Ltd.

Reprinted October 1977
Reprinted May 1978

Library of Congress Cataloging in Publication Data:

Dowrick, D. J.
  Earthquake resistant design.

'A' Wiley–Interscience Publication.'
  1. Earthquakes and building. I. Title.
TA658.44.D67   624'.176   76-26171

ISBN 0 471 99433 2

Printed and bound in Great Britain by The Pitman Press, Bath

MW 10/78 5640

Foreword

Earthquakes are one of nature's greatest hazards to life on this planet; throughout historic time they have caused the destruction of countless cities and villages on nearly every continent. They are the least understood of the natural hazards and in early days were looked upon as supernatural events. Possibly for this reason earthquakes have excited concern which is out of proportion to their actual hazard. Certainly the average annual losses due to wind and flood exceed those due to earthquakes in many parts of the world, and all of these represent lesser life hazards than are accepted daily in our streets and highways. Nevertheless, the totally unexpected nearly instantaneous devastation of a major earthquake has an unique psychological impact which demands serious consideration by modern society.

The hazards imposed by earthquakes are unique in many respects, and consequently planning to mitigate earthquake hazards requires a unique engineering approach. An important distinction of the earthquake problem is that the hazard to life is associated almost entirely with man-made structures. Except for earthquake triggered landslides, the only earthquake effects that cause extensive loss of life are collapses of bridges, buildings, dams and other works of man. It is this fact that has led to the great emphasis placed on earthquake prediction in one of the world's great seismic regions—The People's Republic of China. With even a few hours of advance notice, people can be evacuated from buildings and houses into open fields where loss of life can be almost completely avoided. Apparently such a prediction was effective in saving hundreds or possible thousands of lives during the Hai-cheng, China earthquake of February 1975.

However, it is evident that even a successful prediction cannot eliminate the earthquake hazard; even if all the people are evacuated safely, the structures which largely determine the standard of living of the community remain, and their destruction could be a disastrous loss to the regional economy. This aspect of earthquake hazard can be countered only by the design and construction of earthquake resistant structures, and therefore a completely successful earthquake prediction programme could not eliminate the need for effective earthquake engineering. On the other hand, with effective application of earthquake engineering knowledge, the collapse of structures and the resulting life hazard can be avoided; this would greatly reduce the value of any earthquake prediction programme.

Earthquake hazard poses an unique engineering design problem in that an intense earthquake constitutes the most severe loading to which most civil engineering structures might possibly be subjected, and yet the probability that any given structure will ever be affected by a major earthquake is very low. The optimum engineering approach to this combination of conditions is to design the structure so as to avoid collapse in the most severe possible earthquake—thus ensuring against loss of life—but accepting the possibility of damage, on the basis that it is less expensive to repair or replace the small number of structures which will be hit by a major earthquake than to build all structures strong enough to avoid damage. Clearly this design concept presents the structural engineer with a most challenging problem: to provide an economical design which is susceptible to earthquake damage, but which is essentially proof against collapse in the greatest possible earthquake.

Another unique feature of the earthquake excitation provides the key to the solution of this design problem. In contrast to the other loads considered in structural design—wind, gravity, hydrodynamic, etc.—the intensity of the earthquake loading depends on the properties of the structure. Thus adequate earthquake resistance may be provided either by the traditional approach of increasing strength, or by the unique seismic design concept of reducing stiffness and thereby reducing the forces to be resisted. This additional approach to earthquake design imposes a greater need for understanding of structural behaviour in earthquake engineering than in any other field of civil engineering design. Seemingly minor changes in the framing system or in design details may have an overwhelming influence on the seismic performance; and merely adding more materials—though it will directly increase costs—will not guarantee satisfactory performance.

It is because understanding is the key to good earthquake design, and because the quality of design has such a profound influence on the earthquake performance of structures, that this earthquake design manual by David Dowrick will occupy an important place in engineering design offices throughout the world.

RAY W. CLOUGH
Berkeley, California

Preface

This document was originally written for the guidance of architects and engineers employed in the international practice of the Ove Arup Partnership. Because the book is not written specifically for application in any one country, it should be of assistance to designers in any part of the world. Much of the text should also be of interest to students of architecture and engineering, the elements of Chapter 4 being particularly recommended at that stage.

In preparing the text, valuable advice has been obtained from generous people in many parts of the world, some of whom are acknowledged below. Because of the enormous scope of the book brevity has been essential, with referencing to source material. With the rapid advice in the understanding of much of the subject matter, it is hoped that the text will be revised from time to time and suggestions for its improvement will be welcomed.

Grateful acknowledgement for their assistance in preparing parts of the text is made to my colleagues C. H. I. Balmond, R. J. Bentley, J. C. Blanchard, A. K. Denney, M. V. Harley, P. Parlour, C. P. Wade and R. T. Whittle. The Ove Arup Partnership and the author wish to thank Professor N. A. Mowbray of Auckland University and Professor R. Park of Canterbury University for their kindness, advice and encouragement during part of the preparation of this document. Amongst the many architects, engineers and seismologists who also gave helpful advice, the author is particularly indebted to Professor T. Paulay, Canterbury University, Mr R. Granwall, Professor R. Shepherd and Professor P. W. Taylor, Auckland University; Mr O. A. Glogau, New Zealand Ministry of Works; Professor G. W. Housner, California Institute of Technology; Dr A. G. Brady, U.S. Geological Survey, San Fransisco; Professor R. V. Whitman, Massachusetts Institute of Technology and Mr J. Lord, Seismic Engineering Associates, Los Angeles.

The author is grateful to the Literary Executor of the late Sir Ronald A. Fisher, F.R.S., to Dr Frank Yates, F.R.S. and to Longman Group Ltd., London, for permission to reprint Table III from their book *Statistical Tables for Biological, Agricultural and Medical Research.* (6th edition, 1974.)

The author is especially grateful to Professor R. W. Clough of Berkeley for his careful scrutiny of the manuscript and his constructive Foreword.

List of contents

Introduction

This document is intended to help architects and engineers carry out good earthquake-resistant design expeditiously. Earthquake-resistant design is such a wide and immature subject that there exists a genuine difficulty in deciding what design criteria and analytical methods should be applied to any one project.

The principal objectives are as follows

(1) To discuss the chief aspects of seismic risk evaluation and earthquake-resistant design.

(2) To evaluate various alternative design techniques.

(3) To give guidance on topics where no generally accepted method is currently available.

(4) To suggest procedures to be adopted in earthquake regions having no official zoning or lateral force regulations.

(5) To indicate the more important specialist literature.

The general principles of this document apply to the whole range of building construction and civil engineering, while the more detailed sections relate to the structural rather than the heavy civil engineering industry.

Whereas an initial attempt has been made to provide guidance in the more important areas of design, the coverage can scarcely be exhaustive, even in the long term.

THE DESIGN PROCESS FOR EARTHQUAKE RESISTANT CONSTRUCTION

Earthquakes provide architects and engineers with a number of important design criteria foreign to the normal design process. As some of these criteria are fundamental in determining the form of the structure it is crucial that adequate attention is given to earthquake considerations at the correct stage in the design. To this end a simplified flow chart of the design process for earthquake resistant structures is shown in the figure on the next page.

Although the real interrelationships between all the factors shown in the diagram are obviously much more complex than indicated, the overall sequence is correct. All factors 2–10 are related when evaluating the level of seismic risk, as the risk depends not only on the possible earthquake loadings but also on the capacity of the construction to avoid damage.

Few clients wish to be involved in deciding the acceptable level of risk, but in any case it is important that the client should be informed of the

1

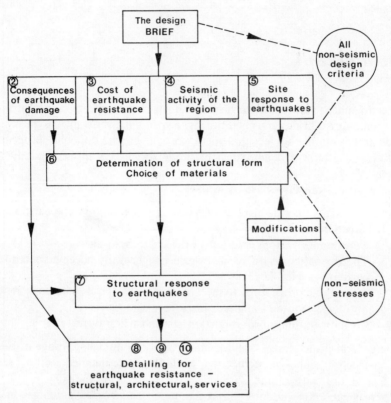

Simplified flow chart for design of earthquake-resistant construction

risks consequent to the available options. Even when the design is done according to a good local code, high risks may still exist.

The various stages of the design process are now discussed following the sequence of numbers in the figure.

Chapter 1

Consequences of earthquakes

1.1 CONSEQUENCES OF EARTHQUAKE DAMAGE

There are two basic results of earthquakes;

(i)   Loss and impairment of human life;
(ii)   Destruction and damage of the constructed and natural environment

Both financially and technically it is only possible to reduce these hazards for large earthquakes. The basic design aims are therefore confined to the reduction of loss of life in any earthquake, either through structural collapse or through secondary damage such as falling debris or fire, and to the reduction of damage and loss of use of the constructed environment.

Obviously some structures demand greater earthquake resistance than others, because of their greater social and/or financial significance. It is important to determine in the design brief not only the more obvious intrinsic value of the structure, but also the survival value placed upon it by the client.

In some countries the greater importance to the community of some types of structure is recognized by statutory requirements, such as in New Zealand where all public buildings are designed for higher earthquake forces than other buildings. Some of the most vital structures to remain functional after destructive earthquakes are dams, hospitals, fire stations, government offices, bridges, radio and telephone services, schools, or in short anything vitally concerned with preventing loss of life in the first instance and with operating emergency services afterwards.

In some cases the client should be made fully aware of the consequences of damage to his structure. It is worth noting that even in earthquake conscious California, it is only since the destruction of three hospitals and some important bridges in the San Fernando earthquake of 1971 that there are likely to be statutory requirements for extra protection of various vital structures throughout the U.S.A.

The choice of an acceptable level of seismic risk is a complex problem, involving consideration of the consequences of earthquake damage, both social and financial, as well as the probable degree of physical risk, i.e. the seismicity of the site. To date little literature has been published on this

3

subject but three recent papers make valuable contributions.[1,2,3] A brief discussion of economic aspects of earthquakes is given in the next section.

1.2 THE COST OF EARTHQUAKE RESISTANCE

During the briefing and budgeting stages of a design, the cost of providing earthquake resistance will have to be considered, at least implicitly, and sometimes explicitly. The cost will depend on such things as the type of project, site conditions, the form of the structure, the seismic activity of the region, and statutory design requirements. The capital outlay actually made may in the end be determined by the wealth of the client and his attitude to the consequences of earthquakes, and insurance to cover losses.

Unfortunately it is not possible to give simple guides on costs, although it would not be misleading to say that most engineering projects designed to the fairly rigorous New Zealand regulations would spend a maximum of 10 percent of the total cost on earthquake provisions, with 5 percent as an average figure.

Where the client simply wants the minimum total cost satisfying local regulations, the usual cost-effectiveness studies comparing different forms and materials will apply. For this a knowledge of good earthquake-resistant forms will of course hasten the determinaton of an economical design, whatever the material chosen.

In many cases, however, a broader economic study of the cost involved in prevention and cure of earthquake damage may be fruitful. In purely economic terms the cost of an earthquake may be examined under three categories;

(i)   cost of life;
(ii)   cost of damage;
(iii)   losses through a facility being out of service.

These costs can be estimated on a probability basis and a cost-effectiveness analysis made to find the relationship between capital expenditure on earthquake resistance on the one hand, and the cost of repairs and loss of income together with insurance premiums on the other hand. Hollings[4] has discussed the earthquake economics of several engineering projects. In the case of a 16 storey block of flats with a reinforced concrete ductile frame it was estimated that the cost of incorporating earthquake resistance against collapse and subsequent loss of life was 1·4 percent of the capital cost of building, while the cost of preventing other earthquake damage was reckoned as a further 5·0 percent, a total of 6·4 percent. The costs of insurance for the same building were estimated as 4·5 percent against deaths and 0·7 percent against damage, a total of 5·2 percent. Clearly a cost-conscious client would be interested in outlaying a little more capital against danger from collapse, thus reducing the life insurance premiums, and he might well consider offsetting the danger of damage mainly with insurance.

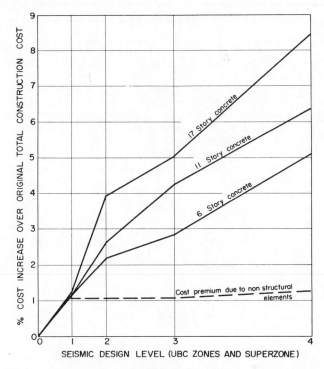

Figure 1.1 Effect on cost of a seismic design of typical concrete apartment buildings in Boston (after Whitman *et al.*[3])

Loss of income due to the building being out of service was not considered in the preceding example. In a hypothetical study of a railway bridge, Hollings showed that up to 18 percent of the capital cost of the bridge could be spent in preventing the bridge going out of service, before this equalled the cost of complete insurance cover.

In a study by Whitman *et al.*,[3] an estimate was made of the costs of providing various levels of earthquake resistance for typical concrete apartment buildings of different heights, as illustrated in Figure 1.1. Until further studies of this type have been done, results such as shown in the figure should be used qualitatively rather than quantitatively.

It is most important that at an early stage the client should be advised of the relationship between strength and risk so that he can agree to what he is buying. Where stringent earthquake regulations must be followed the question of insurance versus earthquake resistance may not be a design consideration: but it can still be important, for example for designing non-structural partitions to be expendable or if a 'fail-safe' mechanism is proposed for the structure. Where there are loose earthquake regulations or none at all, insurance can be a much more important factor, and the client may wish to spend little on earthquake resistance and more on insurance.

REFERENCES

1. Wiggins, J. H., and Moran, D. F. *Earthquake safety in the city of Long Beach, based on the concept of balanced risk*, J. H. Wiggins Co., Palos Verdes Estates, California, Sept. 1971.
2. Grandori, G., and Benedetti, D. 'On the choice of acceptable risk', *Earthquake Engineering, Proc. 4th European Symposium on Earthquake Engineering, London, 1972*, Bulgarian National Committee on Earthquake Engineering, Sofia, 1973, pp. 321–330.
3. Whitman, R. V., Biggs, J. M., Brennan, J., Cornell, C. A. Neufville, R. de, and Vanmarcke, E. H. 'Seismic design analysis', *Structures Publication No. 381*, Massachusetts Institute of Technology, March, 1974, 33 pages.
4. Hollings, J. P. 'The economics of earthquake engineering', *Bulln. New Zealand Society for Earthquake Engineering*, Vol. 4., No. 2, April, 1971, pp. 205–221.

Chapter 2

Seismic activity of a region—seismic risk

2.1 SEISMIC ACTIVITY

2.1.1. Introduction

As the seismic risk for a given project obviously depends on the seismic activity of the region, an early evaluation of this topic will have to be made. Background information may be obtained from various sources such as local officials, engineers, seismologists, local building regulations, published papers or reference books. A rough preliminary indication of seismicity may be obtained from Table 5.4 of this manual, while Figure 2.1 indicates the most seismically active parts of the world, although the available seismic design data in many areas are often inadequate or insufficiently evaluated for safe design, making it desirable to carry out a basic seismicity study of the area concerned. In this chapter, the following types of seismicity study will be described;

(1)   Regional geological evidence;
(2)   Preparation of maps of seismic events;
(3)   Strain–release studies;
(4)   Statistical estimates of design parameters such as return periods for magnitude and acceleration.

As some reference to seismological data may be necessary, some basic definitions are given prior to discussing the seismicity studies themselves.

2.1.2 Definitions of terms used in seismology

Unfortunately, the literature relating to earthquakes is afflicted by the lack of precise definitions of fundamental seismological terms. Hence, definitions of intensity, magnitude, seismicity and seismology, have been set out below. Definitions of other terms may be found in seismology textbooks such as that by Richter.[1]

Intensity is a subjective measure of the effects of an earthquake. It refers to the degree of shaking at a specified place. Over the years, various scales have been devised by different people, notably by Mercalli and also one

7

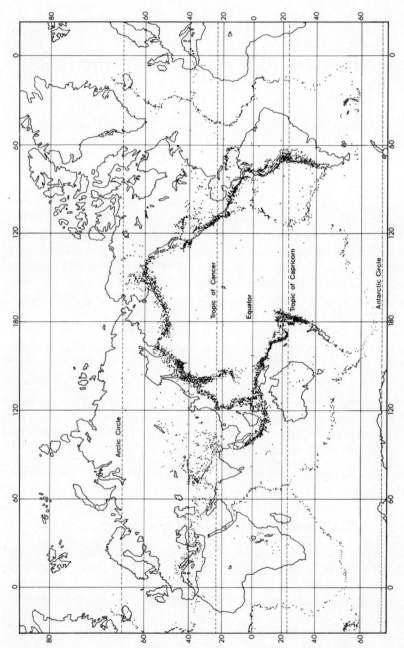

Figure 2.1 Seismicity map of the world. The dots indicate the distribution of seismic events in the mid 20th Century (after Barazangi and Dorman[2])

by Rossi and Forel. The most widely adopted is the Modified Mercalli scale (commonly denoted M.M) which has twelve grades denoted by the Roman numerals I–XII. A detailed description of this intensity scale is given in Appendix B.1.

Magnitude is a quantitative measure of the size of an earthquake, which is independent of the place of observation. It is calculated from amplitude measurements on seismograms, and is on a logarithmic scale expressed in ordinary numbers and decimals. The most commonly used magnitude scale is named after Richter and is denoted by M. It is defined as

$$M = \log A - \log A_0$$

where A is the maximum recorded trace amplitude for a given earthquake at a given distance as written by a standard instrument, and A_0 is that for a particular earthquake selected as standard. Seismologists measure magnitudes in terms of different types of ground motion, such as body waves, M_b, or shear waves, M_s. These values of magnitude are near enough equal for practical engineering purposes. The greatest earthquake magnitude yet recorded is $M \approx 8.9$. Magnitude may be calculated from the amplitude caused by various types of elastic waves (surface or body), depending on the depth of the earthquake focus and its distance from the recording point. Various attempts have been made to relate magnitude to the total energy released by an earthquake. This has led Gutenberg to form a slightly more scientific definition of magnitude called *unified magnitude* and denoted m.

Seismicity is most strictly defined as 'the frequency per unit area of earthquakes in a given region'. The term is often misused and may be more generally thought of as signifying 'the seismic activity of a given region'. This latter definition indicates that seismicity gives some indication of the amount of seismic energy released in a particular area, and in this sense the term seismicity is more meaningful for engineers than the more limited first definition given above.

Seismology may be defined as the science and study of earthquakes, and their causes, effects and related phenomena.

2.2 REGIONAL GEOLOGY

2.2.1 Introduction

Geological evidence of the seismic activity of a region is a valuable tool in the evaluation of seismic risk.[3-8] It is helpful in estimating the likely magnitude, locations and frequency of seismic events. Also, by indicating the types of fault movement prevalent on a given fault, some of the characteristics of the ground motions in the fault vicinity may be anticipated.

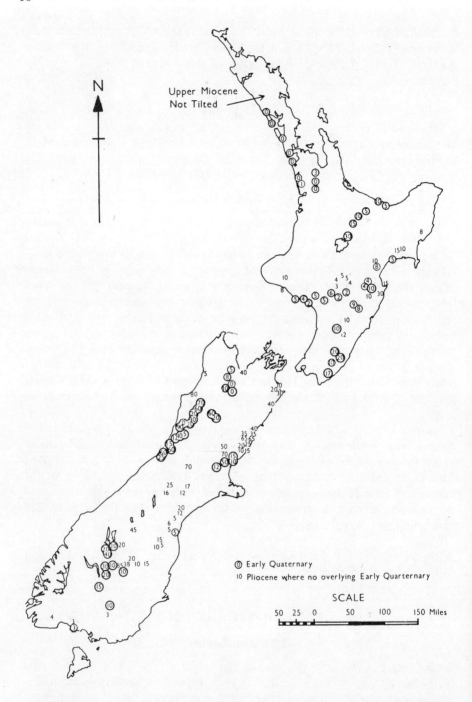

Figure 2.2 Map of New Zealand showing dip angles on Early Quaternary and Pliocene beds (after Clark *et al.*[3])

Regional earthquake geology involves a study of tectonic deformations. The term tectonic refers to rock structure resulting from deep-seated crustal and sub-crustal forces in the earth. The object of a study of the tectonic deformations will be to determine their nature, position, age and movement history. The main geological features to be studied are warping, tilting, faulting, and tectonic structure, of which faulting generally receives most attention. Evidence for one or other type of deformation will not be available for all earthquakes, though in some regions, such as New Zealand, it appears that most major earthquakes involve tilting, warping and faulting.

2.2.2 Tilting

Tilt is helpful in determining the amount and recency of crustal movement in a region, and is measured by the slope of beds which are known to have been originally deposited almost horizontally. The most seismically active regions of the world are in belts of late Tertiary and Quaternary deformation, and by dating sloping beds the age of the activity may be estimated. On the map of New Zealand (Figure 2.2) are plotted the slopes of tilted strata of two periods of geological time.

2.2.3 Faulting

Three main features of faulting are relevant to earthquake engineering: location, activity and type.

2.2.3.1 Fault Location

In most seismically active areas faulting is the main source of information regarding seismic risk. This is partly because faults are relatively easy to describe and sensitive to the measurement of movement, and partly because they provide the focus of energy release in most earthquakes. Even so, maps of fault lines must always be considered incomplete, as old or recent faulting may be difficult or impossible to detect on account of overlying soft soils (or water) which are incapable of maintaining a fault plane displacement. Nevertheless evidence for some hidden faulting may be inferred by interpolating or extrapolating outcropping faults (Figure 2.3).

2.2.3.2 Faulting activity

Uppermost in the engineer's mind is the question 'will this fault move during the lifetime of my project?' In some faults there is evidence that continuous creep movement is taking place, and although this *may* mean that no large earthquake will occur on that particular fault while strain-energy is being gently released, it is clear that few structures should be built across the fault. In most cases the best answer the geologist may be able to give

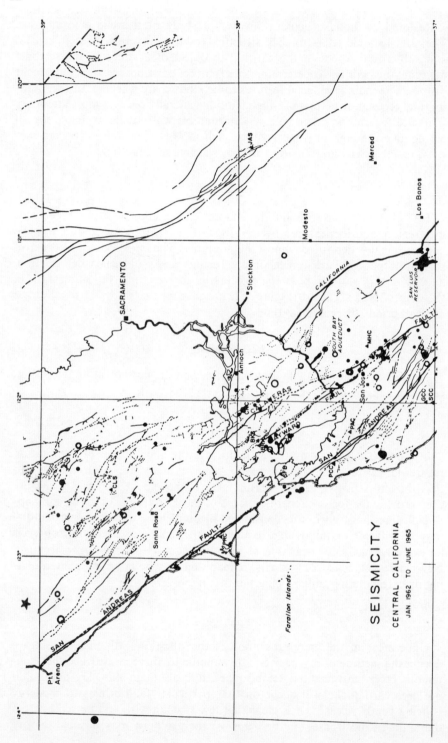

SEISMICITY
CENTRAL CALIFORNIA
JAN 1962 TO JUNE 1965

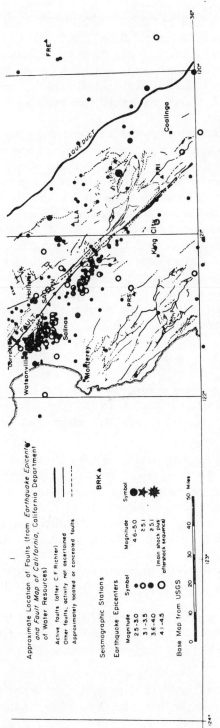

Figure 2.3 Map of California showing faults and locations of earthquakes which occurred between January 1962 and June 1965 (after Bolt[4]. Reproduced by permission of Prentice-Hall Inc., Englewood Cliffs, New Jersey)

is to estimate when the most recent significant movements occurred. For faults which have not been *known* to move in historical times, this is done by dating the youngest soil deposit displaying a fault displacement by examining a section through the fault zone. Unfortunately such sections may not be readily available, although cliffs and slip faces, or quarries and road cuttings may provide free points of inspection. Sometimes special trenches may be cut across a fault zone for this purpose.

Faults are sometimes classified as active or inactive for engineering purposes. Some faults may be unarguably called *active*, where several movements have been recorded in recent times such as on the San Andreas fault in California. In such cases the average return period of earthquakes on a given length of fault line may be used as an earthquake design criterion as discussed later in this chapter. For less frequently active faults the division between the classification of active and inactive is arbitrarily drawn, and is dependent on the ability to date past fault movements. It is technically convenient, for example, to consider faults which have moved within the last 35,000 years as 'active', because this falls within the geologically describable (and datable) 'recent' period. In the siting of nuclear power plants, faults are also considered to be active if they have moved twice in 500,000 years. This again is an expedient time interval depending on dating techniques.

2.2.3.3 Types of Fault

It appears that the characteristics of strong ground motion in the general vicinity of the causative fault can be strongly influenced by the type of faulting. Housner[5] suggests that four types of fault should be considered in the study of destructive earthquakes;

(a) low-angle, compresssive, underthrust faults (Figure 2.4a). These result from tectonic sea-bed plates spreading apart and thrusting under the adjacent continental plates, a phenomenon common to much of the circum-Pacific earthquake belt;

(b) compressive, overthrust faults (Figure 2.4b): compressive forces cause shearing failure forcing the upper portion upwards, as occurred in San Fernando California in 1971;

(c) extensional faults (Figure 2.4c): this is the inverse of the previous type, extensional strains pulling the upper block down the sloping fault plane;

(d) Strike-slip faults (Figure 2.4d): relative horizontal displacement of the two sides of the fault takes place along an essentially vertical fault plane, such as occurred at San Francisco in 1906 on the San Andreas fault.

Few pure examples of the above occur, most earthquake fault movements having components parallel and normal to the fault trace. A useful discussion of faulting has been given by Bonilla.[6]

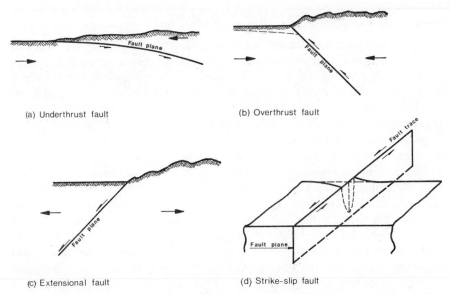

(a) Underthrust fault

(b) Overthrust fault

(c) Extensional fault

(d) Strike-slip fault

Figure 2.4 The main fault types to be considered in the study of strong ground motion characteristics (after Housner[5])

2.2.4 Tectonic structures

Further general information about seismicity may be derived from the relationship of the site to tectonic structure. Mogi[7] has pointed out that the majority of large shallow earthquakes occur in ocean-facing slopes of deep-sea trenches, or in local depressions or troughs or ends of depressions. The magnitude and frequency of earthquakes in a given area may be derived in broad terms from the size and strength of the fault blocks[8]. The larger and stronger the block, the larger is the maximum size of earthquake which can be generated along the boundaries of that block. Also the greater the rate of tectonic movement and the less the competency of the tectonic structures, the more rapid is the build up of the stress needed for a fault movement, and the more frequent will be the occurrence of the maximum magnitude of earthquake for that structure.

2.2.5 Conclusions

From the foregoing we see that useful information on the seismic activity of a region may be obtained from studies of crustal tilting, warping, faulting and tectonic structure. In estimating seismic risk for a given site this largely qualitative information should be used in conjunction with quantitative estimates of earthquake magnitude and frequency as outlined in the following sections of this chapter. Obviously no geological study of seismicity would be complete without reference to maps of known seismic events (Section 2.3).

2.3 MAPS OF SEISMIC EVENTS

For design purposes, much more detailed information of seismicity is obviously required than is shown on Figure 2.1. Larger scale seismicity maps of many areas are given in various publications, principally the reference works by Gutenberg and Richter,[9] Karnik[10] and Lomnitz.[11] A useful set of large maps covering the whole world has been published by the U.S. Department of Commerce[12]: these maps show the position of all earthquakes of magnitude $M \gtrsim 4.0$ which were recorded from 1962–1969 inclusive, but the maps give no indication of the magnitude of individual events.

The type of map of seismic events of most immediate help in the design of a particular structure is as shown in Figure 2.5. This map indicates the locations in plan, the order of depths, and the magnitudes of all recorded earthquakes of magnitudes $M \gtrsim 5.0$ which have occurred within 300 km of the site (Djakarta) since 1900. For preparing such a map, the data of seismic events is most conveniently obtained in the form of computer print–out from organizations such as the Seismology Unit of the Institute of Geological Sciences in Edinburgh. By specifying the geographical location (latitude and longitude, or place name if well known) of the site, a list of seismic events within a certain distance of that site can be obtained for a modest fee. The Seismology Unit currently (1975) provides the following listing of data;

(i)	dates of seismic events since c. 1900 A.D. within a radius of 500 km (or 300 km if requested);
(ii)	latitude and longitude;
(iii)	distance from site;
(iv)	focal depth;
(v)	magnitude (M);
(vi)	intensity (I) observed and calculated;
(vii)	source of the data (recording agency);
(viii)	plot of locations;
(ix)	calculated ground accelerations and velocities.

Each earthquake may be listed several times with the data computed by various recording sources. All the data must be cautiously evaluated as its quality varies considerably according to each agency's technical expertise at the time of each event. Generally speaking, the earlier the event, the less accurate the data. It is unlikely that the above listing will be complete for each item, particularly the intensity. Nevertheless, the information obtained is of great value in forming an assessment of the recent seismic activity of any part of the world.

The type of seismicity map shown on Figure 2.5 is conveniently prepared on a size A2 drawing sheet, to a scale of about 25 km:10 mm. The choice of symbols poses something of a problem as there is no international convention on symbols. Various systems have been used by seismologists such as Gutenberg and Richter or the U.S. Coast and Geodetic Survey, but these

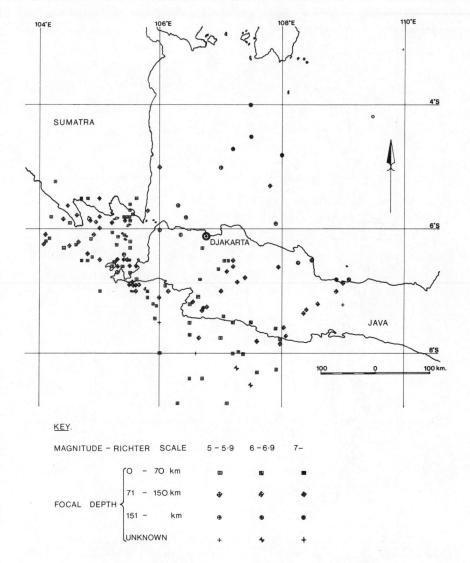

Figure 2.5 Seismic event map for Djakarta (1900–1972)

all suffer from the disadvantage that different magnitudes of earthquakes are shown by symbols which differ only in size. Such symbols cannot be easily distinguished by eye and hence are not as rapidly interpreted as they could be.

It is therefore recommended that for engineering purposes, the symbols used on Figure 2.5 should be used. Magnitudes less than about 5·0 are generally of little direct design significance, as such earthquakes cause little structural damage. Therefore, events of $M < 5·0$ have been excluded from the

Figure 2.6 Mean annual strain release map of southern America, for period 1920–1971 (after Carmona and Castano[14])

notation. However in areas of low seismicity, it may be worth plotting events where $M \geqslant 4$, in order to emphasize the pattern of seismic activity and hence help delineate the zones of greater risk. Earthquakes occurring at great depths cause little damage, and for that reason all events occurring at depths greater than 150 km have been grouped together under one symbol.

2.4 STRAIN–RELEASE STUDIES

The strain released during an earthquake is taken to be proportional to the square root of its energy release. The relationship between energy E,

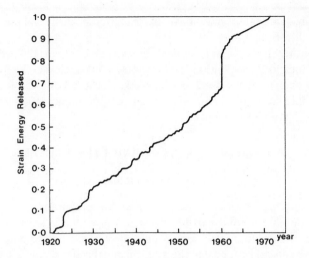

Figure 2.7 Rate of strain energy release in portion of south America shown in Figure 2.6 (after Carmona and Castano[14])

(in ergs) and magnitude M, for shallow earthquakes has been given by Richter[1] as

$$\log E = 11\cdot4 + 1\cdot5\,M \qquad (2.1)$$

The strain release U for a region can be summed and represented by the equivalent number of earthquakes of $M = 4$ in that region, $N(U4)$. The equivalent number of earthquakes $N(U4)$ is divided by the area of the region to give a measurement of the strain release in a given period of time for that region which can be used for comparisons of one region with another (Figure 2.6), or one period of time with another (Figure 2.7). Further examples of strain energy studies may be found in references 13 and 15.

Large shocks constitute the main increments to a cumulative strain release plot. For example, a shock of $M = 5\cdot5$ is about thirteen times stronger than one of $M = 4$. It therefore requires extensive low seismic activity to equal the energy release of a large shock. Nevertheless, in the study of relative strain release rates, comprehensive information is required of low magnitude activity. The summation of many low energy shocks in one region may be comparable to that of a few large shocks in another region. Because the recording of low magnitude events is still in its early years, this method of estimating seismicity may be either impossible for some regions or of limited value due to lack of data. However, where suitable records exist strain release plots can be very illuminating.

A plot of strain release against time is a step function to which an upper bound curve can be drawn, giving an indication of the trend in energy release for that region (Figure 2.7). Obviously, if a flattening of the curve tends to be asymptotic to a constant strain value over a significant time, then the faults in the region may have at least temporarily taken up a more stable configuration.

On the other hand, a mechanical blockage of strain release may have occurred which only a pending large shock could release. Therefore strain release curves cannot be used on their own for earthquake prediction, but should be used in conjunction with magnitude-frequency plots and a knowledge of local fault movements.

2.5 EARTHQUAKE PROBABILITY STUDIES

2.5.1 Introduction

Given a suitable set of data such as used for preparaing seismicity maps, various probability studies can be made using standard statistical methods. One of the most valuable of such studies would be to estimate the largest earthquake likely to occur near the site during the life of the structure being designed. If the frequency of lesser earthquakes can also be realistically predicted, a rational earthquake loading philosophy may be developed as long as the ground accelerations related to such earthquake magnitudes can also be determined.

Methods of studying magnitude-frequency relationships and acceleration return periods are outlined in the following sections. *It cannot be too heavily emphasized that the validity of the results of any probability analysis is conditioned by the quantity and quality of the data used.* Scientific data on earthquakes are relatively scarce, particularly in some areas. Accurate recording instruments are of only recent invention and long term predictions from short term data make little sense. Nevertheless, probability studies intelligently used in relation to general historical seismicity information and strain-release studies, can be valuable in forming design loading judgements.

2.5.2 Magnitude-frequency relationships

Detailed studies of magnitude distribution in many individual areas such as California, New Zealand, Japan, the Mediterranean and the USSR, as well as general studies of the world as a whole, have been carried out by various investigators. Foremost of these are Gutenberg and Richter[9] who derived an empirical relationship between magnitude and frequency of the form

$$\log N = A - bM \tag{2.2}$$

where N is the number of shocks of magnitude M and greater than M per unit time, and A and b are seismic constants for any given region (Figure 2.8).

A varies significantly from study to study while b varies from about $0.5–1.5$ over various regions of the earth. Values of A and b derived by Kaila and

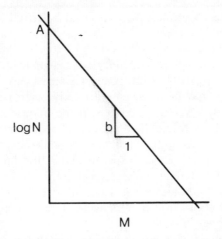

Figure 2.8 Magnitude-frequency relationship for earthquake occurrence

Narain[16] are given in Table 2.1. An authoritative discussion of magnitude-frequency relationships in various parts of the world has been given by Evernden[17], whose values of b should be compared with those in Table 2.1.

Various researchers have tried to relate A and b; Kaila and Narain[16] obtained the relationship

$$A = 6\cdot35b - 1\cdot41 \tag{2.3}$$

Table 2.1. Values of A and b based on 14 years of shallow earthquake data, for various regions normalized to a $2° \times 2°$ grid (Extracted from Kaila and Narain[16])

Region	Boundary				A	b
Japan	26N	40N	132E	150E	6·86	1·22
New Guinea	13S	1N	132E	148E	7·83	1·35
New Zealand	48S	37S	164E	180E		1·04
W. Canada	47N	65N	142E	115W	5·05	1·09
W. United States	25N	47N	135W	105W	5·94	1·14
E. United States	25N	47N	105W	51W	5·79	1·38
Central America	10N	25N	120W	85W	7·36	1·45
Colombia-Peru	18S	6N	85W	60W	5·60	1·11
N. Chile	37N	18S	78W	60W	4·78	0·88
S. Chile	63S	37S	78W	60W	4·46	0·92
Mediterranean	30N	50N	20W	48E	5·45	1·10
Iran-Turkmenia	15N	42N	48E	65W	6·02	1·18
Java	13S	5S	90E	118E	5·37	0·94
E. Africa	40S	30N	20E	48E	3·80	0·87

Note: Great care must be taken to use a sufficiently large sample of earthquakes over as long a period of time as possible in order to obtain reasonably meaningful values of b.

In order to derive this equation, values of A and b derived for various regions of the earth were plotted against each other, and the least squares line was found. The correlation coefficient or goodness of fit was 0·90.

The slope b of the least squares line has a significant seismic meaning. A decrease in b over a period of time indicates an increase in the proportion of large shocks. This may be caused by a relative increase in the frequency of large shocks, or by a relative decrease in the frequency of small shocks. Some investigators have found that periods of maximum strain release in the earth's crust have been preceded and accompanied by a marked decrease in b. From uniaxial compression experiments in the laboratory, Scholz[18] found that the magnitude-frequency relationship for microfractures in a given rock is characterized by b decreasing when the stress level is raised. Consequently regional variations in b may indicate variations in the level of compressive stress in the earth's crust.

For any given region if enough data is available a plot of M against $\log N$ can be made, and the 'best' line for equation (2.1) can be determined using a linear regression analysis as described in Section 2.6.3. Records of events that give magnitude $M < 4$ go back only a few years, and it is usual to neglect these values as they may give a misleading bias to the relationship

$$\log N = A - bM \qquad (2.2)$$

Having evaluated equation (2.2) for the site in question, the value of M for a given return period is found directly and may be used to determine the peak ground motions by methods such as outlined in the following section.

In obtaining equation (2.2) for a given site, some allowance for variations in the distance D of the events from the site should be made. This may be done most simply by inspection of the map of events, taking geological evidence of active fault lines into account. More formal allowance for the effect of D may be made by taking D as an extra variable in the statistical model, such as in the technique outlined by Cornell[19] and by Rascón and Cornell[20]. Cornell uses extreme value probability distributions, and presents convenient methods for allowing for different geometrical arrangements of earthquake sources, e.g. linear distributions along faults, or general distributions when events have an equal likelihood of occurrence anywhere within a given area. For more general accounts of statistical methods of assessing earthquake risk, reference should be made to Newmark and Rosenblueth[21] or Lomnitz[11].

2.5.3 Peak ground motions

In determining the peak ground motions at a given site for a given return period, the attenuation of ground motion with distance from the origin of the earthquake is of great importance. It is common to express the peak ground motion, acceleration or velocity, in terms of the magnitude, M and

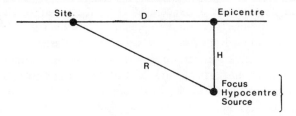

Figure 2.9 Geometric relationship between focus and site

either the epicentral distance D, or the focal distance R (Figure 2.9). A great number of such expressions have been published over the years, based on different amounts of data, of different quality from measurements made in different parts of the world. There is a very wide scatter in the predictions of such equations and only the most recent ones, which give statistical evaluations of the scatter in the data used, can be used with much confidence.

One of the big problems is whether the motion is appropriate for bedrock or for softer overlying soils. Seed et al.[22] have given a valuable description of rock motions in earthquakes, while Donovan[23] gives a clear picture of the scatter in acceleration data for rock-like as well as soil sites even for a single event, the San Fernando earthquake, as shown in Figure 2.10. For

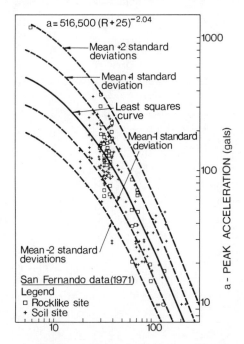

Figure 2.10 Attenuation of peak ground acceleration. Least squares and standard deviation curves for San Fernando California 1971 earthquake (after Donovan[23])

example, at a hypocentral distance of 30 km, the mean acceleration minus two standard deviations was about 55 gals (i.e. cm/s^2) while the mean acceleration plus two standard deviations was about 280 gals.

There has been much uncertainty about whether attenuation rates vary significantly in different parts of the world. With the recent trend of installing more reliable recording networks in different parts of the world this issue will soon be clarified, but some recent work suggests that attenuation equations based on the abundant Californian data are more widely applicable than was previously thought.

Perhaps the most broadly based relationship between magnitude, distance and peak acceleration a, is that of Donovan[23]

$$a = \frac{1080 \, e^{0 \cdot 5M}}{(R + 25)^{1 \cdot 32}} \tag{2.4}$$

where a is in cm/s^2 and R is the distance from the hypocentre in km (Figure 2.9). The equation is the expression for the mean of 678 acceleration values of Western U.S.A., Japan and Papua New Guinea, and represents a conservative estimate of mean peak acceleration on sites with 6 m or more of soil overlying the rock.

Another attenuation relationship based on recent corrected and statistically evaluated data is that of Esteva[24], who gives the expressions for peak ground acceleration and peak ground velocity v, as

$$a = \frac{5600 \, e^{0 \cdot 8M}}{(R + 40)^2} \tag{2.5}$$

$$v = \frac{32 \, e^M}{(R + 25)^{1 \cdot 7}} \tag{2.6}$$

These two expressions are based on Californian data and are valid for focal distances in excess of 15 km. It should be noted that attenuation equations are generally inappropriate for the epicentral area, i.e. within a distance of about 15–20 km from the epicentre. This area needs special consideration, and the understanding of it is still very limited. For example there are great problems in relating peak ground acceleration to maximum earthquake intensity as discussed by Donovan[23] and Ambraseys.[25] The enormous scatter is shown in Figure 2.11.

Our ability to reliably predict peak ground displacement is rather less than for acceleration or velocity, mainly because estimates of ground displacement are mostly derived from double integrations of acceleration records, and numerical errors add to the uncertainties. A rough estimate for ground displacement d, may be made from the empirical relationship given by Newmark and Rosenblueth[21]:

$$5 \lesssim \frac{ad}{v^2} \lesssim 15 \tag{2.7}$$

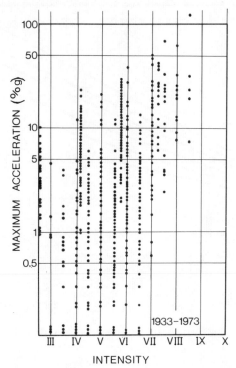

Figure 2.11 Peak ground accelerations plotted against the corresponding intensity as observed in earthquakes occurring between 1933 and 1973 (after Ambraseys[25])

in which the value 5 is appropriate to large epicentral distances, say 100 km, and the value 15 is for small epicentral distances.

In addition to acceleration, velocity and displacement, the frequency content of ground motion is vital in the study of earthquakes. It is thought

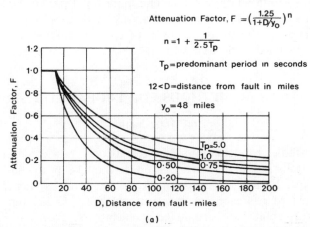

Attenuation Factor, $F = \left(\frac{1.25}{1+D/y_0}\right)^n$

$n = 1 + \frac{1}{2.5T_p}$

T_p = predominant period in seconds

$12 < D$ = distance from fault in miles

y_0 = 48 miles

(a)

Figure 2.12a Attenuation factors for earthquake ground motions (after Benioff[26])

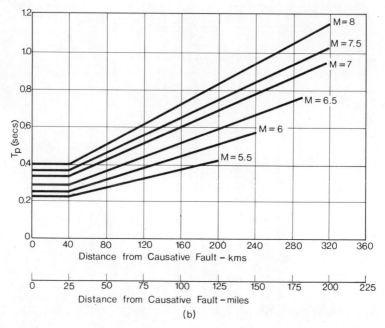

Figure 2.12b Predominant periods for maximum accelerations in rock (after Seed *et al.*[22])

that the frequency content, especially in the epicentral area is a function of the earthquake source mechanism. There is some evidence to suggest[22] that the predominant period of vibration associated with peak accelerations in rock is a function of earthquake magnitude (Figure 2.12a), although this is not statistically well supported as yet. Figure 2.12(a) also shows the tendency for the predominant period to lengthen with distance from the epicentre, as the shorter period vibrations attenuate more rapidly than the long period shaking. This effect is shown in another way in Figure 2.12(b). The significance for design of the various features of ground shaking discussed above are further discussed in Section 3.4.

2.6 STATISTICAL METHODS FOR PROBABILITY STUDIES

2.6.1 Introduction

Probability studies by definition involve statistics. A brief outline of some elementary statistical operations used in earthquake probability problems is given below. Further explanation of these and related matters may be found in standard references on statistical methods.[27,28] For further reading on statistics as applied to earthquake problems, reference may be made to Benjamin and Cornell,[29] Newmark and Rosenblueth[21] and Lomnitz.[11]

2.6.2 Definitions of some statistical terms

The following terms may be usefully defined here before their use in the subsequent text.

Variance σ^2, is the mean square deviation from the mean. If a set of values (a finite population) consists of n observations x_i, whose mean is μ, the *deviation* of each observation is $x_i - \mu$, and the variance is written

$$\sigma^2 = \frac{\sum_{i=1}^{n}(x_i - \mu)^2}{n} \tag{2.8}$$

If we are considering only a limited number (a sample) out of a population, then the variance is defined as S^2 such that

$$S^2 = \frac{\sum_{i=1}^{n}(x_i - \bar{x})^2}{n-1} \tag{2.9}$$

where $\bar{x}$ is the mean of the sample. It is important to choose the correct variance, either equation (2.8) or (2.9), except of course where n is large and the difference between the equations becomes negligible.

Standard deviation σ is the square root of the variance and is therefore written

$$\sigma = \sqrt{\frac{\sum_{i=1}^{n}(x_i - \mu)^2}{n}} \tag{2.10}$$

Distributions. A set of observations may be arranged in various frequency distribution forms. Earthquakes are generally considered to be randomly occurring phenomena and hence have what is commonly termed a *normal distribution* described by the so-called Gauss function

$$p = Ce^{-h^2X^2} \tag{2.11}$$

where

X = deviation from the mean

p = probability of occurrence of this deviation

$C = \dfrac{1}{\sigma\sqrt{(2\pi)}}$, a constant equal to the maximum height of the curve (Figure 2.13)

$h = \dfrac{1}{\sigma\sqrt{2}}$, the precision constant determining the spread of the curve.

Hençe equation (2.11) can be written

$$p = \frac{e^{-X^2/2\sigma^2}}{\sigma\sqrt{(2\pi)}} \qquad (2.12)$$

The standard deviation σ determines the horizontal spread of the distribution curve, and for many purposes it is convenient to use σ as a unit of deviation from the mean, i.e. let

$$Z = \frac{1}{\sigma\sqrt{(2\pi)}} e^{-Z^2/2} \qquad (2.13)$$

Call $F(Z)$ the area under the curve between the mean and $Z = Z$. The areas under the normal distribution curve $F(Z)$ corresponding to deviations in steps of one standard deviation are written on Figure 2.13.

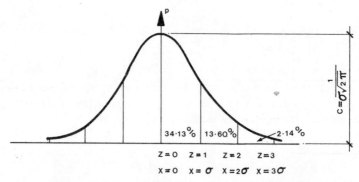

Figure 2.13 Area under the normal probability curve

Significance level represents the probability of drawing an erroneous conclusion. When observations are normally distributed, $\{1-2F(Z)\}$ represents the probability of a value falling outside the range. For example, when $Z = 1.96$, $\{1-2F(Z)\} = 0.05$, and it is said that the level of significance is 5 percent.

Confidence level represents the probability of drawing a correct conclusion. It is described by $2F(Z)$ expressed as a percentage. For example, when $Z = 1.96$, $2F(Z) = 0.95$, i.e. there is a 95 percent probability of a value falling within the range $\mu \pm 1.96\sigma$, and the confidence level is 95 percent.

Confidence limits are defined as $(\mu - Z\sigma)$ and $(\mu + Z\sigma)$, the interval between them being the *confidence interval*.

Degrees of freedom v may be defined as the number of *independent* observations that can be hypothesized. For example, in determining a least squares line, two dependent observations exist, namely the slope and the intercept. Hence if n observations (of M, say) exist, then there are $v = n - 2$ degrees of freedom.

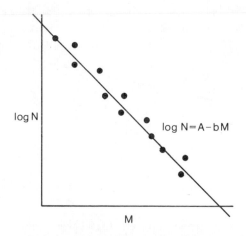

Figure 2.14 Typical plot of magnitude against frequency of occurrence

2.6.3 Establishing relationships from data of seismic observations

From a set of earthquake data for a given site a typical relationship that we may wish to obtain is the magnitude-frequency relationship as discussed in Section 2.5.2. If the magnitude M is plotted against $\log N$, where N is the number of earthquakes of magnitude M or greater per year, a scatter of points is obtained (Figure 2.14).

First let us make the usual assumption of the linear relationship discussed in Section 2.5.2, i.e.

$$\log N = A - bM \tag{2.2}$$

As well as obtaining the 'best' values of A and b, it is desirable to be able to

(i)   evaluate how well the line fits the data, and also estimate
(ii)   what confidence we can have in the 'best' fitting line.

In order to do both of these things, the statistical operations discussed in Sections 2.6.4 and 2.6.5 should be carried out.

But first we must fit the 'best' straight line to the data by the method of least squares. This line is called the *linear regression line*, the principle of which is that the most probable position of the line is such that the sum of the squares of deviations of all points from the line is a minimum.

Considering the line shown in Figure 2.15,

$$y = a + bx \tag{2.14}$$

the deviations ϵ_i are measured in the direction of the y-axis. The underlying assumption is that x has either negligible or zero error (being assigned), while y is the observed or measured quantity. The observed y is thus a random value from a population of values of y corresponding to a given x.

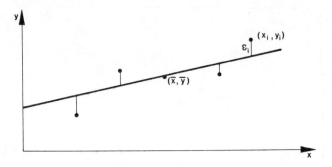

Figure 2.15 Fitting the 'best' straight line through data points

The regression line as defined above is found when the sum of the squares of the deviations is a minimum, i.e. when

$$P = \sum \epsilon_i^2$$

is a minimum, i.e.

$$P = \Sigma\{y_i - (a + bx_i)\}^2$$

is minimized.
By differentiating, P is found to be a minimum for values of a and b such that the equation of the line of best fit is

$$y = \frac{\Sigma x\, \Sigma y^2 - \Sigma y\, \Sigma xy}{n\Sigma x^2 - (\Sigma x)^2} + \frac{n\Sigma xy - \Sigma x\, \Sigma y}{n\Sigma x^2 - (\Sigma x)^2} x \qquad (2.15)$$

Equation (2.15) is the regression line of y upon x.
If the properties of the variables are reversed such that y is assigned and x is observed then the regression line of x upon y is found to be described by equation (2.16).

$$x = \frac{\Sigma x\, \Sigma y^2 - \Sigma y\, \Sigma xy}{n\Sigma y^2 - (\Sigma y)^2} + \frac{n\Sigma xy - \Sigma x\, \Sigma y}{n\Sigma y^2 - (\Sigma y)^2} y \qquad (2.16)$$

The two equations (2.15) and (2.16) coincide at the centroidal point $(\bar{x}, \bar{y})$ of the data, but generally will have differing slopes (Figure 2.16).

2.6.4 Estimating goodness of fit—correlation

Having fitted the 'best' straight line to the data by the regression analysis outlined above, the 'goodness of fit' of the line to the data should be ascertained. Although the regression line has the most appropriate slope and passes through the centroid of the data, it may still be an inappropriate description of the data, such as in the case illustrated in Figure 2.17.

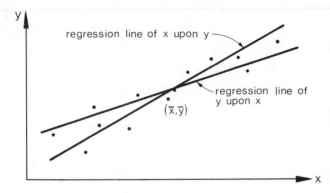

Figure 2.16 Regression lines based on the two alternative variables, x and y.

A convenient test for 'goodness of fit' is to measure what is termed the *correlation* between the two variables. It can be seen from equation (2.15) that if there is no correlation between y and x (if y is independent of x) the coefficient of x is zero, i.e.

$$\frac{n\Sigma xy - \Sigma x \, \Sigma y}{n\Sigma x^2 - (\Sigma x)^2} = 0 \tag{2.17}$$

Similarly, from equation (2.16) if there is no correlation between x and y the coefficient of y is zero, i.e.

$$\frac{n\Sigma xy - \Sigma x \, \Sigma y}{n\Sigma y^2 - (\Sigma y)^2} = 0 \tag{2.18}$$

If there is no correlation between the two variables being studied, the product of the slopes given by equations (2.17) and (2.18) is zero, i.e.

$$r^2 = \frac{n\Sigma xy - \Sigma x \, \Sigma y}{n\Sigma x^2 - (\Sigma x)^2} \cdot \frac{n\Sigma xy - \Sigma x \, \Sigma y}{n\Sigma y^2 - (\Sigma y)^2} = 0 \tag{2.19}$$

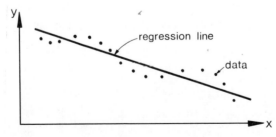

Figure 2.17 Illustration of difference between a 'best' straight line and 'goodness of fit'

Conversely, when there is perfect correlation, i.e. all the points lie exactly on each of the two regression lines, the lines coincide; their slopes are the same and hence from equation (2.19):

$$r^2 = 1 \qquad (2.20)$$

A measure of the correlation of two variables is thus given by r, the *correlation coefficient* and

$$r = \frac{n\Sigma xy - \Sigma x \, \Sigma y}{\sqrt{[\{n\Sigma x^2 - (\Sigma x)^2\} \{n\Sigma y^2 - (\Sigma y)^2\}]}} \qquad (2.21)$$

or simply, writing $X = x - \bar{x}$, and $Y = y - \bar{y}$,

$$r = \frac{\Sigma XY}{\sqrt{[\Sigma X^2 \Sigma Y^2]}} \qquad (2.22)$$

The correlation coefficient lies in the range $0 \leqslant r \leqslant 1$. If r is positive, y increases with increasing x, and if r is negative, y decreases with increasing x. If $r = \pm 1$ perfect correlation exists, i.e. all the data points lie exactly on one straight line.

In order to assess the meaning of a calculated value of r, Table 2.2 should be consulted. If the calculated (absolute) value of r is equal to or greater than the appropriate value of r in Table 2.2, we conclude that correlation exists with a level of significance (Section 2.6.2) equal to or better than that implied by the table. The level of significance represents the probability of our having drawn a wrong conclusion, i.e. it represents the probability that the relationship of the points to a straight line arose by chance and not because there was a real linear relationship. In most seismicity studies we would choose a 5 percent level of significance. If the calculated value of r is less than the appropriate value of r in Table 2.2, the correlation between the data and the best fitting line is worse than that desired, i.e. there is smaller probability that we have found a valid straight line.

Example 2.1—Correlation

Consider the data points plotted in Figure 2.18. The regression line for this data is as shown in the figure. How well does this line correlate with the data?

Firstly calculate the correlation coefficient using equation (2.22).

$$r = \frac{\Sigma XY}{\sqrt{[\Sigma X^2 \Sigma Y^2]}}$$

$$= \frac{2(9 + 25 + 63 + 63 + 120 + 99)}{2(2^2 + 3^2 + 5^2 + 7^2 + 9^2 + 10^2 + 11^2) \times 2(2^2 + 3^2 + 5^2 + 7^2 + 9^2 + 12^2)}$$

$$= \frac{379}{389}$$

i.e.

$$r = 0.974$$

To enter Table 2.2, the number of degrees of freedom equals the number of observations minus 2, that is

$$v = n - 2$$
$$= 14 - 2 = 12$$

For a 1 percent level of significance and $v = 12$, from Table 2.2, $r = 0.661$. As the calculated value of r exceeds this value, then the line fits the data with a level of significance better than 1 percent, i.e. the data fits the regression line very well.

The data is linearly distributed but with quite large scatter from the line. The meaning of this scatter is studied in the following section.

2.6.5 Confidence limits of regression estimates

Having found the regression line through the observed data and having determined the 'goodness of fit' by the correlation factor, we must still discover what confidence we can have in this estimate of the equation to the line. A commonly used confidence test is the so-called t–test. For the desired level of significance and the appropriate number of degrees of freedom we find t such that the true value of some parameter y lies in the range

$$y \pm tS_y$$

The probability of being wrong is equal to the level of significance of the value of t. S_y is the standard deviation of the observed values of y. Values of t are tabulated in Table 2.3. The confidence limits are thus defined as $(y + tS_y)$ and $(y - tS_y)$. For a regression line we can determine the confidence limits for such parameters as the mean $\bar{y}$, the slope b, or a future individual value of y_i related to a specific value of x. These confidence limits are illustrated in the following example.

Example 2.2—Confidence limits

We will refer to the data of the previous example as plotted in Figure 2.18.

(a) *Confidence limits for the mean,* $\bar{y}$. The variance of y

$$S_y^2 = \frac{\epsilon_i^2}{v} = \frac{10 \times 2^2}{12} = \frac{10}{3}$$

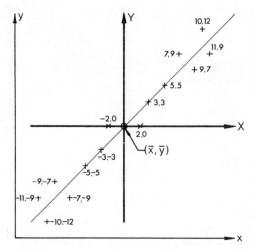

Figure 2.18 Data points and regression line for Example 2.1

The variance of $\bar{y}$

$$S_{\bar{y}}^2 = \frac{S_y^2}{n} = \frac{10}{3 \times 14} = \frac{5}{21}$$

From Table 2.3 taking $v = 12$ and a 5 percent level of significance

$$t = 2 \cdot 179$$

Therefore the confidence limits for $\bar{y}$ are

$$\bar{y} \pm tS_y$$

$$= \bar{y} \pm 2 \cdot 179 \sqrt{(\tfrac{5}{21})}$$

$$= \bar{y} \pm 1 \cdot 063$$

There is a 95 percent probability that $\bar{y}$ lies in this range. Since the regression line must pass through the centroidal point, an error in $\bar{y}$ leads to a constant error in y for all points on the line as shown in Figure 2.19(a).

(b) *Confidence limits for the slope, b.* The variance of the slope b

$$S_b^2 = \frac{S_y^2}{\Sigma(x - \bar{x})^2} \tag{2.23}$$

$$= \frac{S_y^2}{\Sigma X^2}$$

$$= \frac{10}{3} \times \frac{1}{2(2^2 + 3^2 + 5^2 + 7^2 + 9^2 + 10^2 + 11^2)}$$

$$S_b^2 = \frac{5}{3 \times 389}$$

As before, for a 5 percent level of significance

$$t = 2\cdot179$$

Therefore the confidence limits for b are

$$b \pm tS_b$$

$$= 1\cdot00 \pm 2\cdot179 \sqrt{\left(\frac{5}{3 \times 389}\right)}$$

$$= 1\cdot00 \pm 0\cdot143$$

Again there is a 95 percent probability that b lies in this range (Figure 2.19b).

$$S_{y_i}^2 = S_y^2\left(1 + \frac{1}{n} + \frac{X^2}{\Sigma X^2}\right) \qquad (2.24)$$

$$= \frac{10}{3}\left(1 + \frac{1}{14} + \frac{11^2}{2 \times 389}\right)$$

$$= 4\cdot09.$$

As before, for a 5 percent level of significance

$$t = 2\cdot179.$$

Therefore the corresponding confidence limits for y_i (or Y_i) at $X = 11$ are

$$Y_i \pm tS_{y_i} = Y_i \pm 2\cdot179\sqrt{4\cdot09}$$

$$= Y_i \pm 4\cdot407$$

Note that the confidence limits for Y_i at $X = -11$ equal those at $X = +11$. Similarly the confidence limits for y_i at $X = 0$ can be found by substituting $X = 0$ in equation (2.24). This gives the confidence limits for y_i (or Y_i) at $X = 0$ as

$$Y_i \pm 4\cdot118$$

Again there is a 95 percent probability that $Y_{(X=0)}$ lies in this range. These confidence limits are illustrated in Figure 2.19(c).

36

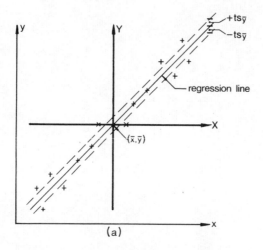

(a)

Figure 2.19a Confidence limits for $\bar{y}$

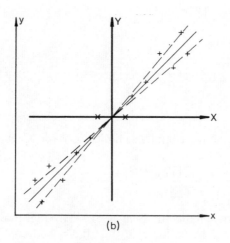

(b)

Figure 2.19b Confidence limits of slope

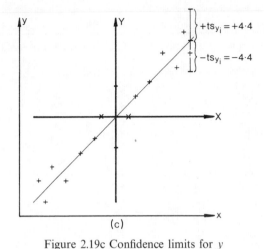

Figure 2.19c Confidence limits for y

Figure 2.19 Confidence limits for regression estimates

The confidence limits found in Example 2.2 show there is appreciable scatter from the regression line, despite the fact that the correlation coefficient $r = 0.974$ indicates that the data fit the straight line very well. Part (c) of Example 2.2 shows how we can estimate a probable maximum value of magnitude M, or ground acceleration a, for a given return period from seismicity data such as shown on Figure 2.14.

For examples of the use of statistics on earthquake acceleration data, see Figure 2.10 and its source reference.

Table 2.2. Values of Correlation Coefficient r. (From Crow, Davis and Maxfield, *Statistical Manual*, Dover, 1960)

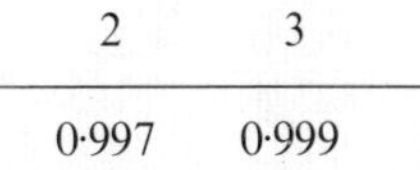

v	5 Percent level of significance				1 Percent level of significance				v
	Total number of variables				Total number of variables				
	2	3	4	5	2	3	4	5	
1	0·997	0·999	0·999	0·999	1·000	1·000	1·000	1·000	1
2	0·950	0·975	0·983	0·987	0·990	0·995	0·997	0·998	2
3	0·878	0·930	0·950	0·961	0·959	0·976	0·983	0·987	3
4	0·811	0·881	0·912	0·930	0·917	0·949	0·962	0·970	4
5	0·754	0·836	0·874	0·898	0·874	0·917	0·937	0·949	5
6	0·707	0·795	0·839	0·867	0·834	0·886	0·911	0·927	6
7	0·666	0·758	0·807	0·838	0·798	0·855	0·885	0·904	7
8	0·632	0·726	0·777	0·811	0·765	0·827	0·860	0·882	8
9	0·602	0·697	0·750	0·786	0·735	0·800	0·836	0·861	9
10	0·576	0·671	0·726	0·763	0·708	0·776	0·814	0·840	10

[*cont.*

Table 2.2—*continued*

	5 Percent level of significance				1 Percent level of significance				
	Total number of variables				Total number of variables				
v	2	3	4	5	2	3	4	5	v
11	0·553	0·648	0·703	0·741	0·684	0·753	0·793	0·821	11
12	0·532	0·627	0·683	0·722	0·661	0·732	0·773	0·802	12
13	0·514	0·608	0·664	0·703	0·641	0·712	0·755	0·785	13
14	0·497	0·590	0·646	0·686	0·623	0·694	0·737	0·768	14
15	0·482	0·574	0·630	0·670	0·606	0·677	0·721	0·752	15
16	0·468	0·559	0·615	0·655	0·590	0·662	0·706	0·738	16
17	0·456	0·545	0·601	0·641	0·575	0·647	0·691	0·724	17
18	0·444	0·532	0·587	0·628	0·561	0·633	0·678	0·710	18
19	0·433	0·520	0·575	0·615	0·549	0·620	0·665	0·698	19
20	0·423	0·509	0·563	0·604	0·537	0·608	0·652	0·685	20
21	0·413	0·498	0·552	0·592	0·526	0·596	0·641	0·674	21
22	0·404	0·488	0·542	0·582	0·515	0·585	0·630	0·663	22
23	0·396	0·479	0·532	0·572	0·505	0·574	0·619	0·652	23
24	0·388	0·470	0·523	0·562	0·496	0·565	0·609	0·642	24
25	0·381	0·462	0·514	0·553	0·487	0·555	0·600	0·633	25
26	0·374	0·454	0·506	0·545	0·478	0·546	0·590	0·624	26
27	0·367	0·446	0·498	0·536	0·470	0·538	0·582	0·615	27
28	0·361	0·439	0·490	0·529	0·463	0·530	0·573	0·606	28
29	0·355	0·432	0·482	0·521	0·456	0·522	0·565	0·598	29
30	0·349	0·426	0·476	0·514	0·449	0·514	0·558	0·591	30
35	0·325	0·397	0·445	0·482	0·418	0·481	0·523	0·556	35
40	0·304	0·373	0·419	0·455	0·393	0·454	0·494	0·526	40
45	0·288	0·353	0·397	0·432	0·372	0·430	0·470	0·501	45
50	0·273	0·336	0·379	0·412	0·354	0·410	0·449	0·479	50
60	0·250	0·308	0·348	0·380	0·325	0·377	0·414	0·442	60
70	0·232	0·286	0·324	0·354	0·302	0·351	0·386	0·413	70
80	0·217	0·269	0·304	0·332	0·283	0·330	0·362	0·389	80
90	0·205	0·254	0·288	0·315	0·267	0·312	0·343	0·368	90
100	0·195	0·241	0·274	0·300	0·254	0·297	0·327	0·351	100
125	0·174	0·216	0·246	0·269	0·228	0·266	0·294	0·316	125
150	0·159	0·198	0·225	0·247	0·208	0·244	0·270	0·290	150
200	0·138	0·172	0·196	0·215	0·181	0·212	0·234	0·253	200
300	0·113	0·141	0·160	0·176	0·148	0·174	0·192	0·208	300
400	0·098	0·122	0·139	0·153	0·128	0·151	0·167	0·180	400
500	0·088	0·109	0·124	0·137	0·115	0·135	0·150	0·162	500
1,000	0·062	0·077	0·088	0·097	0·081	0·096	0·106	0·116	1,000

The critical value of r at a given level of significance, total number of variables, and degrees of freedom v, is read from the table. If the computed $|r|$ exceeds the critical value, then the null hypothesis that there is no association between the variables is rejected at the given level. The test is an equal-tails test, since we are usually interested in either positive or negative correlation. The shaded portion of the figure is the stipulated probability as a level of significance.

Table 2.3. Distribution of t (Table 2.3 is taken from Table III of Fisher and Yates: *Statistical Tables for Biological, Agricultural and Medical Research*, published by Longman Group Ltd., London (previously published by Oliver and Boyd, Edinburgh), and by permission of the authors and publishers)

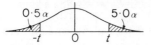

Degrees of freedom ν	Probability α			
	0·10	0·05	0·01	0·001
1	6·314	12·706	63·657	636·619
2	2·920	4·303	9·925	31·598
3	2·353	3·182	5·841	12·941
4	2·132	2·776	4·604	8·610
5	2·015	2·571	4·032	6·859
6	1·943	2·447	3·707	5·959
7	1·895	2·365	3·499	5·405
8	1·860	2·306	3·355	5·041
9	1·833	2·262	3·250	4·781
10	1·812	2·228	3·169	4·587
11	1·796	2·201	3·106	4·437
12	1·782	2·179	3·055	4·318
13	1.771	2·160	3·012	4·221
14	1·761	2·145	2·977	4·140
15	1·753	2·131	2·947	4·073
16	1·746	2·120	2·921	4·015
17	1·740	2·110	2·898	3·965
18	1·734	2·101	2·878	3·922
19	1·729	2·093	2·861	3·883
20	1·725	2·086	2·845	3·850
21	1·721	2·080	2·831	3·819
22	1·717	2·074	2·819	3·792
23	1·714	2·069	2·807	3·767
24	1·711	2·064	2·797	3·745
25	1·708	2·060	2·787	3·725
26	1·706	2·056	2·779	3·707
27	1·703	2·052	2·771	3·690
28	1·701	2·048	2·763	3·674
29	1·699	2·045	2·756	3·659
30	1·697	2·042	2·750	3·646
40	1·684	2·021	2·704	3·551
60	1·671	2·000	2·660	3·460
120	1·658	1·980	2·617	3·373
∞	1·645	1·960	2·576	3·291

This table gives the values of t corresponding to various values of the probability α (level of significance) of a random variable falling inside the shaded areas in the figure, for a given number of degrees of freedom ν available for the estimation of error. For a one-sided test the confidence limits are obtained for $\alpha/2$.

2.7 DEFINING THE DESIGN EARTHQUAKE

Studies of the seismic activity of a region as discussed in the previous sections of this chapter supply the material for defining the design earthquake for a given project. The characteristics of the design earthquake may be used in conjunction with the dynamic characteristics of the site to determine the dynamic design criteria for the project as discussed in Chapter 3.

An adequate definition of a design earthquake is very elusive, even prior to consideration of site conditions, because of difficulties in defining past earthquake behaviour and difficulties in predicting future seismic events. The main variables derived or implied in this chapter for use in defining the design earthquake are: magnitude, return period, epicentral distance, focal depth, fault positions, fault types, and associated dependent variables such as peak ground acceleration, peak ground velocity, peak ground displacement, duration of strong shaking, dominant period of shaking and fault length activated.

Data on the above aspects of earthquakes are variable, inaccurate, and scarce. This means that the interpretation of the data must be highly subjective, and the use of mean values or some other value such as the 95 percent confidence level may be open to argument. A considerable amount of idealization is necessary.

In order to illustrate the definition of design earthquakes for a given site, reference will be made to Figure 2.20. Assume that studies of the earthquake history of the region have suggested the use of two design earthquakes, A and B, with the characteristics tabulated in Figure 2.20. It is quite common practice to consider two different design earthquakes with magnitudes and return periods as suggested above; normally the larger, less frequent, earthquake would be considered the worst design condition for use as ultimate loading, while the smaller, more frequent, earthquake would be used as a criterion for control of non-structural damage. However in the situation illustrated in Figure 2.20, the associated fault types would render this use of the design earthquakes inappropriate.

Because the sloping plane for earthquake B outcrops near the site, the intensity of ground shaking at the site due to earthquake B may be as intense as at positions closer to the epicentre E_B. If the fault trace BB' had been undetected or not allowed for at the time of the design, the intensity of ground motion at the site would be underestimated assuming normal attenuation from an epicentre 30 km away.

For a brief qualitative discussion of current techniques in regional seismicity evaluation, see the section entitled 'regional seismicity' in a paper by Rosenblueth.[30] A more detailed account of this subject has been given by Lomnitz.[11]

As an example of the type of information obtained by the simple methods of this chapter, consider the seismicity of Djakarta as illustrated in Figure 2.5. A probability study of the events within 300 km of Djakarta recorded

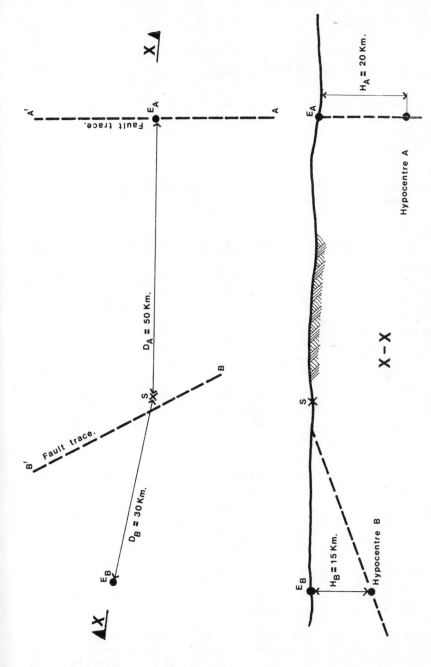

Figure 2.20 Hypothetical example of site relationship to two design earthquakes A and B with epicentres E_A and E_B respectively

between 1900 and 1972 indicated that for a 50 year return period there was a 95 percent probability that the greatest magnitude of event would be $M = 8.0$, and that for a 25 year return period the maximum magnitude of event would be $M = 7.6$. But Figure 2.5 indicates no larger shallow events recorded near to Djakarta. As can be seen from Figure 2.1 the tectonic discontinuity of the Sunda Arc lies adjacent to Java, the main concentration of events being near the South coast and many events being fairly deep seated. Without a detailed knowledge of the local fault lines, a reasonable fifty year design earthquake for Djakarta might therefore be a shallow event of magnitude 7.5 with an epicentral distance of 100 km.

REFERENCES

1. Richter, C. F., *Elementary seismology*, Freeman, San Francisco, 1958.
2. Barazangi, M., and Dorman, J., 'World seismicity map of ESSA coast and geodetic survey epicentre data for 1961–67', *Bulln. Seismological Society of America*, **59**, 369–380 (1969).
3. Clark, R. H., Dibble, R. R., Fyfe, H. E., Lensen, G. J., and Suggate, R. P., 'Tectonic and earthquake risk zoning in New Zealand', *Proc. 3rd World Conference on Earthquake Engineering, New Zealand*, **1**, I–107 to I–124 (1965).
4. Bolt, B. A., 'Causes of earthquakes', in *Earthquake engineering*, (Ed. R. L. Wiegel), Prentice-Hall, Englewood Cliffs, N.J., 1970, Chap. 2.
5. Housner, G. W., 'Important features of earthquake ground motion', *Proc. 5th World Conference on Earthquake Engineering, Rome*, **1**, CLIX–CLXVIII (1973).
6. Bonilla, M. G., 'Surface faulting and related effects', in *Earthquake Engineering*, (Ed. R. L. Wiegel), Prentice-Hall, Englewood Cliffs, New Jersey, 1970.
7. Mogi, K., 'Relationship between the occurrence of great earthquakes and tectonic structures', *Bulletin of Earthquake Research Institute, University of Tokyo*, **47**, Part 3, 429–451 (May, 1969).
8. Gubin, I. E., 'Earthquakes and seismic zoning', *Bulletin International Institute of Seismology and Earthquake Engineering*, **4**, pp 107–126 (1967).
9. Gutenberg, B., and Richter, C. F., *Seismicity of the earth*, Hafner, New York, 1965.
10. Karnik, V., *Seismicity of the European area*, Reidel, Dordrecht, Part 1, 1969, and Part 2, 1971.
11. Lomnitz, C., *Global tectonics and earthquake risk*, Elsevier, Amsterdam, 1974.
12. United States Dept. of Commerce, National Earthquake Information Center, *World Seismicity 1961–1969*, (15 maps), Washington, D.C., 1970.
13. Algermissen, S. T., 'Seismic risk studies in the United States', *Proc. 4th World Conference on Earthquake Engineering, Santiago*, Vol. **1**, Part. A-I, 14–27 (1969).
14. Carmona, J. S., and Castano, J. C., 'Seismic risk in South America to the south of 20°', *Proc. 5th World Conference on Earthquake Engineering, Rome*, **2**, 1644–1653 (1973).
15. Brooks, J. A., 'Seismicity of the Territory of Papua and New Guinea', *Proc. 3rd World Conference on Earthquake Engineering, Auckland*, **1**, III–15 to III–26 (1965).
16. Kaila, K. L. and Narain, H., 'A new approach for the preparation of quantitative seismicity maps', *Bulln. Seismological Society of America*, **61**, No. 5, 1275–1291 (Oct. 1971).
17. Evernden, J. F., 'Study of regional seismicity and associated problems', *Bulletin Seismological Society of America*, **60**, No. 2, April, 1970, 393–446 (April, 1970).
18. Scholz, C. H., 'The frequency-magnitude relationship of microfracturing in rock and its relationship to earthquakes', *Bulln. Seismological Society of America*, **58**, No. 2, 399–415 (April 1968).

19. Cornell, C. A., 'Engineering seismic risk analysis', *Bulln. Seismological Society of America*, **58**, No. 5, 1538–1606 (Oct. 1968).
20. Rascón, O. A., and Cornell, A. C., 'A physically based model to simulate strong earthquake records on firm ground', *Proc. 4th World Conference on Earthquake Engineering, Chile*, **1**, A–1, 84–96 (1969).
21. Newmark, N. M., and Rosenblueth, E., *Fundamentals of earthquake engineering*, Prentice-Hall, 1971.
22. Seed, H. B., Idriss, I. M., and Kiefer, F. W., 'Characteristics of rock motions during earthquakes', *Jnl. Soil Mechanics and Foundations Division, ASCE*, **95**, No. SM5, 1199–1218 (Sept. 1969).
23. Donovan, N. C., 'A statistical evaluation of strong motion data including the February 9, 1971 San Fernando earthquake', *Proc. 5th World Conference on Earthquake Engineering, Rome*, **1**, 1252–1261 (1973).
24. Esteva, L., 'Geology and predictability in the assessment of seismic risk', *Proc. 2nd Int. Conf. Assoc. Eng. Geologists, Sao Paolo* (1974).
25. Ambraseys, N. N., 'Dynamics and response of foundation materials in epicentral regions of strong earthquakes', *Proc. 5th World Conference on Earthquake Engineering, Rome*, **1**, CXXVI–CXLVIII (1973).
26. Benioff, H., Unpublished report to A. R. Golze, Chief Engineer, Department of Water Resources Consulting Board for Earthquake Analysis, U.S.A. November, 1962.
27. Neville, A. M., and Kennedy, J. B., *Basic statistical methods for engineers and scientists*, International Textbook Co., Pennsylvania, 1964.
28. Weatherburn, C. E., *A first course in mathematical statistics*, Cambridge University Press, 1968.
29. Benjamin, J. R., and Cornell, C. A., *Probability, statistics and decision for civil engineers*, McGraw-Hill, New York, 1970.
30. Rosenblueth, E. 'Analysis of risk', *Proc. 5th World Conference on Earthquake Engineering, Rome* **1**, CIL–CLVIII (1973).

Chapter 3

Site response to earthquakes

3.1 INTRODUCTION

The evaluation of the response of a site to earthquakes is one of the most important and difficult parts of the design process. For this to be done it is first necessary to define the design earthquake (or earthquakes) by methods such as those described in Chapter 2, and then to judge how the site may respond to such an earthquake. In this chapter the effects of local geology and soil conditions will be considered. This includes not only the determination of the individual dynamic properties of different soil types, but also the means of determining the overall seismic response of construction sites. Finally various means of selecting ground motion input for dynamic response analyses of soil and structures are discussed.

3.2 LOCAL GEOLOGY AND SOIL CONDITIONS

In many earthquakes the local geology and soil conditions have had a profound influence on site response. The term 'local' is a somewhat vague one, generally meaning local compared to the total terrain traversed between the earthquake focus and the site. On the assumption that the gross bedrock vibration will be similar at two adjacent sites, local differences in geology and soil produce different surface ground motions at the two sites. Factors influencing the local modifications to the underlying motion are the topography and nature of the bedrock and the nature and geometry of the depositional soils. Thus the term 'local' may involve a depth of a kilometre or more, and an area within a horizontal distance of several kilometres from the site.

Soil conditions and local geological features affecting site response are numerous and some of the more important are now discussed with reference to Figure 3.1.

(i)   The greater the horizontal extent (L_1 or L_2) of the softer soils the less the boundary effects of the bedrock on the site response. Mathematical modelling is influenced by this, as discussed in Section 3.4.3.1.

44

45

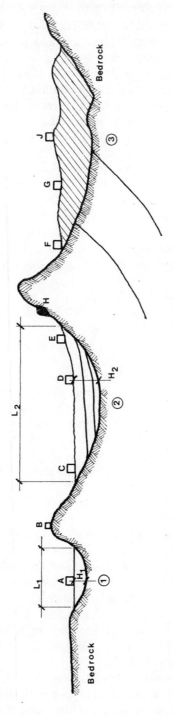

Figure 3.1 Schematic diagram illustrating local geology and soil features

(ii)   The *depth* (H_1 or H_2) *of soil overlying bedrock* affects the dynamic response, the natural period of vibration of the ground increasing with increasing depth. This helps determine the frequency of the waves filtered out by the soils and is also related to the amount of soil-structure interaction that will occur in an earthquake (Section 5.5.3.1). The Mexico City earthquake of 1957 witnessed extensive damage to long-period structures in the area of the city sited on deep (> 1000 m) compressible alluvium.[1] The natural tendency for long-period ground motions to be amplified in the structural response was intensified in this earthquake because the epicentral distance was quite large at 230 km. Another notable example of an earthquake where the fundamental period of structures which were damaged appeared closely related to depth of alluvium, was that in 1967 at Caracas.[2] Again long-period structures were damaged in areas of greater depth of alluvium.

(iii)   The *slope of the bedding planes* (valleys 2 and 3 in Figure 3.1) of the soils overlying bedrock obviously affects the dynamic response; but it is less obvious how to deal with non-horizontal strata (end of Section 3.4.3.1).

(iv)   *Changes of soil types horizontally* across a site (sites F and G in Figure 3.1) affect the response locally within that site, and may profoundly affect the safety of a building straddling the two soil types.

(v)   The *topography of both the bedrock and the deposited soils* has various effects on the incoming seismic waves such as reflection, refraction, focusing and scattering. Unfortunately many of these effects will always remain suppositional; for instance, while focusing effects in bedrock (valleys 1 and 2 in Figure 3.1) may be amenable to calculation, how are the response modifications at sites G and J to be reliably predicted due to these effects in valley 3?

It may well be that geological features such as hidden irregularities in the bedrock topography explain the otherwise unexplained differences of response observed at two nearby sites in the 1971 San Fernando earthquake.[3] At this time at two locations on the campus of the California Institute of Technology, the peak acceleration recorded at one site was 21 percent g while only 11 percent g was recorded at the other; whereas the local soil profiles at both locations were considered identical.

(vi)   Another topographical feature affecting response is that of *ridges* (Site B in Figure 3.1) where magnification of basic bedrock motion may occur (Section 3.4.2).

(vii)   *Slopes of sedimentary deposits* may of course completely fail in earthquakes. In steep terrain (Site H in Figure 3.1) failure may be in the form of avalanches. This occurred in the Northern Peru

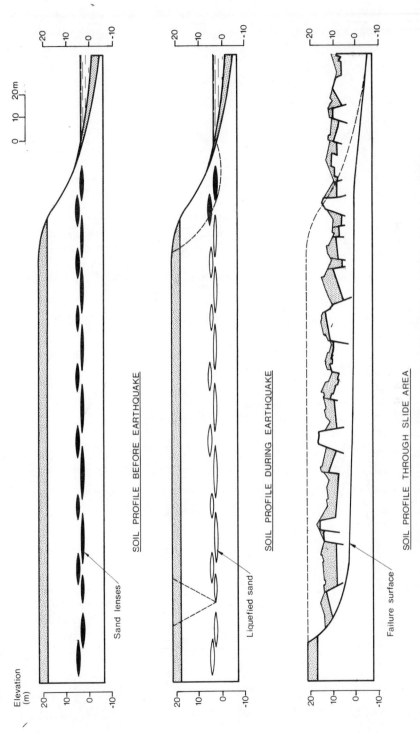

Elevation
(m)

0 10 20m

SOIL PROFILE BEFORE EARTHQUAKE

Sand lenses

SOIL PROFILE DURING EARTHQUAKE

Liquefied sand

SOIL PROFILE THROUGH SLIDE AREA

Failure surface

Figure 3.2 Conceptual development of Turnagain Heights landslide, Anchorage, Alaska, due to liquefaction of sand lenses (after Seed[5])

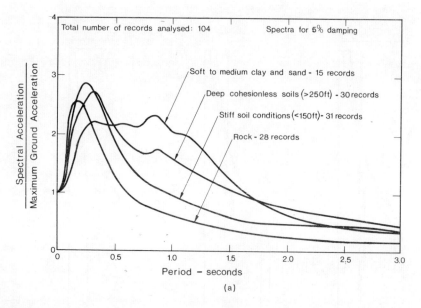

Figure 3.3a Average acceleration spectra for different sites conditions (after Seed *et al.*[8])

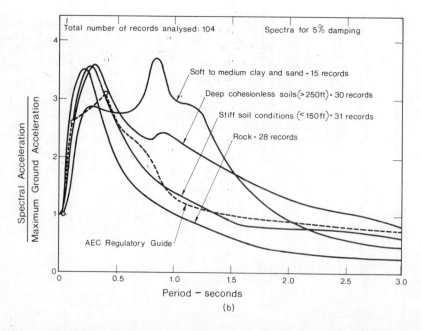

Figure 3.3b 84 percentile acceleration spectra for different site conditions (after Seed *et al.*[8])

earthquakes of May 31, 1970, in which whole towns were buried and about 20 000 people were killed[4] by one particular avalanche which travelled 18 km at speeds of 200–400 km/h.

(viii)   Spectacular soil failures can also occur in *gentle slopes*, as seen in the 1964 Alaskan earthquake[5] and again in the 1968 Tokachi-Oki earthquake.[6] The slope failures in the Alaskan earthquake were mostly related to liquefaction of layers of soil. For instance landslides occurred even in basically clay deposits (Figure 3.2) where liquefaction occurred in thin lenses of sand contained in the clay. In the Tokachi–Oki earthquake, some of the slope failures resulted from upper soil layers sliding on a slippery (wet) supporting layer of clay. This 'greasy back' situation could occur as illustrated in Figure 3.1, Site *E*.

(ix)   The *water content* of the soil is an important factor in site response. This applies not only to sloping soils as mentioned above, but liquefaction may also occur in flat terrain composed of saturated cohesionless soils (Section 5.5.2.2). Classical examples of failures of this type occurred in the Alaskan and Tokachi–Oki earthquakes referred to above, and in the much studied 1964 Niigata earthquake.[7]

(x)   Finally, the seismic response of a site and structures on it is of course a function of the local *soil types* and their condition. This is illustrated by the very different response spectra for different soils shown in Figure 3.3. The dynamic properties of individual soils are described in terms of mechanical properties such as shear modulus, damping, density and compactability as discussed in Section 5.5.2.

3.3 SITE INVESTIGATIONS AND SOIL TESTS

3.3.1 Introduction

For any construction project it is normal to carry out some investigations of the site, generally using fairly standardized operations in the field and in the laboratory such as drilling boreholes and carrying out triaxial tests. In this section only those investigating techniques related to the seismic response of soils are discussed.

The scope of the site investigations will depend on the site and on the budget and importance of the project, but in general it will be desirable to examine *to some degree* the factors relating to local geology and soil conditions discussed in Section 3.2. In Tables 3.1 and 3.2 the main variables in seismic site response have been related to some means for evaluating them. It is not proposed that these tables are exhaustive, but the field and laboratory test methods listed have been chosen because of their availability, reliability, or economy. For some parameters such as radiation damping and Poissons's ratio, no suitable tests for their evaluation exist.

Table 3.1. List of the main seismic soil factors with the most suitable tests used in their evaluation

		Field tests	Laboratory tests
Settlement of dry sands		Penetration resistance	Relative density
Liquefaction		Penetration resistance;	Relative density;
		Groundwater conditions	Particle size
Dynamic response parameters	Shear modulus	Shear wave velocity	Resonant column or cyclic triaxial
	Damping		Resonant column or cyclic triaxial
	Mass Density		Density
	Fundamental soil period	Vibration test	

Table 3.2. List of the best field and laboratory tests related to the evaluation of the seismic response of soils

Field determinations and tests	Related to
Soil distribution and layer depth	Response calculations
Depth to bedrock	Response calculations
Groundwater conditions	Response calculations and liquefaction
Penetration resistance	Settlement and liquefaction
Shear wave velocity	Shear modulus
Fundamental period of soil	Response calculations

Laboratory tests	
Particle size distribution	Liquefaction
Relative density	Liquefaction and settlement
Cyclic triaxial	Shear modulus and damping
Resonant column	Shear modulus
Unit mass	Response calculations

For the description of the dynamic behaviour of soils see Section 5.5, where the main dynamic design parameters such as shear modulus and damping are defined. The application of the results of the site investigation to soil response and design problems may be found in various parts of the text, namely Sections 3.4, 4.5, 5.5 and 5.6.

3.3.2 Field determination and tests of soil characteristics

A brief description of the nature, applications and limitations of those site investigations pertaining to seismic behaviour of soils as listed in Tables 3.1 and 3.2 now follows.

3.3.2.1 Soil distribution and layer depth

Standard borehole drilling and sampling procedures are satisfactory for determining layer thicknesses for most seismic response analysis purposes as well as for normal foundation design. In the upper 15 metres of soil, sampling is usually carried out at about 0·75 or 1·5 metre intervals; from 15–30 metres depth, a 1·5 metre interval may be desirable; while below 30 metres depth, 1·5 or 3·0 metres may be adequate depending on the soil complexity. If the site may be prone to liquefaction or slope instabilities, thin layers of weak materials enclosed in more reliable material may need to be identified, requiring more frequent or continuous sampling in some cases.

The depth to which the deepest boreholes are taken will depend as usual on the nature of the soils and of the proposed construction. For instance, for the design of a nuclear power plant on deep alluvium, detailed knowledge of the soil is required to a depth of perhaps 200 m, while general knowledge of the nature of subsoil will be necessary down to bedrock or rock-like material.

3.3.2.2 Depth to bedrock

For use in response calculation a knowledge of the depth to bedrock or rock-like material is essential. Beyond the ordinary borehole depth of 50–100 m, bedrock may be determined from geophysical refraction surveys, preferably checked by reference to information from geological records, artesian water or oil boreholes where available. In areas of deep overburden, for seismic response purposes the depth at which bedrock or equivalent bedrock is reached may have to be defined fairly arbitrarily, for example on some sites it may be reasonable to say that equivalent bedrock is material for which the shear wave velocity at low strains (0·0001 percent) is $v_s \geqslant 750$ m/s, where such material is not underlain by materials having significantly lower shear wave velocities.[21] In California on a typical site, effective bedrock would be found within 30 m of the surface, while on virtually all Californian sites it would be found within 150 m depth. The order of accuracy of bedrock depth determination as currently required for seismic response calculations is as follows:

Bedrock depth (m)	Approximate accuracy (m)
0–30	1·5
30–60	1·5–3·0
60–150	6–15
150–300	15–30
> 300	60

The large errors permissible in the measurement of deeper bedrock reflects the great approximations made in soil response analyses at the present time.

3.3.2.3 Groundwater conditions

Adequate standard borehole installations are available for accurately measuring groundwater conditions at any site. For response calculations this information is used indirectly through effective confining pressures as they affect both shear modulus and damping of the soil. Those sites which are most susceptible to liquefaction have their water table within three metres of the surface, while sites with water tables within about eight metres of ground level may also be potentially liquefiable depending on other soil parameters.

3.3.2.4 Penetration resistance tests

The penetration resistance test is really an indirect means of determining the relative density or degree of compaction of granular deposits. It is therefore an important factor in the study of settlement and liquefaction of soils in earthquakes. Because it can be carried out simply, frequently and cheaply as part of routine subsoil investigations, it is probably preferable to the direct laboratory test for determining relative density.

Two basic types of penetrometer are in common use for penetration tests, namely hollow tube samplers and cone penetrometers. Both types may be either driven by a falling weight (dynamic method) or driven by a static load into the undisturbed soil at the bottom of the borehole as drilling proceeds. In America and some other countries the preferred method is the standard penetration resistance test (SPT) which is a dynamic method having the advantage of sample recovery. The static cone tests, particularly the Dutch cone, have found favour in some countries because of the greater consistency of results deriving from the simple static load application. This advantage is offset, however, because the cone test does not recover samples, so that no visual examination of the material being tested is possible.

When using the results of penetration tests for assessing the condition of granular soils, they may in some cases be used directly (for example in

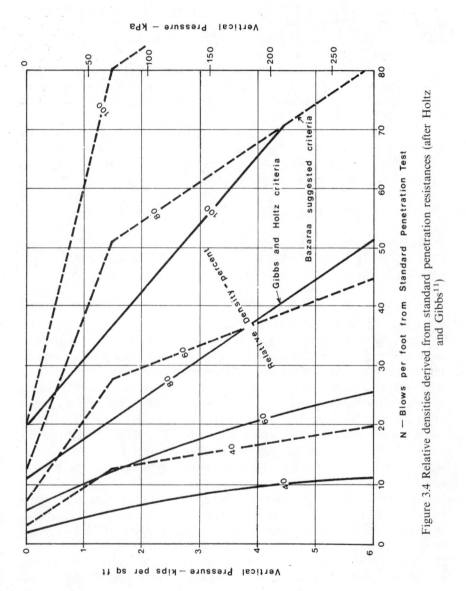

Figure 3.4 Relative densities derived from standard penetration resistances (after Holtz and Gibbs[11])

Figure 5.21) or else indirectly, i.e. after conversion to relative density (Table 5.5). As the various penetrometer tests yield different numerical results for the same soils, the exact type of equipment used in each case must be known and appropriate conversions made where necessary for assessing results. For example Schmertmann[10] related the static cone penetration resistance, Q_c (kg/cm^2), to standard penetration resistance (blows/foot) for fine sands, but the relationship has been found to vary with grain size.

It is particularly important to bear in mind the large scatter of results obtained using all penetration tests; therefore penetrometer readings should be used to establish trends of soil compaction rather than be considered as absolute values.

For conversion of results of the American standard penetration test (ASTM designation: D1586–67) to relative density values, Figure 3.4 shows the correlation given by Holtz and Gibbs[11] and Bazaraa.[12] The Holtz and Gibbs criteria appear to be the more widely accepted.[13]

3.3.2.5 Field determination of shear wave velocity

Although the shear wave velocity is used directly in response analyses (Section 5.5.3.1), it may be thought of mainly as a means of determining the shear modulus G of a soil (Section 5.5.2.3) from the relationship

$$G = \rho v_s^2 \tag{3.1}$$

where ρ is the mass density of the soil.

In the geophysical method of determining v_s low energy waves are propagated through the soil deposit, and the shear wave velocity is measured directly. Three techniques using boreholes are illustrated in Figure 3.5.

In each case waves are generated by an explosive charge or a hammer and the time of first arrival of the shear wave travelling from energy source to geophone is recorded. Difficulties in interpreting results arise from uncertainties in separating the first arrival of shear waves from the faster travelling longitudinal waves. Unfortunately these latter P-waves are not suitable for shear modulus calculations as they are greatly influenced by the presence of groundwater, whereas shear waves are not.

The cross-hole technique shown in Figure 3.5 measures shear wave velocities horizontally between adjacent boreholes, and is clearly well suited to response calculations of reasonably homogeneous or thick strata. With thinly bedded deposits, various routes may be taken by waves between source and geophone and the interpretation of arrival times is more problematical and should be viewed with caution. When using the up-hole and down-hole techniques of Figure 3.5 the different wave types can be distinguished more easily, but care must be taken to deal with misleading local borehole effects. For example, where casing has to be used in a borehole, the waves transmitted by the casing may disguise the slower and weaker signals in the soil and experienced resolution of the results is required.

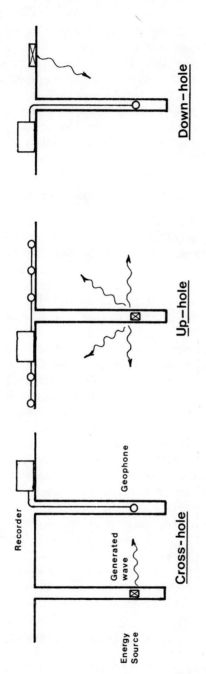

Figure 3.5 Geophysical methods of evaluating shear wave velocity

The above geophysical methods of determining v_s are the most applicable field procedures because they involve a large mass of soil, they can be carried out in most soil types, and they permit v_s to be determined as a function of depth. Furthermore their cost is reasonable and in many countries the necessary equipment is available. Because these tests are only feasible at low levels of soil strain of 10^{-5}–10^{-3} percent, compared with design earthquake strains of about 10^{-3}–10^{-1} percent, values of shear modulus calculated from these values of v_s will be scaled down for seismic response purposes (Section 5.5.2.3). It is also normal to compare values of G computed in this manner with values determined from laboratory tests as discussed in Section 3.3.3.

3.3.2.6 Field determination of fundamental period of soil

A knowledge of the predominant period of vibration of a given site is helpful in assessing a design earthquake motion (Section 3.4) and the vulnerability of the proposed construction to earthquakes (Sections 4.2.6 and 5.5.3). Many attempts have been made to measure the natural period of vibration of different sites; the vibrations measured have generally been microtremors, some arising from small earthquakes[9] or those induced artificially such as by explosive charges, pile driving, passing trains or nuclear test explosions.[14,15]

For an important or seismically vulnerable project, a vibration test may well be warranted, but problems of interpretation of results arise as such tests involve much lower magnitudes of soil strain than occur in design earthquakes. If a local correlation between soil periods in strong motion earthquakes and periods recorded during microtremors does not exist, cautious comparisons with strong motion results on similar soils in different areas will have to be made. In the case of vibration tests carried out for the Parque Central Development in Caracas,[14] the measured periods were increased by 50 percent in order to convert the microtremor behaviour into strong ground motion. This adjustment factor was derived through comparison of studies of the 1967 Caracas earthquake with the site tests.

It should be noted that the fundamental period of the soil will generally be between about 0·2 and 4·0 seconds, depending on the stiffness and depth of the soils overlying bedrock.

A review of various microtremor recording techniques, and a detailed discussion of a particular method used in New Zealand are given in two papers by Parton and Taylor.[15,16]

3.3.3 Laboratory tests relating to dynamic behaviour of soils

A brief description of the nature, applications, and limitations of the laboratory tests relating to the dynamic behaviour of soils, as summarized in Tables 3.1 and 3.2, is set out below.

3.3.3.1 Particle size distribution

This soil property is related to the liquefaction of saturated cohesionless soils as discussed in Section 5.5.2.2. As the test for its determination is a standard laboratory procedure, it will not be described here. Although a number of classifications of grain size and standard sieves exist, correlations are straightforward, so that use of any scale of sizes can easily be applied to the liquefaction potential graph shown in Figure 5.22 which incorporates an American sieve grading.

3.3.3.2 Relative density test

The *in situ* relative density or degree of compaction is helpful in determining the likely settlement of dry sands and the liquefaction potential of saturated cohensionless soils is earthquakes (Sections 5.5.2.1 and 5.5.2.2). As this property has a significant influence on the dynamic modulus,[13] it indirectly relates to response analyses. Relative density for the void ratio must also be assessed in order to reproduce field conditions in samples which are recompacted in the laboratory for cyclic loading tests. As is well known by soils engineers, larger scatter occurs in the results of relative density tests, the chief reason being the virtual impossibility of retrieving reliable undisturbed samples of granular deposits.

The relative density may be found from either

$$D_r = \frac{e_{max} - e}{e_{max} - e_{min}} = \frac{\rho_{max}(\rho - \rho_{min})}{\rho(\rho_{max} - \rho_{min})} \tag{3.2}$$

where e_{max} and e_{min} are the maximum and minimum void ratios, e and ρ are the natural (*in situ*) void ratio and unit mass respectively, and ρ_{max} and ρ_{min} are the maximum and minimum unit mass.

In the laboratory e, the void ratio of the undisturbed sample is first determined by measuring the appropriate quantities in

$$e = \frac{G\rho_w}{\rho_d} - 1 \tag{3.3}$$

where G is the specific gravity of the solids, ρ_w is the unit mass of water and ρ_d is the dry unit mass of the sample.

The minimum mass density may be found by pouring oven-dry material gently through a funnel into a mould, using a method such as the American one designated ASTM:D2049—69. For reasonably clean sands this method is reliable.

More difficulty is experienced in determining the maximum density ρ_{max} with equal consistency, different methods of compaction giving modestly different results. Vibratory compaction techniques seem better for uniform sands with few fines, while impact methods seem better for sands with more fines.

Vibration and impact techniques generally used in America comply with ASTM tests designated D2049—69 and D1557—70 respectively.

If the percentage passing the 200-mesh sieve exceeds approximately 15 percent, laboratory determination of relative density is of doubtful validity. In this case more reliance will have to be made upon the penetration resistance tests as a measure of relative density as discussed in Section 3.3.2.

3.3.3.3 Cyclic triaxial test

This test is one of the best laboratory methods at present available for determining the shear modulus and damping of cohesive and cohesionless soils for use in dynamic response analyses (Section 5.5.2.3). In this test cyclically varying axial compression is applied to a cylindrical specimen (Figure 3.6), and the compressive stress–strain characteristics are measured directly.[17,18] The compressive modulus E so obtained is converted to the shear modulus G using the relationship

$$G = \frac{E}{2(1 + v)}$$

where v is Poisson's ratio. The damping ratio may also be obtained from this test from the resulting hysteresis diagram as illustrated in Figure 5.23. Depending on the range of strains produced in the test, any desired level of strain may be chosen for plotting the hysteresis loops.

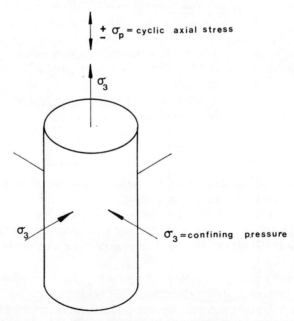

Figure 3.6 Cyclic triaxial test

As well as having the facility for applying a variety of stress conditions, the cyclic triaxial test has the advantages that it can be applied to all types of soils except gravel, that the test equipment is widely available and precise in its control, and that testing is comparatively cheap. The disadvantages of this test are related to its inability to reproduce the stress conditions found in the field, i.e. that the cyclic shear stresses are not applied symmetrically in the test, that zero shear stresses are applied in the laboratory with isotropic rather than anisotropic consolidation and also that the test involves deformations in the three principal stress directions, whereas in earthquakes the soil in many cases is thought to be deformed mainly unidirectionally in simple shear.

Cyclic shear tests are carried out at high strains (10^{-2}–5 percent) equal to and larger than the strains occurring in strong earthquakes; since geophysical test involve low strains, values of G at intermediate strains may be determined by interpolating between G values found from these different methods, but as there is no overlap between the strains occurring in these two tests cross-checking between the field and laboratory methods is not possible. It is also to be noted that in the use of this test to determine soil damping characteristics, no field method of evaluating damping is as yet available for comparison, and hence any values of damping coefficient obtained should be treated with appropriate caution.

3.3.3.4 Resonant column test

This test provides a good alternative to the cyclic tri-axial test for the laboratory determination of shear modulus of most soils. A cylindrical column of soil is vibrated at small amplitudes on one end, either torsionally or longitudinally (Figure 3.7), varying the frequency until resonance occurs.

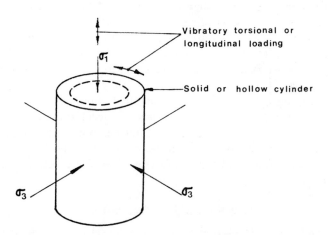

Figure 3.7 Resonant column test

Wilson and Dietrich[19] proposed that the shear or compression modulus for a solid cylinder may be found from

$$G \text{ or } E = 1\cdot59 \times 10^{-8} f^2 h^2 \rho \, (\text{MN/mm}^2) \qquad (3.4)$$

where h is the height of the soil cylinder (mm), ρ is the unit mass of soil (Mg/m^3), and f is the resonant frequency of torsional vibration in cycles per second when determing G, or the resonant frequency of longitudinal vibration in cycles per second when determining E.

It will be seen that by determining E and G separately from these tests, a value of Poisson's ratio v may be determined, but as this test involves low strain and no suitable extrapolation method exists, such values of v are not suitable for most earthquake engineering purposes. Although this test has the disadvantage of being carried out at low strains (10^{-2}–10^{-4} percent), it has the advantages of simplicity, cheapness of equipment, and applicability to most soil types.

Details of the equipment used in this test may be found elsewhere.[19,20]

3.4 DYNAMIC DESIGN CRITERIA FOR A GIVEN SITE

3.4.1 Introduction

In order to determine the effect of an earthquake on a given soil deposit or structure it will be necessary to choose an appropriate dynamic design criterion, which may be in the form of an input ground motion (accelerogram) or its equivalent response spectrum (see Figures 3.3 and 3.14 and Section 5.4). The basic information required in making this choice consists of defining the design earthquake as determined in Chapter 2, and knowing the geometry and dynamic properties of the soils at the site (Sections 3.2 and 3.3). As with other aspects of design the degree of detail entered into selecting dynamic input will depend on the size and vulnerability of the project.

In this section simple and complex methods of formulating dynamic design input are discussed, *but because there are still so many imponderables in this topic only the simpler methods will be warranted in most cases.* With the rapid growth in the use of dynamic analysis, an outline of the more sophisticated means of choosing dynamic loading is justified, although it is to be hoped that rapid improvements in these techniques will occur, especially with experience derived from the design problems of nuclear power stations.

In the following discussion it is convenient first to consider sites at which bedrock is at the surface and then to discuss sites where bedrock is overlain by softer deposits.

3.4.2 Sites with surface bedrock

At sites where the bedrock is at the surface, the selection of dynamic design criteria for a structure is limited to choosing a bedrock motion or response spectrum corresponding to the design earthquakes. Unfortunately this apparently straightforward task is at present made difficult because very few strong ground motions have as yet been recorded on rock sites. Accelerograms of this type have been recorded in the U.S.A. as follows:

Helena, Montana, Oct. 31, 1935

$$M = 6 \cdot 0 \, I_{max} = VIII$$

Golden Gate Park, San Francisco, Mar. 22, 1957

$$M = 5 \cdot 3 \, I_{max} = VII$$

Pacoima Dam, California, Feb. 9, 1971

$$M = 6 \cdot 6 \, I_{max} = XI$$

Although the above is not a complete list, insufficient recordings of bedrock motion have been made for a definitive study to be possible. Care is required in the application of individual existing recordings to other sites; for example at Pacoima Dam, peak horizontal acceleration was the very large value of $1 \cdot 17 \, g$. Part of this enormous surface acceleration may be explained by the amplification occurring due to the cracking of the rock below the instrument station, and there may also have been magnification of base bedrock motion arising from the location of the recording instrument on a steep ridge of the valley.[3,22] In any case at the present time it is difficult to see whether this particular ground motion record should be directly used elsewhere.

As so few bedrock strong ground motions have been recorded to date, it may be necessary to derive dynamic design criteria from random vibration theory as discussed in the next subsection.

3.4.2.1 Simulated accelerograms at surface bedrock

For most design purposes it can be assumed that ground motion is a random vibratory process, and that accelerograms can be mathematically simulated with random vibration theory. This will be most true at distances from the causative fault sufficient to ensure that the details of the fault displacement are not significant in the ground shaking. Because of the scarcity of actual bedrock recordings, at present the modelling of simulated earthquakes is necessarily based on the more numerous accelerograms recorded on softer soils. This is considered reasonable as there is much to suggest that the main difference between bedrock and soft-soil motions is one of frequency content; this difference can be dealt with in the simulation process.

By far the most common pattern of ground motion is one of an abrupt transition from zero to maximum shaking, followed by a portion of more or less uniformly intense vibration, and finally a rather gradual attenuation (Figure 3.15). In the terminology of random vibration theory, the middle portion may be considered as a stationary random process, whereas the initial and final phases, being transitional, are nonstationary. At the present time no other type of ground motion is seriously considered as a model for deriving earthquake loads. A short sharp jolt such as experienced at Agadir (1960) and Skopje (1964) is the only type of earthquake, essentially different from the random process described above, that has so far been identified.

Accelerograms have been simulated with mathematical models of varying complexity, including

(i)   nonstationary processes of various types[23,24,25]

(ii)   stationary processes such as the white noise type[26,27,28] and the Gaussian white noise type.[29,30]

(i) *Nonstationary random processes.* As only the middle portion of strong ground motion can be reasonably described as a stationary process, it has been argued that the shaking of smaller earthquakes and the tail of larger shocks are most accurately modelled by nonstationary processes. As some of the latter become mathematically very complex, it is fortunate that one of the simplest types of nonstationary models seems satisfactory. This consists of a stationary process multiplied by a nonstationary envelope function. The stationary process is usually derived from a segment of band-limited or filtered white noise;[31] this means that the frequency content has been limited to a prescribed band, or that the ensemble average Fourier spectrum has a prescribed shape which may be deduced from actual records.[32]

Various forms of the stationary process have been taken such as the Gaussian random process used by Jennings *et al.*[25] and Ruiz and Penzien,[33] or the filtered Poisson process used by Amin and Ang.[24] In these cases the nonstationarity was achieved with an amplitude envelope as mentioned in the preceding paragraph. Figure 3.8 illustrates the amplitude envelopes used by Jennings *et al.* in modelling their four earthquakes, which ranged in magnitude from about 5·0 to 8·0.

For modelling ground motion *in the vicinity of the causative faults* Housner[31] suggests characteristics related to the types of faulting discussed above in Regional Geology (Section 2.2).

(ii) *Stationary random processes.* Some workers consider that as so few records of bedrock motion exist, it is reasonable to carry out strong motion studies using simple stationary processes in the form of banded white noise segments to simulate accelerograms.[26,28,34] Parton[34] found good agreement when comparing the velocity response spectra for three white noise segments with Housner's averaged response spectra (Figure 3.9). Any inaccuracy involved in not modelling the tail of the earthquake may be considered tolerable in relation to other important simplifications which are usually made in analysis, such as the assumption of elastic material behaviour. In this way the use of segments of white noise for the motion of bedrock overlain by softer deposits is arguably even more justifiable, as the simplifying assumptions in dynamic soil analyses are gross compared to those in structures.

63

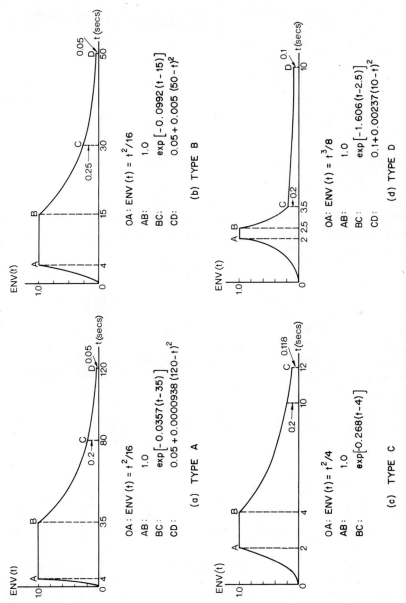

OA : ENV (t) = $t^2/16$

AB :   1.0

BC : $\exp\left[-0.0357(t-35)\right]$

CD : $0.05 + 0.0000938\,(120-t)^2$

(a) TYPE A

OA : ENV (t) = $t^2/16$

AB :   1.0

BC : $\exp\left[-0.0992(t-15)\right]$

CD : $0.05 + 0.005\,(50-t)^2$

(b) TYPE B

OA : ENV (t) = $t^2/4$

AB :   1.0

BC : $\exp\left[-0.268(t-4)\right]$

(c) TYPE C

OA : ENV (t) = $t^3/8$

AB :   1.0

BC : $\exp\left[-1.606(t-2.5)\right]$

CD : $0.1 + 0.00237(10-t)^2$

(d) TYPE D

Figure 3.8 Envelope functions for nonstationarity of simulated earthquakes based on typical Californian earthquakes (after Jennings et al.[25] by permission of the California Institute of Technology)

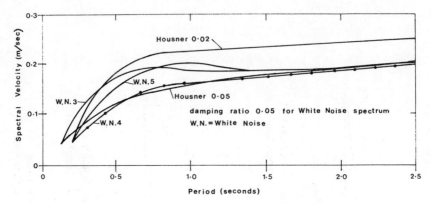

Figure 3.9 Comparison of Housner's averaged curves with response spectra from three white noise segments (after Parton[34])

Band-limited white noise has the practical advantages that it is simple to generate and use in analysis, and programs for its generation are readily available in computer centres.

The three main characteristics of an accelerogram requiring consideration in the simulation process are peak acceleration, duration and frequency content. These are functions of the design earthquake, particularly its magnitude and distance as discussed in Chapter 2. As well as taking account of the preceding factors, simulated accelerograms should be integrated to check that realistic peak velocities and displacements occur.

In determining the duration of the simulated earthquake a white noise segment is clearly equivalent to the region of strongest shaking of a real

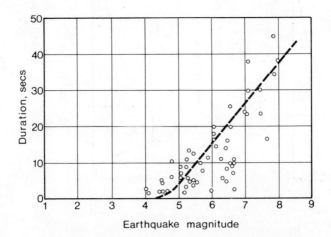

Figure 3.10 Duration of strong shaking as a function of earthquake magnitude (after Donovan[35]. Reproduced by permission of Dames & Moore)

earthquake. As the tails of most earthquakes subside gradually the cut-off point of strong shaking is clearly fairly arbitrary, depending on the definition of 'strong'. Donovan[35] has recently confirmed with more data a working relationship (formerly proposed by Housner[36]) between the duration of strong shaking and earthquake magnitude as shown in Figure 3.10. In elastic analyses the duraton of excitation is relatively unimportant, except that in order to determine peak response the duration should be several times the fundamental period of the system.

When considering relatively short epicentral distances, the nominally uniform frequency content of white noise seems at least as appropriate as more refined models of bedrock motion. At great epicentral distance the attenuation of high frequency vibration may be worth consideration. This trend is illustrated in Figure 2.12(b) which shows the dominant period of bedrock vibration increasing with epicentral distance. Some scaling of the response spectra resulting from the simulated accelerograms may be warranted to ensure that peak acceleration response occurs at periods corresponding roughly to Figure 2.12(b). But at present the data is so sparse and its scatter so great that the wisdom of this procedure is in doubt, particularly for shorter epicentral distances. In any case it is recommended that a number of different white noise accelerograms be used in any response analysis, and particularly in response spectrum analyses the averaged spectrum should be used (Section 3.4.4).

3.4.3 Sites with subsurface bedrock

In deriving dynamic design criteria for a site with subsurface bedrock, the effect of the softer overlying soil layers on the basic bedrock motion must be taken into account, as illustrated in Figure 3.3.

The dynamic input used will depend on the analytical problem posed by the site and the structure, and on the technical and financial resources available. In order to illustrate this, consider the three analytical situations indicated in Figure 3.11. Site (i) represents the general site evaluation problem; here the stability of the overburden in earthquakes is to be determined with regard to phenomena such as settlement, liquefaction or landslides, in relation to the feasibility of future construction on this site or the safety of adjacent sites.

For the dynamic response analysis of this site, an accelerogram or response spectrum must be applied at B1 to the soil system between B1 and S1. This necessitates the choice of a suitable bedrock motion, and it is proposed that this should be done in the same manner as described in the preceding section concerning surface bedrock. Attempts have also been made to compute bedrock motion from surface motion at another site, as discussed in the following section.

As even fewer strong motion recordings have been made on subsurface bedrock than surface bedrock, the same justifications of simulated earthquakes apply in each case. When sufficient recordings from near-by subsur-

66

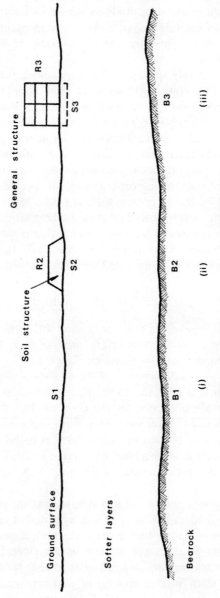

Figure 3.11 Schematic diagram of sites with subsurface bedrock

face and surface bedrock motions have been made, some basic difference may be proven, but it seems likely that the difference will be slight.

Sites (ii) and (iii) in Figure 3.11 represent any site with any structure. The dynamic analysis of a structure on such a site may be carried out in either of two ways. Firstly, the total soil and structure system from bedrock to the top of the structure may be analysed together by applying bedrock motion at B2, B3 and determining the responses of the whole system including that of the structure R2, R3. This is the ideal means of analysis, as full allowance for interaction between *in situ* soils and constructed soils is included. The dynamic input at bedrock is chosen as for Site (i). Or secondly, the structure may be analysed by applying a dynamic input at its base (S2, S3) or at some arbitrary distance below ground surface. The dynamic input appropriate for application at S2, S3 may be either: (1) surface motion accelerograms or spectra derived specially for the site by computing the modifications caused by the overlying soils on the bedrock motions input at B2, B3, or (2) surface motion accelerograms or spectra derived without specific dynamic analysis of the soil layers. These two alternative methods of deriving surface motion dynamic design criteria are now discussed in turn.

3.4.3.1 Surface motion criteria derived from bedrock motion

Surface motion may be estimated from bedrock motion by a dynamic response analysis of the site. Referring to Figure 3.11 (ii) and (iii), both the input at bedrock (B2, B3) and the output at the surface (S2, S3) may be in the form of accelerograms or response spectra. The bedrock motion is chosen as for surface bedrock (Section 3.4.2).

At the present time the most common and probably most practical technique for modelling the dynamic behaviour of the soil above bedrock is that of the vertical shear beam model, which is so-called because of the use of shear wave theory. Several types of errors or limitations apply to the shear beam model as discussed below.

(a) Errors arise in representing a three-dimensional problem by a one-dimensional model.[37]

(b) Nearly all shear beam models assume linear material behaviour as a crude approximation to the real non-elastic behaviour.

(c) Errors arise from the use of viscous rather than hysteretic damping.

(d) Errors from the use of approximate mathematical solutions.

(e) The shear beam model is valid only for sites where ground motion is dominated by shear waves propagating vertically through the soil. It is argued[38] that this is reasonably true at many sites related to the earthquake focus as illustrated in Figure 3.12, but is more likely to be valid at special sites like Mexico City rather than at locations where normal thicknesses of overburden exist. The shear waves will be approximately vertical for deep focus earthquakes, but this will

68

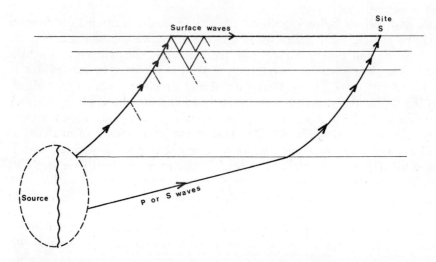

Figure 3.12 Schematic relationship of source, travel paths and site as assumed in one-dimensional shear wave studies

not be true at sites near the source of shallow earthquakes. At some distance from shallow-focus earthquakes the significant seismic waves may approach local soils horizontally; for this case it has been suggested that the shear beam model should be used only if the half-wave length $\lambda/2$ of incoming waves is large compared to the lateral extent of the soil layers (Figure 3.13).

(f)   For the shear beam model to be applicable the boundaries of the site must be essentially horizontal, allowing the soil profile to be treated as a series of semi-infinite layers.

(g)   Finally the effect of the presence of the proposed structure (or other structures) is not readily included in the computation of surface motion. For further discussion of the soil-structure interaction problem see Section 5.5.3.

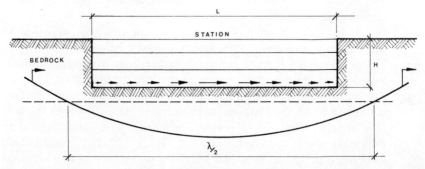

Figure 3.13 Waves arriving horizontally from shallow focus earthquake. Shear-beam site response theory applicable only if $\lambda/2 \geqslant L$ (after Tsai[38])

Two main types of vertical shear beam model are in use, firstly the lumped mass methods and secondly continuous solutions in the frequency domain. The chief characteristics of each of these methods are now discussed briefly.

(i) In *the lumped mass model* the soil profile is idealized with discrete mass concentrations interconnected by stiffness elements which represent the structural properties of the soil. A time-step modal analysis is commonly used because of the consequent computer economies and because of the familiarity of modal superposition to earthquake engineers. In modal analysis it is necessary to assume linear material behaviour and viscous damping. As the damping of soils is more nearly hysteretic it is common to use an *equivalent* viscous damping which assumes constant damping in all modes, rather than true viscous damping in which the critical damping ratio would increase in proportion to the natural frequency of each mode.

Further allowances for the non-linearity of soil behaviour and hysteretic damping have been made by Seed and Idriss,[39] who used an iterative procedure to adjust the soil properties according to the level of strain. Even so they make the considerable simplification of averaging the properties of all layers in the soil profile. Further discussion on the problems involved in determining suitable soil properties for use in modal solutions may be found in Section 5.5.2.3 and in a paper by Whitman *et al.*[40]

(ii) A *continuous solution in the frequency domain* provides an 'exact' alternative to the lumped mass treatment of the vertical shear beam model. In this method the transfer of the bedrock motion to the surface is derived by consideration of the equation of motion of one-dimensional wave propagation in a continuous medium,

$$\rho \frac{\partial^2 u}{\partial t^2} = G \frac{\partial^2 u}{\partial x^2} + \eta \frac{\partial^3 u}{\partial t \, \partial x^2} \tag{3.5}$$

where ρ is the density of a semi-infinite soil layer, G = shear modulus, η the viscosity constant, and $u(x, t)$ the displacement of a point in the soil layer.

Transfer functions may be derived which modify input bedrock harmonic motions into corresponding surface motions in terms of the elastic properties of the intervening layers and of the bedrock layer itself. By multiplying the Fourier spectrum corresponding to the time-dependent bedrock motion by the transfer function, the surface Fourier spectrum is found. This Fourier spectrum may then be converted into the surface accelerogram. Fuller discussion of the continuous solution to the shear beam problem has been given by Roesset[41] and Schnabel *et al.*[42] Computer programs involving Fourier analysis and transfer functions[43] are simple and may be more economical than those using the lumped mass solution, if output at only a few points is desired. The continuous solution has the advantages that it readily handles many soil layers with different properties including the bedrock layer, and any linear damping may be used.

An interesting feature of the transfer function technique is the facility with which bedrock motions can be estimated from surface motion recorded at a given site. This is a useful source of input bedrock data at another site. The main problem with transferring surface motion at one site to surface motion at another site lies in the incorporation of *two* sets of errors implicit in modelling ground motion transfer downwards through one soil profile as well as upwards through another.

An example of a site response study[44] in which surface ground response spectra were computed from bedrock motion is illustrated in Figure 3.14. An artificial accelerogram for a magnitude 7·0 earthquake at an epicentral distance of 50 km was generated using a nonstationary process, and applied to each of the soil profiles shown. The surface accelerograms (not shown) were computed using the continuous solution mentioned above, and the corresponding response spectra are shown at the top of Figure 3.14 for two values of damping. An indication of the effect of the soil layers on the site response can be obtained by comparing the output response spectra with the response spectra for the input motion shown at the bottom of Figure 3.14.

Finally some mention should be made of those sites which in no way conform to the assumptions of the shear beam model. Clearly many sites have soil conditions which are not amenable to mathematical analysis, particularly irregular or heterogeneous deposits. Some situations such as valleys filled with reasonably uniform alluvium, or sloping but uniform layers, may be capable of analysis by finite elements.[45] The cost of the latter technique and the difficulty in interpreting the results will be a deterrent to its use except for research studies. It would appear that for the time being there are enough problems in deriving usable results from response analyses of more regularly distributed soils using the shear beam theory, without embarking on more difficult mathematical studies. For sites for which dynamic response analysis is inappropriate the surface ground motions may be determined as suggested in the next section.

3.4.3.2 Surface motion criteria derived without site response analysis

Dynamic analysis of the overlying softer soils is not always carried out for sites with subsurface bedrock, in order to determine dynamic design criteria for a structure. A soil response analysis may be too expensive or the site may not be amenable to present day analytical techniques. Earthquake accelerograms or response spectra appropriate to surface ground motion on a given site may be derived by:

(i)   choosing from records of similar real earthquakes, or
(ii) generating simulated accelerograms to match the design earthquake, or

(iii)   construction of response spectra from relationships between design earthquake magnitude, focal distance, spectral velocity and acceleration.[46,47] This method should be approached with great caution because of the difficulties involved in choosing suitable factors for amplification and ductility, or allowing for local soil conditions as illustrated in Figure 3.3, or

(iv)   deriving averaged or smoothed spectra from a set of earthquake records with various levels of confidence.[48]

The first two of the above methods are discussed briefly below.

(i)   In choosing from amongst real earthquake records (Section 3.4.5) it will be desirable to match as nearly as possible the design conditions of magnitude, epicentral distance, focal depth and soil profile with those of the real earthquakes. Close matching of magnitude and distance is desirable for minimizing scaling errors. Unfortunately not all of the above factors may be known or be readily available for the real events, and a second best criterion is that the records should come from a geologically similar area. Unfortunately it will not be possible to fulfil this latter condition in seismic areas which have few if any strong motion recordings of their own and little in common geologically with regions rich in accelerograms such as Japan and California. But even when soil conditions are reasonably matched, it is well known that each individual earthquake record has strong features characteristic only of that particular earthquake and site. To rely *only* on records of similar earthquakes is probably the least satisfactory method of choosing dynamic input for structural analysis.

Major difficulties arise over the choice of a real earthquake, with suitable peak accelerations, as some small earthquakes have much greater peak accelerations than large earthquakes. This produces misleading results in elastic response analyses, unless factors are applied to take account of the duration of strong shaking (Sections 5.1, 5.2).

(ii)   When simulating surface ground motion accelerograms at sites where softer soils overlie bedrock, the approach is similar to that for simulated accelerograms for surface bedrock (Section 3.4.2.1). The difference lies in the allowance made for the softer soils. When filtering the white noise, deducing the prescribed shape of the average Fourier spectrum from actual records may have more relevance than for surface bedrock, as most surface records have been made on softer soils. For the same reason nonstationary envelope functions such as shown in Figure 3.8 are also likely to be more relevant to simulated accelerograms for soft surface layers. And because of the filtering effects of softer layers, nonstationary processes are even

72

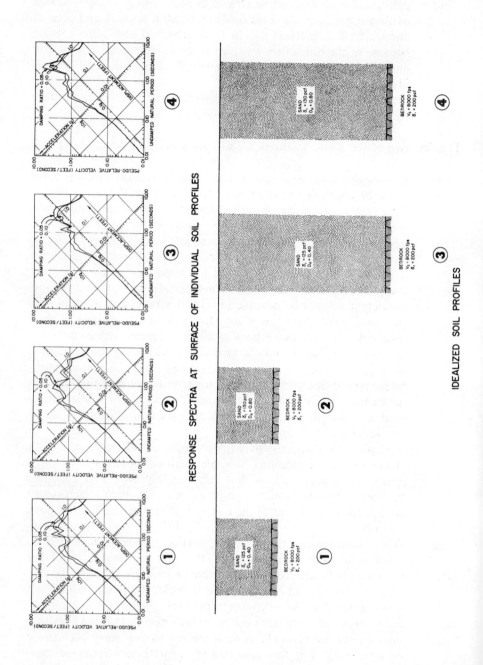

RESPONSE SPECTRA AT SURFACE OF INDIVIDUAL SOIL PROFILES

IDEALIZED SOIL PROFILES

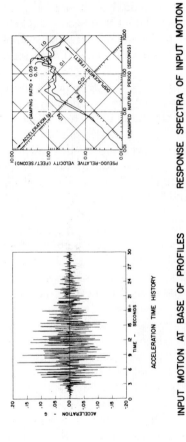

ACCELERATION TIME HISTORY

INPUT MOTION AT BASE OF PROFILES
(FROM SET OF GENERATED TIME HISTORIES)

RESPONSE SPECTRA OF INPUT MOTION

Figure 3.14 Site response analysis summary showing ground motion criteria at bedrock and at surface after one-dimensional shear wave propagation through soil layer (after Valera and Donovan[44])

more appropriate to the surface of softer layers than to bedrock motion.

3.4.4 Application of ground motion criteria to structures

Dynamic response analysis of soils or any type of structure may be carried out using accelerograms or response spectra as input (Chapter 5). Whichever form of dynamic input is used a number of earthquakes should be used or implied. Because of the random nature of earthquakes it seems unlikely that any single seismic event can be shown to be safely representative of the design risk, without choosing an uneconomically powerful ground motion.

The use of accelerograms in a time-dependent analysis is analytically more powerful than response spectrum analysis and may be significantly more informative about the dynamic response of the structure. Individual accelerograms may induce local response peaks (Figure 5.17) in elastic analyses which may be difficult to interpret or justify. In non-elastic analyses this difficulty is largely overcome, but a number of accelerograms should be taken in all cases. It is common practice to take three or four accelerograms in a given study, these often being a mixture of real events (sometimes scaled) and simulated records.

When using response spectra as input, either several response spectra from individual events (e.g. Figure 5.17) or a single spectrum which is the average of several events (Figures 5.18 or 3.15) should be taken. This will help to allow for the randomness of earthquakes, and smoothed average spectra will eliminate undue influence of local peaks in response. Figure 3.15 indicates the scatter of response from five simulated earthquakes; it is worth noting that the standard deviation in the response spectra is likely to have been much greater if real rather than simulated accelerograms had been used.

It is strongly argued by many engineers[44] that where surface motions have been computed from bedrock motions as in Section 3.4.3.1, these surface motions should be used for structural analysis in the form of averaged response spectra rather than accelerograms. It is considered that so many simplifying assumptions are made in site response analyses, that very sophisticated use of the computed surface accelerograms can scarcely be justified for practical design purposes.

3.4.5 Sources of accelerograms and response spectra

Earthquake engineers experienced at working outside basic code requirements have developed sources of information of their own, through government and university organizations specializing in seismology and earthquake engineering. As the problem of availability of information varies so widely from place to place and as the situation is changing so rapidly, this section will simply discuss a few of the chief sources of data presently existing.

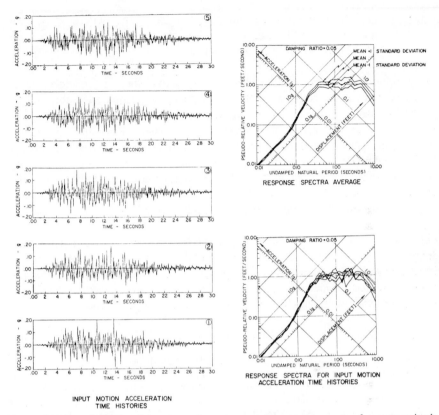

Figure 3.15 Set of simulated input accelerograms and response spectra for a magnitude 7·0 earthquake with a closest distance to the causative fault of 50 km (after Valera and Donovan[44])

(i) *Accelerograms of real earthquakes.* The most thoroughgoing source of accelerograms at the present time is the series of records published by the California Institute of Technology.[49,50] These accelerograms are available in uncorrected and corrected forms and have been digitized for use in computers. They are also available in punched card form. To date this series has covered only earthquakes recorded in the U.S.A. Various other institutions throughout the world also have digitized accelerograms.

(ii) *Accelerograms of simulated earthquakes.* Many earthquake engineering research organizations throughout the world have computer programs for generating artificial earthquakes. Eight simulated earthquakes have been described by Jennings et al.,[25] and the computer cards from these are available from the California Institute of Technology. Two computer programs called PSEQGN and SIMEAR for the generation of simulated earthquakes are available as listings and as decks, from the National Information Services Earthquake Engineering, address NISEE, 729 Davis Hall, University of California, Berkeley,

76

California 94720. White noise segments can readily be generated at most computer centres. When filtering white noise, reference may be made to Fourier amplitude spectra of actual earthquakes.[32]

(iii) *Response spectra of real earthquakes.* Response spectra are more readily available than accelerograms as they are easily described in diagram form in the literature. The classical averaged response spectra (Figure 5.18) developed by Housner,[52] are widely referred to elsewhere. Response spectra can be computed from earthquake accelerograms by computer programs such as SPECEQ which is available through NISEE at address given in part (ii) above. A considerable early collection of American response spectra was prepared by Alford *et al.,*[53] but this has been largely superseded by the series being produced by the California Institute of Technology.[49,50,51]

(iv) *Response spectra of simulated earthquakes.* As mentioned in (ii) above, Jennings *et al.*[25] have described eight simulated earthquakes, giving response spectra as well as accelerograms. Response spectra may readily be computed from simulated accelerograms from computer programs such as SPECEQ, which are available through NISEE as described in part (ii) above.

REFERENCES

1. Rosenblueth, E., 'Earthquake of 28th July, 1957 in Mexico City, *Proc. 2nd World Conference on Earthquake Engineering, Japan,* **1,** 359–379 (1960).
2. Seed, H. B., Whitman, R. V., Dezfulian, H., Dobry, R., and Idriss, I. M., 'Soil conditions and building damage in 1967 Caracas earthquake', *Jnl. Soil Mechanics and Foundations Division, ASCE,* **98,** No. SM8, Aug. 1972, 787–806 (Aug. 1972).
3. Housner, G. W., and Jennings, P. C., 'The San Fernando California earthquake', *Int. Jnl. of Eathquake Engineering and Structural Dynamics,* **1,** 5–32 (1972).
4. Cluff, L. S., 'Peru earthquake of May 31, 1970; engineering geology observations', *Bulln. Seismological Society of America,* **61,** No. 3, 5111–533 (June 1971).
5. Seed, H. B., 'Landslides during earthquake due to soil liquefaction', *Jnl. of the Soil Mechanics and Foundations Division, ASCE,* **94,** No. SM5, 1053–1122 (Sept. 1968).
6. Suzuki, Z., (ed), *General report on the Tokachi-Oki earthquake of 1968,* Keigaku Publishing Co., Ltd., Tokyo, 1971.
7. Seed, H. B., and Idriss, I. M., 'Analysis of soil liquefaction, Niigata earthquake', *Jnl. Soil Mechanics and Foundations Division, ASCE,* **97,** No. SM9, 1249–1274 (1971).
8. Seed, H. B., Ugas, C., and Lysmer, J., 'Site dependent spectra for earthquake resistant design', *Report No. EERC 74-12, Earthquake Engineering Research Center,* University of California, Berkeley, Nov. 1974.
9. Espinosa, A. F., and Algermissen, S. T., 'Soil amplification studies in areas damaged by the Caracas earthquake of July 29, 1967,' *Proc. Microzonation Conference, Seattle, Washington,* **II,** 455–464 (1972).
10. Schmertmann, J. H., 'Static cone to compute static settlement over sand' *Journal of the Soil Mechanics and Foundations Division, ASCE,* **96,** No. SM3, 1011–1043 (May, 1970).
11. Holtz, W. G., and Gibbs, H. J., Discussion of 'Settlement of spread footings on sand', (by D. J. D'Appolonia, E. D'Appolonia, and R. F. Brissette, *Journal of the Soil Mechanics and Foundations Division, ASCE,* May, 1968), *Journal of the Soil Mechanics and Foundations Division, ASCE,* **95,** No. SM3, 900–905 (May 1969).

12. Bazaraa, A. R. S. S., 'Use of the standard penetration test for estimating settlements of shallow foundations on sand', *Ph.D. thesis*, University of Illinois, 1967.

13. Shannon and Wilson, Inc., and Agbabian-Jacobsen Associates., *Soil behaviour under earthquake loading conditions—state of the art evaluation of soil characteristics for seismic response analyses*, prepared for the U.S. Atomic Energy Commission by Shannon and Wilson, Inc., Seattle, and Agbabian-Jacobsen Associates, Los Angeles, 1972.

14. Ravara, A., Pereira, J., Oliveira, C., and Lourtie, P., 'Estudos estruturais dos edifícios de Parque Central—2°- Relatóno: Análise dinâmica dos edifícios de apartamentos', *LNEC Report*, Lisbon, 1971.

15. Parton, I. M., and Taylor, P. W., 'Analysis of microtremor recordings', *Bulletin of the New Zealand Society of Earthquake Engineering*, **6**, No. 3, 96–109 (Sept. 1973).

16. Parton, I. M., and Taylor, P. W., 'Microtremor recording techniques', *Bulletin of the New Zealand Society of Earthquake Engineering*, **6**, No. 2, 87–92 (June 1973).

17. Seed, H. B., and Lee, K. L., 'Pore-water pressures in earth slopes under seismic loading conditions', *Proc. Fourth World Conference on Earthquake Engineering, Chile*, **3**, A5, 1–11 (1969).

18. Parton, I. M. and Smith, R. W. M., 'Effect of soil properties on earthquake response', *Bulletin of the New Zealand Society for Earthquake Engineering*, **4**, No. 1, 73–93 (March 1971).

19. Wilson, S. D., and Dietrich, R. J., 'Effect of consolidation pressure on elastic and strength properties of clay', *Proc. ASCE Research Conf. on Shear Strength of Cohesive Soils, University of Colorado*, 1960, pp. 419–435, discussion pp. 1086–1092.

20. Drnevich, V. P., Hall, J. R., and Richart, F. E., 'Effects of amplitude of vibration on the shear modulus of sand', *Proc. International Symposium on Wave Propagation and Dynamic Properties of Earth Materials, New Mexico*, 1967, pp. 189–192.

21. Seismology Committee, SEAOC., *Recommended lateral force requirements and conmentary*, Structural Engineers Association of California, 1974.

22. Reimer, R. B., Clough, R. W., and Raphael, J. M., 'Evaluation of the Pacoima dam accelerogram', *Proc. Fifth World Conference on Earthquake Engineering, Rome*, **2**, 2328–2337 (1973).

23. Bolotin, V. V., 'Statistical theory of a seismic design of structures' *Proc. Second World Conference on Earthquake Engineering, Tokyo*, **2**, 1365–1374 (1960).

24. Amin, M., and Ang, A. H. S., 'A nonstationary stochastic model for strong-motion earthquakes', *Structural Research Series No. 306*, University of Illinois, Department of Civil Engineering, April, 1966.

25. Jennings, P. C., Housner, G. W., and Tsai, N. C., *Simulated earthquake motions*, Earthquake Engineering Research Laboratory, California Institute of Technology, April, 1968.

26. Bycroft, G. N., 'White noise representation of earthquakes', *Jnl. of the Engineering Mechanics Division, ASCE*, **86**, No. EM2, 1–16 (April 1960).

27. Housner, G. W., 'Characteristics of strong-motion earthquakes', *Bulletin of the Seismological Society of America* **37**, No. 1, 19–31 (Jan. 1947).

28. Werner, S. D., *A study of earthquake input motions for seismic design*, prepared for U.S. Atomic Energy Commission by Agbabian-Jacobsen Associates, Los Angeles, June, 1970.

29. Tajimi, H., 'A statistical method of determining the maximum response of a building structure during an earthquake', *Proc. Second World Conference on Earthquake Engineering, Tokyo*, **2**, 781–797 (July 1960).

30. Housner, G. W., and Jennings, P. C. 'Generation of artificial earthquakes', *Journal of the Engineering Mechanics Division, ASCE*, **90**, No. EM1, 113–150 (Feb. 1964).

31. Housner, G. W., 'Important features of earthquake ground motion', *Proc. Fifth World Conference on Earthquake Engineering, Rome*, **1**, CLIX–CLXVIII (1973).

32. California Institute of Technology, *Analyses of strong motion earthquake accelerograms, Vol. IV—Fourier amplitude spectra*, Earthquake Engineering Research Laboratory, California Institute of Technology, Pasadena. (Issued serially).

33. Ruiz, P., and Penzien, J., 'Stochastic seismic response of structures', *Journal of the Engineering Mechanics Division, ASCE*, **97**, No. EM2, 441–456 (April 1971).

34. Parton, I. M., 'Site response to earthquakes, with reference to application of microtremor measurements', *Report No. 80*, School of Engineering, University of Auckland, May, 1972.

35. Donovan, N. C. 'Earthquake hazards for buildings', *Engineering Bulletin*, No. 46, Dames and Moore, Los Angeles, 3–20 (Dec. 1974).

36. Housner, G. W., 'Intensity of earthquake ground shaking near a causative fault', *Proc. Third World Conference on Earthquake Engineering, New Zealand*, **1**, III–94 to III–115 (1965).

37. Newmark, N. M., Robinson, A. R., Ang. A. H. S., Lopez, L. A., and Hall, W. J., 'Methods for determining site characteristics', *Proc. of Int. Conf. on Microzonation for Safer Construction, Research and Applications, Seattle*, **1**, 113–129 (1972).

38. Tsai, N. C., 'Influence of local geology on earthquake ground motion', *Ph.D. Thesis*, California Institute of Technology, 1969.

39. Seed, H. B., and Idriss, I. M., 'Influence of soil conditions on ground motions during earthquakes', *Journal of the Soil Mechanics and Foundations Division, ASCE*, **95**, No. SM1, 99–137 (Jan. 1969).

40. Whitman, R. V., Roesset, J. M., Dobry, R., and Ayestaran, L., 'Accuracy of modal superposition for one-dimensional soil amplification analysis', *Proc. Int. Conf. on Microzonation, Seattle*, **2**, 483–498 (1972).

41. Roesset, J. M., 'Fundamentals of soil amplification', in *Seismic Design of Nuclear Reactors*, (Ed. R. J. Hansen) M.I.T. Press, 1972, pp. 183–244.

42. Schnabel, P. B., Seed, H. B., and Lysmer, J., 'Modification of seismograph records for effects of local soil conditions', *Report EERC No. 71–8*, Earthquake Engineering Research Center, University of California, Berkeley, Calif., 1971.

43. Schnabel, P. B., Seed, H. B., and Lysmer, J. 'Shake. A computer program for earthquake response analysis of horizontally layered sites', *Report No. EERC 72–12*, Earthquake Engineering Research Center, University of California, Berkeley, Calif., 1972.

44. Valera, J. E., and Donovan, N. C., 'Incorporation of uncertainties in the seismic response of soils', *Proc. Fifth World Conference on Earthquake Engineering, Rome*, **1**, 370–379 (1973).

45. Idriss, I. M., and Seed, H. B., 'Response of earth banks during earthquakes', *Journal of the Soil Mechanics and Foundations Division, ASCE*, **93**, No. SM3, 61–82 (May 1967).

46. Newmark, N. M., and Rosenblueth, E., *Fundamentals of earthquake engineering*, Prentice-Hall, Englewood Cliffs, New Jersey, 1971.

47. Newmark, N. M., and Hall, W. J. 'A rational approach to seismic design standards for structures, *Prof. Fifth World Conference on Earthquake Engineering, Rome*, **2**, 2266–2275 (1973).

48. Newmark, N. M., Blume, J. A., and Kapur, K. K., 'Seismic design spectra for nuclear power plants', *Jnl. of the Power Division, ASCE*, **99**, No. PO2, 287–303 (Nov. 1973).

49. California Institute of Technology, *Strong motion earthquake accelerograms, Digitized and plotted data, Vol. I—Uncorrected accelerograms*, Earthquake Engineering Research Laboratory, California Institute of Technology, Pasadena. (Issued serially).

50. California Institute of Technology, *Strong motion earthquake accelerograms, Digitized and plotted data, Vol. II—Corrected accelerograms and integrated ground velocity and displacement curves*, Earthquake Engineering Research Laboratory, California Institute of Technology. (Issued serially).

51. California Institute of Technology, *Analyses of strong motion earthquake accelerograms, Vol. III—Response spectra*, Earthquake Engineering Research Laboratory, California Institute of Technology, Pasadena. (Issued serially).
52. Housner, G. W., 'Behaviour of structures during earthquakes', *Journal of the Engineering Mechanics Division, ASCE*, **85,** No. EM4, 109–129 (Oct. 1959).
53. Alford, J. L., Housner, G. W., and Martel, R. R., *Spectrum analyses of strong-motion earthquakes*, California Institute of Technology, Pasadena, 1951. (Revised 1964).

Chapter 4

Determination of structural form

4.1 INTRODUCTION

In earthquake regions it is of paramount importance that the structural form is sound. This chapter is addressed to architects as well as engineers because *the structural engineer cannot make a poor structural form behave satisfactorily in an earthquake.*

For the design team to provide the client with the most appropriate structure, the form should not be fixed until adequate background information is available. As illustrated in the diagram in the introduction the design team should know sufficient about the consequences of earthquake damage, the economic factors in resisting that damage and the degree of risk to different types of structure on the site in question, to enable a wise choice of structural form to be made.

4.2 THE FORM OF THE SUPERSTRUCTURE

4.2.1 Introduction

There is of course no universal ideal form for a particular type of structure, but there are certain guiding principles to be borne in mind. Briefly, the structure should

(i)   be simple;
(ii)   be symmetrical;
(iii)   not be too elongated in plan or elevation;
(iv)   have uniform and continuous distribution of strength;
(v)   have horizontal members which form hinges before the vertical members;
(vi)   have its stiffness related to the sub-soil properties.

An earthquake will relentlessly seek out every structural weakness, whether previously acknowledged or not. The above rules give the engineer the best chance of understanding the earthquake behaviour of the structure and even if that has been far from perfect, wise detailing of the right structure (in any material) is the best guarantee of success.

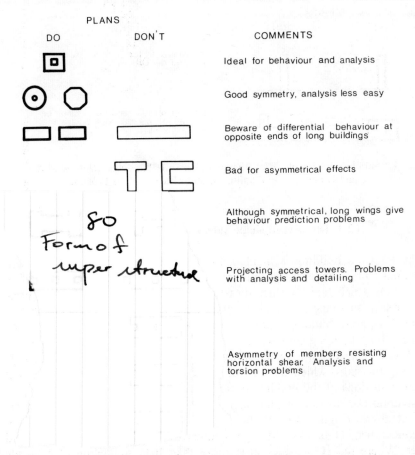

Figure 4.1 Simple rules for plan layouts of aseismic buildings. (*Only* with dynamic analysis and careful detailing should these rules be broken)

4.2.2 Simplicity and symmetry

Earthquakes repeatedly demonstrate that the simplest structures have the greatest chance of survival. There are two main reasons for this. Firstly, our ability to understand the overall earthquake behaviour of a structure is markedly greater for a simple one than it is for a complex one, and secondly, our ability to understand structural details is considerably greater for simple details than it is for complicated ones.

Symmetry is desirable for much the same reasons. It is worth pointing out that symmetry is important in both directions on plan (Figure 4.1), a point too often neglected. Lack of symmetry produces torsional effects which are difficult to assess properly and which can be very destructive.

82

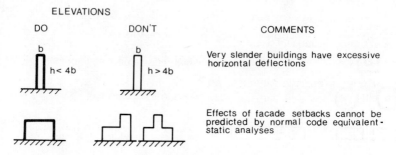

Figure 4.2 Simple rules for elevation shapes of aseismic buildings. (*Only* with dynamic analysis and careful detailing should these rules be broken)

4.2.3 The overall shape should not be too elongated

The longer a building is in plan, the more chance there is of different earthquake movements being applied simultaneously to the two ends of it, a situation which may produce disastrous results. If for a given plan area, a squarish plan shape is not satisfactory for architectural reasons, then two or more separate buildings may be the answer. This is sometimes done by slicing a long building into sections with movement-gaps between. But this can only be considered a partial solution because of the difficulty of properly detailing the gaps, which have to be 100 mm or more in width to prevent adjacent sections of the building battering each other.

An important aspect of plan layout is the general undesirability of re-entrant angles (Figure 4.1). While T– and L–shaped plans are doubly to be condemned, H–blocks although symmetrical should not be encouraged either. Where the H provides little more than light modelling of a facade with a small set-back, this plan type could be adopted with reasonable confidence, so long as the effects of the discontinuities in the horizontal members on the stepped facades can be properly understood and detailed against. External lift and stairwells provide similar dangers which tend to act on their own in earthquakes, with force concentrations, torsions and out of balance forces which are difficult to predict without complex and expensive dynamic analyses.

For the elevation it seems reasonable to suggest a limited slenderness for most buildings: Height/width $\not> 3$ or 4 (Figure 4.2). The more slender a building the worse the overturning effects of an earthquake and the greater the earthquake stresses in the outer columns, particularly the overturning compressive forces which can be very difficult to deal with.

On some sites the ground conditions may be such that the foundations will strongly influence the overall proportions and the layout of the vertical structure, for both practical and economic reasons. This aspect of the design is considered in more detail in Section 4.5.

4.2.4 Uniform and continuous distribution of strength

This concept is closely related to that of simplicity and symmetry. The structure will have the maximum chance of surviving an earthquake if;

(a) the load bearing members are uniformly distributed;
(b) all columns and walls are continuous and without offsets from roof to foundation;
(c) all beams are free of offsets;
(d) columns and beams are coaxial;
(e) reinforced concrete columns and beams are nearly the same width;
(f) no principal members change section suddenly;
(g) the structure is as continuous (redundant) and monolithic as possible.

In qualification of the above recommendations it can be said that while they are not mandatory they are well proven, and the less they are followed the more vulnerable and expensive the structure will become.

While it can readily be seen how these recommendations make structures more easily analysed and avoid undesirable stress concentrations and torsions, some further explanation may be warranted. The restrictions to architectural freedom implied by the above, sometimes make their acceptance difficult. Perhaps the most contentious is that of uninterrupted vertical structure, especially where cantilevered facades and columns supporting shear walls are fashionable. But sudden changes in lateral stiffness up a building are *not* wise (Figure 4.3), firstly because even with the most sophisticated and expensive computerized analysis the earthquake stresses cannot be determined adequately, and secondly, in the present state of knowledge we probably could not detail the structure adequately at the sensitive spots even if we knew the forces involved. The damage to the Sheraton–Macuto Hotel in the 1967 Caracas earthquake[1] illustrates this point, which is further discussed in Section 5.8.2.

This leads naturally into a discussion of the so-called 'soft storey' concept. In principle it is advantageous to isolate a structure from excessive ground movements by some sort of spongy layer. It has been proposed that a basically stiff structure could be protected from short-period vibrations by making the bottom storey columns relatively flexible (Figure 4.3). Unfortunately, many modern buildings of this type have not performed well in earthquakes. Recent studies have shown the soft storey concept to have theoretical as well as practical problems, and leading engineers in the U.S.A., New Zealand and elsewhere are advising against it at the present time. Chopra *et al.*[2] found that a very low yield force level and an essentially perfectly plastic yielding mechanism are required in the first storey, and that the required displacement capacity of the first storey mechanism is very large.

Item (e) above recommends that in reinforced concrete structures, contiguous beams and columns should be of similar width. This promotes good detailing and aids the transfer of moments and shears through the junctions

84

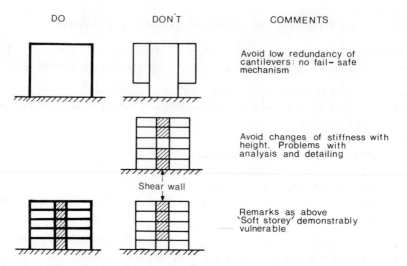

DO	DON'T	COMMENTS
		Avoid low redundancy of cantilevers: no fail-safe mechanism
		Avoid changes of stiffness with height. Problems with analysis and detailing
		Remarks as above 'Soft storey' demonstrably vulnerable

Shear wall

Figure 4.3 Simple rules for vertical frames in aseismic buildings

of the members concerned. Very wide, shallow beams have been found to fail near junctions with normal sized columns, and at present there is a large area of ignorance in the behaviour of such junctions (Figure 4.4).

The remaining main point worth elaborating is item (g) above, which says that a structure should be as redundant as possible. The earthquake resistance of an economically designed structure depends on its capacity to absorb apparently excessive energy input, mainly in repeated plastic deformations of its members. Hence the more continuous and monolithic a structure is made, the more plastic hinges and shear and thrust routes are available for energy absorption. This is why it is so difficult to make precast concrete structures work for strong earthquake motions.

Making joints monolithic and fully continuous is not only important for energy absorption; it also eliminates a frequent source of serious local failure due to high local stresses engendered solely by the very large movements and rotations caused by earthquakes. This problem can arise in such places as the connection of major beams to slabs or minor beams, and beams to columns or corbels.

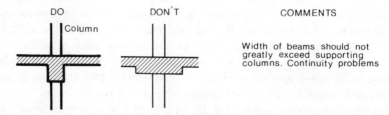

DO	DON'T	COMMENTS
Column		Width of beams should not greatly exceed supporting columns. Continuity problems

Figure 4.4 Simple rule for widths of beams and columns in aseismic reinforced concrete buildings

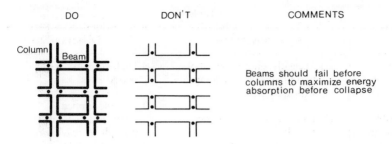

Figure 4.5 Simple rule for relationship between column and beam strengths in aseismic frames

4.2.5 Horizontal and vertical members

In framed building structures it is a fundamental earthquake requirement that horizontal members should fail before vertical members. It is a very important life-saver in that it postpones complete collapse of a structure. Beams and slabs generally do not fall down even after severe damage at plastic hinge positions, whereas columns will rapidly collapse under their vertical loading once sufficient spalling has taken place. This means, for example, that continuous spandrels on light columns are not appropriate in earthquake regions (Figure 4.5). If deep downstand or upstand beams are desirable for architectural reasons their effective depth would have to be substantially reduced by a deep movement gap on each side of all columns (Figure 8.7).

4.2.6 Stiff structures versus flexible

In the past there has been much unresolved discussion of this aspect of design. It revolves around the fact that if the local ground motion is largely in frequencies close to the natural frequency of the building, then the structure will take the maximum punishment. If the local sub-soil will filter out much of the high frequency ground motion (as in Mexico City) then a stiff structure should be subjected to lower seismic forces than a flexible structure, and vice versa (Figure 3.3).

But notwithstanding the above question of resonance with the ground, there still are two schools of thought around the world, the stiff structure school, and those who favour flexible structures. For example in the early 1970's it was alleged that in San Francisco the flexible philosophy currently predominated, while in Los Angeles the stiff school was in vogue, although both cities have a similar range of soil conditions. The chief arguments for and against each form of structure are given in Table 4.1.

One difficulty in weighing up the pros and cons of flexible and stiff structures comes from the lack of clear definitions of the terms 'flexible' and 'stiff'. Perhaps the best that can be said at present is that one structure

Table 4.1

	Advantages	Disadvantages
Flexible Structures	(1) Specially suitable for short period sites, for buildings with long periods (2) Ductility arguably easier to achieve (3) More amenable to analysis	(1) High response on long period sites (2) Flexible framed reinforced concrete is difficult to reinforce (3) Non-structure may invalidate analysis (4) Non-structure difficult to detail
Stiff Structures	(1) Suitable for long period sites (2) Easier to reinforce stiff reinforced concrete (i.e., with shear wall) (3) Non-structure easier to detail	(1) High response on short period sites (2) Appropriate ductility not easy to knowingly achieve (3) Less amenable to analysis

is stiffer than, or less flexible than, another. We shall now consider some current structural forms.

Fully flexible structures may be exemplified by many modern beam and column buildings, where non-structure has been carefully separated from the frame. No significant shear elements exist, actual or potential: all partitioning and infill walls are isolated from frame movements, even the lift and stair shaft walls are completely separated. The cladding is mounted on rocker and roller brackets (of non-corrosive material). This type of completely ductile frame is currently fairly popular in Japan, New Zealand and California. Apart from the points listed in Table 4.1 it has further disadvantages. Floor to floor lateral drift and permanent set may be excessive after a moderate earthquake. In reinforced concrete the joint detailing is very difficult. There is no hidden redundancy (extra safety margin) provided by non-structure as in traditional construction.

Modified flexibility is deliberately sought by some engineers by incorporating limited shear walls in a framed structure, producing what is still a relatively flexible longer-period structure. One approach to this has been Muto's in Japan, where currently all buildings over fourteen storeys have steel frames. Muto introduced slitted concrete shear panels into the steel frames, to reduce horizontal sway in typhoons and moderate earthquakes, and to absorb energy in strong earthquakes (Section 5.8.2).

Many engineers believe that reinforced concrete shear walls should be included in more framed buildings. This would (a)　reduce lateral drift,

(b)　reduce reinforced concrete joint detailing problems,

(c)　help to ensure that plasticity develops uniformly over the structure,

(d)　prevent column failure in sway due to the $P \times \Delta$ effect (i.e. secondary bending resulting from the product of the vertical load and the lateral deflection).

In conclusion it can be said that in many situations either a stiff or a flexible structure can be made to work, but the advantages and disadvantages of the two forms need careful consideration when choosing between them.

4.3 CHOICE OF STRUCTURAL MATERIALS

4.3.1 Seismic strength of materials

In the determination of the form of a structure the choice of material is often an important factor. Sometimes the structural material will be Hobson's Choice, dictated by availability, or political or economic considerations. Whether a fuller choice is possible or not, the following design criteria exist.

Purely in terms of earthquake resistance the best materials have the following properties;

(i)　high ductility;

(ii)　high strength/weight ratio;

(iii)　homogeneity;

(iv)　orthotropy;

(v)　ease in making full strength connections.

Generally the larger the structure the more important the above properties are. By way of illustration the applicability of the major structural materials to buildings is given in Table 4.2 below. The term 'good reinforced masonry' refers to properly detailed hollow concrete block as discussed in Section 6.6.4.

Most fully precast concrete systems are *not* suitable for earthquake resistance, because of the difficulty of achieving a monolithic, continuous and ductile structure.

The order of suitability shown in Table 4.2 is of course far from fixed as it will depend on many things such as the qualities of materials as locally available, the type of structure and the skill of the local labour in using them.

All these factors being equal, there is arguably little to choose between steel and *in situ* reinforced concrete for medium-rise buildings, as long as they are both well designed and detailed. For tall buildings steel-work is generally preferable, though each case must be considered on its merits. Timber performs well in low-rise buildings almost solely because of its high strength/weight ratio, but must be detailed with great care. Further discussion of the use of different materials is given elsewhere.[3] Underdeveloped countries

Table 4.2

	Type of building		
	High-rise	Medium-rise	Low-rise
Best	(1) Steel	(1) Steel	(1) Timber
	(2) *In situ* rein- forced concrete	(2) *In situ* rein- forced concrete	(2) *In situ* rein- forced concrete
Structural		(3) Good precast concrete*	(3) Steel
materials		(4) Prestressed	(4) Prestressed
in approximate		concrete	concrete
order of		(5) Good rein- forced masonry*	(5) Good rein- forced masonry*
suitability			(6) Precast concrete
			(7) Primitive
Worst			reinforced masonry

*These two materials only just qualify for inclusion in the medium-rise bracket. Indeed many earthquake eigineers would not use either material. In Japan masonry is not permitted for buildings of more than three storeys.

have special problems in selecting building materials, from the points of view of cost, availability and technology. Further discussion of these factors has been made by Flores.[4]

4.3.2 Seismic response of structural materials

It is worth bearing in mind while choosing materials that if a flexible structure is required then some materials, such as masonry, are not suitable. On the other hand steelwork is used essentially to obtain flexible structures, although if greater stiffness is desired diagonal bracing or reinforced concrete shear panels may sometimes be incorporated in steel frames. Concrete of course can readily be used to achieve almost any degree of stiffness. See also Section 5.1 and Chapter 6.

A word of warning should be given here about the effect of non-structural materials on the structural response of buildings. The nonstructure, mainly in the form of partitions, may enormously stiffen an otherwise flexible structure and hence must be allowed for in the structural analysis. This subject is discussed in more detail in Section 4.4.

4.4 THE EFFECT OF NON–STRUCTURE

In considering the form of a structure it is important to be aware that some items which are normally non-structural become structurally very responsive in earthquakes. This means anything which will interfere with the free deformations of the structure during an earthquake. In buildings the principal elements concerned are cladding, perimeter infill walls, and internal partitions. Where these elements are made of very flexible materials, they will not affect the structure significantly. But very often, it will be desirable for non-structural reasons to construct them of still materials such as precast concrete or blocks or bricks. Such elements can have a significant effect on the behaviour and safety of the structure. Although these elements may be carrying little vertical load, they can act as shear walls in an earthquake with the following important effects. They may;

(a)   reduce the natural period of vibration of the structure, hence changing the intake of seismic energy and changing the seismic stresses of the 'official' structure;

(b)   redistribute the lateral stiffness of the structure, hence changing the stress distribution;

(c)   cause premature failure of the structure usually in shear or by pounding;

(d)   suffer excessive damage themselves, due to shear forces or pounding.

The more flexible the basic structure is, the worse the above effects will be; and they will be particularly dangerous when the distribution of such 'non-structural' elements is asymmetric or not the same on successive floors. Stratta and Feldman[5] have discussed some of the effects of infill walls during the Peruvian earthquake of May 1970.

In attempting to deal with above problems, either of two opposite approaches may be adopted. The first approach is knowingly to include those extra shear elements into the official structure as analysed, and to detail accordingly. This method is appropriate if the building is essentially stiff anyway, or if a stiff structure is desirable for low seismic response on the site concerned. It means that the shear elements must be effectively tied into the structure, particularly the columns, and that the shear elements themselves will probably require aseismic reinforcement. Thus 'non-structure' is made into real structure. For notes on the analysis of such composite structures, see Section 5.8.

The second approach is to prevent the non-structural elements from contributing their shear stiffness to the structure. This method is appropriate particularly when a flexible structure is required for low seismic response. It can be effected by making a gap against the structure, up the sides and along the top of the element. The non-structural element will need restraint at the top (with dowels, say) against overturning by out-of-plane forces. If the gap has to be filled, a really flexible material must be used, i.e. *not* 'Flexcell'. Some advice on the detailing of infill walls is given in Sections 6.6.5 and 8.2.

Unfortunately, neither of the above solutions is very satisfactory, as the fixing of the necessary ties, reinforcement, dowels, or gap treatments is time-consuming, expensive and hard to supervise properly. Also, flexible gap fillers will not be good for sound insulation.

Finally the client should be warned not to permit construction of solid infill walls without taking structural advice about the earthquake effects.

4.5 THE FORM OF THE SUBSTRUCTURE

Although the form of the substructure must have a strong influence upon the seismic response of structures, little comparative work has been done on this subject. The following notes briefly summarize what appears to be good practice at the present time.

The basic rule regarding the earthquake resistance of substructure is that *integral action* in earthquakes should be obtained. This requires adequate consideration of the dynamic response characteristics of the superstructure and of the subsoil. If a good seismic-resistant form has been chosen for the superstructure (Section 4.2) then at least the plan form of the substructure is likely to be sound, i.e.;

(i)   vertical loading will be symmetrical;
(ii)   overturning effects will not be too large;
(iii)   the structure will not be too long in plan.

As with non-seismic design, the nature of the subsoil will determine the minimum depth of foundations. In earthquake areas this will involve consideration of the following factors;

(a)   transmission of horizontal base shears from the structure to the soil;
(b)   provision for earthquake overturning moments (e.g. tension piles);
(c)   differential settlements (Figure 4.6);
(d)   liquefaction of the subsoil;
(e)   the effects of embedment on seismic response.

The effects of depth of embedment are not fully understood at the present time (Section 5.5.3.3), but some allowance for this effect can be made in soil-structure interaction analyses (Section 5.5.3), or when determining at what level to apply the earthquake loading input for the superstructure analysis.

Three basic types of building foundations may be listed as;

(1)   discrete pads;
(2)   continuous rafts;
(3)   piled foundations.

Piles of course, may be used in conjunction with either pads or rafts. Continuous rafts or box foundations are good aseismic forms only requiring adequate depth and stiffness. Piles and discrete pads require more detailed consideration in order to ensure satisfactory integral action which deals with so many of the structural requirements implied in (i) to (iii) and (a) to (e) above. Integral action should provide sufficient reserves of strength to deal with some of the differential ground movements which are not explicitly

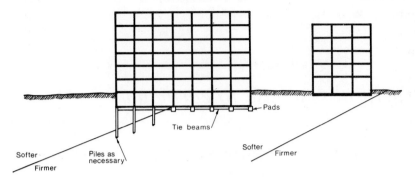

Figure 4.6 Typical structures founded on two types of soil, requiring precautions against differential seismic movements

designed for at the present time. Where a change of soil type occurs under a structure (Figure 4.6), particular care may be necessary to ensure integral substructure action.

This discussion of substructure form is applicable to structures on softer soils only, as structures on rock are naturally integral per media of the rock itself. For a more detailed discussion of foundation design see Section 5.5.4.

REFERENCES

1. Sozen, M. A., Newmark, N. M., and Housner, G. W., 'Implications on seismic structural design of the evaluation of damage to the Sheraton-Macuto', *Proc. 4th World Conference on Earthquake Engineering, Chile*, **III**, J-2, 137–150 (1969).
2. Chopra, A. K., Clough, D. P., and Clough, R. W., 'Earthquake resistance of buildings with a "soft" first storey', *Earthquake Engineering and Structural Dynamics*, **1**, No. 4, 347–355 (June 1973).
3. Dowrick, D. J., 'Modern construction techniques for earthquake areas', *Earthquake Engineering, Proc. 4th European Symposium on Earthquake Engineering, London, 1972*, published by Bulgarian National Committee on Earthquake Engineering, Sofia, 287–300 (1973).
4. Flores, R., 'An outline of earthquake protection criteria for a developing country', *Proc. 4th World Conference on Earthquake Engineering, Chile*, **III**, J4, 1–14 (1969).
5. Stratta, J. L., and Feldman, J., 'Interaction of infill walls and concrete frames during earthquakes', *Bulletin of the Seismological Society of America*, **61**, No. 3, 609–612 (June 1971).

Chapter 5

Structural response to earthquakes

5.1 SEISMIC RESPONSE OF STRUCTURAL MATERIALS

5.1.1 Introduction

This chapter is principally concerned with the determination of seismic stresses and deformations necessary for the detailed design of structures. The chosen earthquake loading (Chapters 2 and 3) is applied to the proposed form and materials of the structure (Chapter 4).

In earthquake conditions the relationship

'Subsoil—Substructure—Superstructure—Non-structure'

ideally should be analysed as a structural continuum. Although in practice this is seldom feasible, each of the parts should be seen as part of the whole when considering boundary conditions.

The problems involved in adequately representing seismic behaviour in structural analysis are numerous, and many compromises will have to be made even in sophisticated analyses. In order to obtain the maximum benefit from even the simplest method of seismic analysis, an understanding of the dynamic response characteristics of structures is essential. For the adequate earthquake resistance of most structures, satisfactory inelastic as well as elastic performance must occur. A brief discussion of the nature of elastic and inelastic dynamic behaviour follows.

5.1.2 Elastic seismic response of structures

Dynamic loading comprises any loading which varies with time, and seismic loading is a complex variant of this. The way in which a structure responds to a given dynamic excitation depends on the nature of the excitation and the dynamic characteristics of the structure, i.e. on the manner in which it stores and dissipates vibrational energy. Seismic excitation may be described in terms of displacement, velocity or acceleration varying with time. When this excitation is applied to the base of a structure it produces a time-dependent response in each element of the structure which may be described in terms of motions or forces.

Perhaps the simplest dynamical system which we can consider is a single-degree-of-freedom system (Figure 5.12) consisting of a mass on a spring which remains in the linear elastic range when vibrated. The dynamic characteristics of such a system are simply described by its natural period of vibration

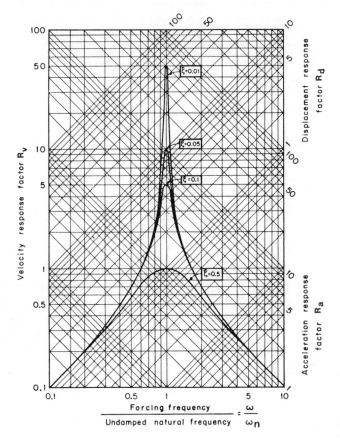

Figure 5.1 Response of linear elastic single-degree-of-freedom system to a harmonic forcing function (after Blake[1])

T, (or frequency ω) and its damping ξ. When subjected to a harmonic base motion described by $u_g = a \sin \omega t$, the response of the mass at top of the spring is fully described in Figure 5.1. The ratios of response amplitude to input amplitude are shown for displacement R_d, velocity R_v, and acceleration R_a, in terms of the ratio between the frequency of the forcing function ω and the natural frequency of the system ω_n.

The significance of the natural period or frequency of the structure is demonstrated by the large amplifications of the input motion at or near the resonance conditions, i.e. when $\omega/\omega_n = 1$. Figure 5.1 also shows the importance of damping particularly near resonance. When the damping $\xi = 0.01$, the resonant amplification of the input motion is fifty fold for this system, but if the damping is increased to $\xi = 0.05$ the resonant amplification is reduced to five times the input motions.

The response of a structure to the irregular and transient excitation of an earthquake will obviously be much more complex than in the simple

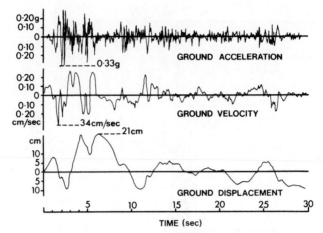

Figure 5.2 1940 El Centro earthquake ground motion:acceleration, velocity and displacement in north-south direction

harmonic steady-state motion discussed above. Consider the ground motion of the 1940 El Centro earthquake, the accelerogram for which is shown in Figure 5.2. If we apply this motion to a series of single-degree-of-freedom structures with different natural periods and damping, we can plot the maximum acceleration response of each of these structures as in Figure 5.3.

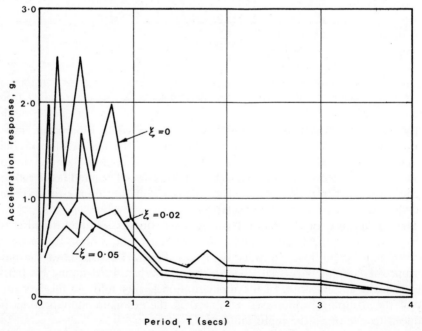

Figure 5.3 Elastic acceleration response spectra of north–south component of the 1940 El Centro earthquake, as derived elsewhere[2]

As with simple harmonic ground motion, the natural period and degree of damping is again evident in Figure 5.3. While no simple periodicity occurs in the ground motion of Figure 5.2, the dominance of the shorter periods is seen from the region of magnified acceleration responses on the left of Figure 5.3. For example a single-degree-of-freedom structure with a period of 0·8 s and damping $\xi = 0·02$ has a maximum acceleration of approximately 0·9 g compared with a peak input ground motion of about 0·33 g. This represents an amplification of 2·7 at $\xi = 0·02$, whereas if the damping is $\xi = 0·05$ the amplification can be seen to reduce to 1·8.

Most structures are more complex dynamically than the single-degree-of-freedom system discussed above. Multi-storey buildings, for example, are better represented as multi-degree-of-freedom structures, with one degree of freedom for each storey, and one natural mode and period of vibration for each storey (Figure 5.19). The response history of any element of such a structure is a function of all the modes of vibration, as well as of its position within the overall structural configuration.

For many multi-degree-of-freedom structures the linear elastic responses can be computed with a high degree of mathematical accuracy. For example, assuming linear elastic behaviour, in the dynamic analysis of a thirty storey building subjected to a ground motion 1·5 times that of Figure 5.2, the maximum horizontal shears at each floor level were computed to be as shown in Figure 5.4. Notice the considerable difference in response between the elastic case assuming 2 percent damping (curve 1) and that for 5 percent damping (curve 3). Further discussion of damping follows in Section 5.1.5.

5.1.3 Inelastic seismic response of structures

For economical resistance against strong earthquakes most structures must behave inelastically. In contrast to the simple linear elastic response model examined in the previous section, the pattern of inelastic stress-strain behaviour is not constant, varying with the member size and shape, the materials used, and the nature of the loading.

The typical stress-strain curves for various materials under repeated and reversed direct loading shown in Figure 5.5 illustrate the chief characteristics of inelastic dynamic behaviour, namely: plasticity, strain hardening, strain softening, stiffness degradation, ductility, and energy absorption.

Plasticity, as exhibited by mild steel (Figure 5.5a), is a desirable property in that it is easy to simulate mathematically and provides a convenient control on the load developed by a member. Unfortunately the higher the grade of steel, the shorter the plastic plateau, and the sooner the *strain hardening* effect shown in Figure 5.5(a) sets in. *Strain softening* is the opposite of strain hardening, involving a loss of stress or strength with increasing strain as seen in Figure 5.5(a), or in the stress-strain envelope for concrete (Figure 5.5c).

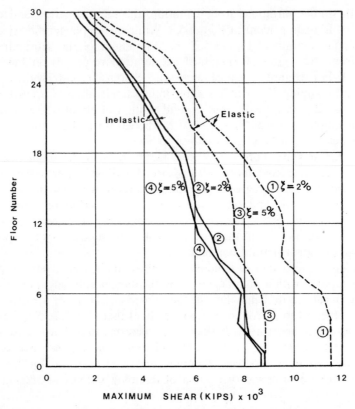

Figure 5.4 Maximum horizontal shear response for Bank of New Zealand Building, Wellington, subjected to 1·5 times El Centro (1940) N–S component (after Albert C. Martin and Associates[3]. Reproduced by permission of the New Zealand National Society for Earthquake Engineering)

In the reversed loading of steel, the *Bauschinger effect* occurs, i.e. after loading past the yield point in one direction the yield stress in the opposite direction is reduced. Another characteristic of the cyclic loading of steel is the increased non-linearity in the elastic range which occurs with load reversal (Figure 5.5b). *Stiffness degradation* is an important feature of inelastic cyclic loading of concrete and masonry materials. The stiffness as measured by the overall stress/strain ratio of each hysteresis loop of Figures 5.5(c) to (f) is clearly reducing with each successive loading cycle.

The *ductility* of a member or structure may be defined in general terms by the ratio

$$\frac{\text{deformation at failure}}{\text{deformation at yield}}$$

In various uses of this definition, 'deformation' may be measured in terms of deflection, rotation or curvature. The numerical value of ductility will also vary depending on the exact combination of applied forces and moments under which the deformations are measured. Ductility is generally desirable in structures because of the gentler and less explosive onset of failure than that occurring in brittle materials. The favourable ductility of mild steel may be seen from Figure 5.5(a) by the large value of ductility in direct tension measured by the ratio $\epsilon_{su}/\epsilon_{sy}$. This ductility is particularly useful in seismic problems because it is accompanied by an increase in strength in the inelastic range. By comparison the high value of compressive ductility for plain concrete, expressed by the ratio $\epsilon_{cu}/\epsilon_{cy}$ in Figure 5.5(c) is far less useful because of the inelastic loss of strength. Steel has the best ductility properties of normal building materials, while concrete can be made moderately ductile with appropriate reinforcement. The ductility of masonry, even when reinforced, is much more dubious. Further discussion of the ductility of the various materials is found elsewhere in this document, particularly in Chapter 6 and Section 5.7.

A high *energy absorption* capacity is often mentioned rather loosely as a desirable property of earthquake-resistant construction. Strictly speaking a distinction should be made between *temporary* absorption and *permanent* absorption of dissipation of energy.

Compare the simple elasto-plastic system represented by OABD in Figure 5.6 with the non-linear mainly elastic system of curves $\hat{OB}$ and $\hat{BE}$; after loading each system to B the total energy 'absorbed' by each system is nearly equal, as represented by the area OABC and $\hat{OBC}$ respectively. But the ratio between temporarily stored strain energy and permanently dissipated energy for the two systems are far from equal. After unloading to zero stress it can be seen that the energy dissipated by the elasto-plastic system is equal to the hysteretic area OABD, while the energy dissipated by the non-linear system is equal to the much smaller hysteretic area $\hat{OBE}$ (shaded in Figure 5.6).

As further elucidation of the seismic energy absorption of structures, consider Figure 5.7 derived from the inelastic seismic analysis of the Bank of New Zealand building, Wellington. A substantial part of the energy is temporarily stored by the structure in elastic strain energy and kinetic energy. After three seconds the earthquake motion is so strong that the yield point is exceeded in parts of the structure and permanent energy dissipation in the form of inelastic strain (or hysteretic) energy begins. Throughout the whole of the earthquake, energy is dissipated by damping, which is of course the means by which the elastic energy is dissipated once the forcing ground motion ceases.

It is evident from the large proportion of hysteretic energy dissipated by this building that considerable ductility is required. A brittle building with the same yield strength, but with no inelastic behaviour ($\epsilon_u/\epsilon_y = 1$), would have begun to fail after three seconds of the earthquake. In other words stronger members would have been necessary in a purely elastic design. This can be seen in another way in Figure 5.4, showing the reduction in storey

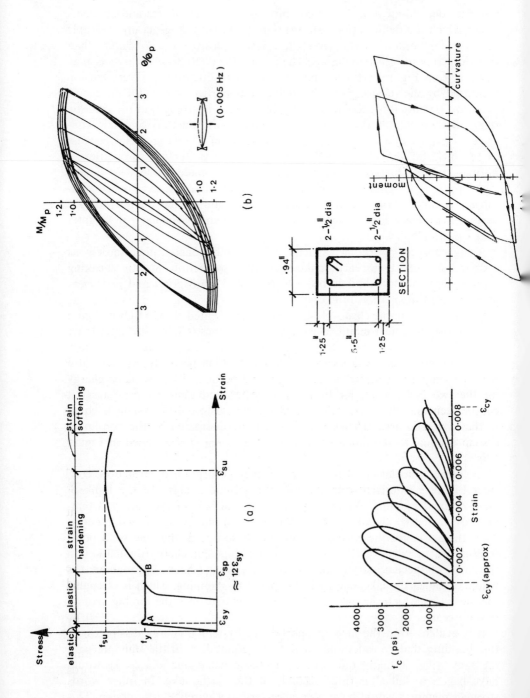

(a)

(b)

(0·005 Hz)

SECTION

2-½" dia

2-½" dia

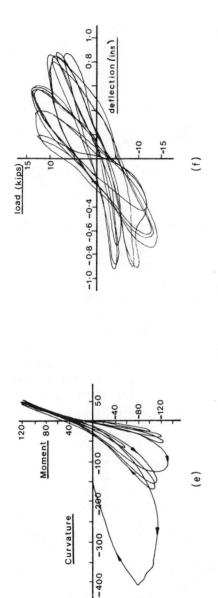

(e)

(f)

Figure 5.5 Elastic and inelastic stress-strain behaviour of various materials under repeated and reversed loading.

(a) Mild steel, monotonic (or repeated axial) loading.
(b) Structural steel under cyclic bending (after Reference 15).
(c) Unconfined concrete, repeated loading (after Reference 4).
(d) Doubly reinforced concrete beam, cyclic loading (after Reference 5).
(e) Prestressed concrete column, cyclic bending (after Reference 6).
(f) Masonry wall, cyclic lateral loading (after Reference 7).

(Part (c) reproduced by permission of the American Concrete Institute)

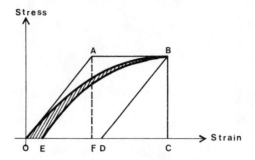

Figure 5.6 Energy stored and dissipated in idealized systems

shears achieved when assuming inelastic behaviour (curve 2) as compared with the elastic case (curve 1).

5.1.4 Mathematical models of inelastic seismic behaviour

When examining the range and complexity of hysteretic behaviour shown in Figure 5.5, the problems involved in establishing usable mathematical stress-strain models for realistic seismic analysis are obvious. A review of work in this field has been made by Park,[8] and the simplest models of hysteresis are shown in Figure 5.8.

Simple models of this type may be reasonable for design purposes when building in steelwork, such as for the building referred to in Figures 5.4 and 5.7; the inelastic dynamic analysis used in this case has also been used elsewhere.[9] However for steel buildings of less convenient form, or when using other materials, the problems of adequately (or economically) predicting inelastic seismic response are daunting at the present time. For further discussion of methods of analysis see Section 5.2.

5.1.5 Level of damping in different structures

The general influence of damping upon seismic response is discussed in Section 5.1.2 above, but when choosing the level of damping for use in the dynamic analyses of a structure, the following factors should be considered.

Damping varies with the materials used, the form of the structure, the nature of the subsoil and the nature of the vibration. Large amplitude post-elastic vibration is more heavily damped than small amplitude vibration, while buildings with heavy shear walls and heavy cladding and partitions have greater damping than lightly clad skeletal structures. There are many reports of differences in damping in different modes of vibration, but no underlying pattern has as yet been established. The overall damping of a structure is clearly related to the damping characteristics of the subsoil, and some allowance may be made for this in more sophisticated analysis involving soil-structure interaction (Sections 5.5.2.3 and 5.5.3).

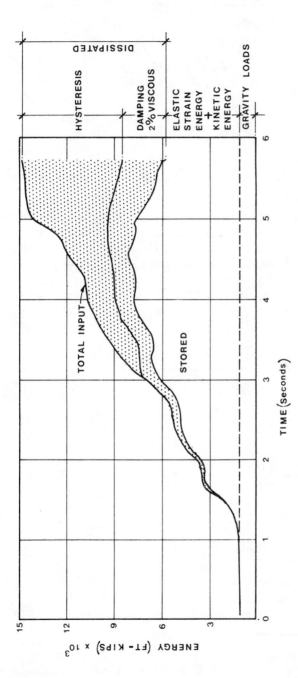

Figure 5.7 Energy expenditure in Bank of New Zealand building, Wellington computed for first part of an earthquake equal to 1·5 times El Centro 1940, north–south component (after Albert C. Martin Associates[3]. Reproduced by permission of the New Zealand National Society for Earthquake Engineering)

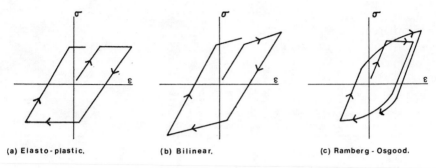

(a) Elasto-plastic.   (b) Bilinear.   (c) Ramberg-Osgood.

Figure 5.8 Idealized hysteresis loops for cyclic behaviour of steel

The many experimentally determined values of damping reported in the literature are generally derived either for individual structural components or for low amplitude vibration of buildings. Hence for whole structures subject to strong ground motion some extrapolation of existing damping data is necessary. Table 5.1 indicates representative values of damping for a range of construction. These values are suitable for normal response spectrum or modal analysis in which viscous damping, equal in all modes, is assumed.

Table 5.1. Typical damping ratios for structures

Type of construction	Damping ξ, percent of critical
Steel frame, welded, with all walls of flexible construction	2
Steel frame, welded or bolted, with stiff cladding, and all internal walls flexible	5
Steel frame, welded or bolted, with concrete shear walls	7
Concrete frame, with all walls of flexible construction	5
Concrete frame, with stiff cladding and all internal walls flexible	7
Concrete frame, with concrete or masonry shear walls	10
Concrete and/or masonry shear wall buildings	10
Timber shear wall construction	15

Note

(1) The term 'frame' indicates beam and column bending structures as distinct from shear structures.

(2) The term 'concrete' includes both reinforced and prestressed concrete in buildings. For isolated prestressed concrete members such as in bridge decks damping values less than 5 percent be appropriate, e.g. 1–2 percent if the structure remains substantially uncracked.

Insufficient evidence exists to warrant any more detailed allowance for differences in structural and non-structural form, and designers will need to use their own judgement to interpret the table.   ·

For further reading on the question of damping of structures see elsewhere.[10-14]

5.2 CRITIQUE OF METHODS OF SEISMIC ANALYSIS

5.2.1 Introduction

The many methods for determining seismic forces in structures fall into two distinct categories;

(i)   equivalent static force analysis;
(ii)   dynamic analysis.

5.2.2 Equivalent static force analysis

These are approximate methods which have been evolved because of the difficulties involved in carrying out realistic dynamic analysis. Codes of practice inevitably rely mainly on the simpler static force approach, and incorporate varying degrees of refinement in an attempt to simulate the real behaviour of the structure. Basically they give a crude means of determining the 'total' horizontal force (base shear) V, on a structure:

$$V = ma$$

where m is the mass of the structure and a is the seismic horizontal acceleration. a is generally in the range $0.05\,g$ to $0.20\,g$. V is applied to the structure by a simple rule describing its vertical distribution. In a building this generally consist of horizontal point loads at each concentration of mass, most typically at floor levels (Figure 5.9). The seismic forces and moments in the structures are then determined by any suitable statical analysis and the results added to those for the normal gravity load cases.

In the subsequent design of structural sections an increase in permissible elastic stresses of 33–50 percent is usually permitted, or a smaller load factor than normal is required for ultimate load design. In regions of high winds

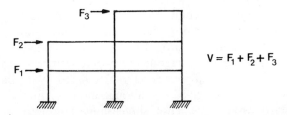

Figure 5.9 Example of frame with equivalent static forces applied at floor levels

and moderate earthquake requirements, the worst design loads of taller structures may well arise from wind rather than earthquake forces. But even so, the form and detail of the structure should still be governed by seismic considerations.

An important feature of equivalent static load requirements in most codes of practice, up till 1972, is the fact that the calculated seismic forces are considerably less than those which would actually occur in the larger earthquakes likely in the area concerned. The forces calculated in more rigorous dynamic analyses based on a realistic earthquake excitation can be as much as ten times greater than those arising from the static load provisions of codes. This state of affairs has been 'justified' by arguing that the force discrepancy will be taken up by plastic behaviour of the structure, which should therefore be detailed to be specially ductile, and some codes do have specific ductility requirements. It is widely recognized that more stringent static load requirements should be laid down, and we can expect the more advanced codes to move in this direction in the 1970s.

5.2.3 Dynamic analysis

For large or complex structures static methods of seismic analysis are not accurate enough and many authorities demand dynamic analyses for certain types and size of structure. Various methods of differing complexity have been developed for the dynamic seismic analysis of structures. They all have in common the solution of the equations of motion as well as the usual statical relationships of equilibrium and stiffness. For any structure with more than three degrees of freedom such analyses are carried out by matrix methods on computers.

The three main techniques currently used for dynamic analysis are;

(i)   direct integration of the equations of motion by step-by-step procedures;
(ii)   normal mode analysis;
(iii) response spectrum techniques.

Direct integration provides the most powerful and informative analysis for any given earthquake motion. A time-dependent forcing function (earthquake accelerogram) is applied and the corresponding response-history of the structure during the earthquake is computed. That is, the moment and force diagrams at each of a series of prescribed intervals throughout the applied motion can be found. Computer programs have been written for both linear elastic and nonlinear inelastic material behaviour, using step-by-step integration procedures. Linear behaviour is seldom analysed by direct integration, unless mode coupling is involved, as normal mode techniques are easier, cheaper, and nearly as accurate. Three-dimensional non-linear analyses have been devised which can take the three orthogonal accelerogram components from a given earthquake, and apply them simultaneously to the structure.[16] In principle, this is the most complete dynamic analysis technique so far devised, and is unfortunately correspondingly expensive to carry out.

Normal mode analysis is a more limited technique than direct integration, as it depends on artificially separating the normal modes of vibration and combining the forces and displacements associated with a chosen number of them by superposition. As with direct integration techniques, actual earthquake accelerograms can be applied to the structure and a stress-history determined, but because of the use of superposition the technique is limited to linear material behaviour. Although modal analysis can provide any desired order of accuracy for linear behaviour by incorporating all the modal responses, some approximation is usually made by using only the first few modes in order to save computation time. Problems are encountered in dealing with systems where the modes cannot be validly separated, i.e. where mode coupling occurs.

The most serious shortcoming of linear analyses is that they do not accurately indicate all the members requiring maximum ductility. In other words the pattern of highest elastic stresses is not necessarily the same as the pattern of plastic deformation in an earthquake structure.[17] For important structures in zones of high seismic risk, non-linear dynamic analysis is sometimes called for.

The response spectrum technique[18,19,20] is really a simplified special case of modal analysis. The modes of vibration are determinnd in period and shape in the usual way and the maximum response magnitudes corresponding to each mode are found by reference to a response spectrum. An arbitrary rule is then used for superposition of the responses in the various modes. The resultant moments and forces in the structure correspond to the envelopes of maximum values, rather than a set of simultaneously existing values. The response spectrum method has the great virtues of speed and cheapness.

Although this technique is strictly limited to linear analysis because of the use of superposition, simulations of non-linear behaviour have been made using pairs of response spectra, one for deflections and one for accelerations.[21,22] The expected ductility factor is chosen in advance and the appropriate spectra are used. This is clearly a fairly arbitrary procedure, and is unlikely to be more realistic than the linear response spectrum method.

Another attempt to study non-linear behaviour by spectral techniques is described by Shepherd and McConnel.[23] They conclude that non-linear response spectrum techniques may be best applied only to structures behaving like a single-degree-of-freedom system, such as bridge piers, as the pattern of hinge points in other systems would be too complicated for predictions by this approximate method.

5.2.4 Selection of method of analysis

In the past there has generally been little choice in the method of analysis, mainly because suitable and economical computer programs have not been readily available. To date most earthquake-resistant structures, even in California, have been analysed with an equivalent static load derived from a code of practice. But this situation is changing. An increasing number of efficient and economical dynamic analysis programs are being written for faster computers, and many design offices have access to such programs. Dynamic analyses are demanded now by some clients, and by the regulations of more countries.

It is difficult to give clear general advice on selecting the means of analysis, as each structure will have its own requirements, technical, statutory, economic and political. Broadly speaking, however, the larger and/or more complex the structure, the more sophisticated the dynamic analysis used. Table 5.2 gives a very simple indication of the applicability of the main methods of analysis.

At the present time, except in very special projects, designers are unlikely to do a three-dimensional dynamic analysis, whether elastic or inelastic, which allows for two orthogonal horizontal components of ground motion simultaneously. In order to make some allowance for the resultant diagonal response, some codes are now stipulating arbitrary means of adding the separately computed orthogonal components. Until further research has been done, there are great problems in doing this without the risk of being too conservative.

It is important to note that the methods of analysis in Table 5.2 become successively more realistic *only if* the appropriate load input is used. For instance a non-linear analysis using inappropriate ground motion could be much less realistic than a response spectrum analysis using suitable spectra. The difficult question of selecting dynamic input is discussed in the following section, and in Section 3.4.

Table 5.2

Type of structure	Method of analysis
Small simple structures ↓ Progressively more demanding structures ↓ Large complex structures	(1) Equivalent static forces (appropriate code) (2) Response spectra (appropriate spectrum) (3) Modal analysis (appropriate dynamic input) (4) Non-linear plane frame (appropriate dynamic input) (5) Non-linear 3-D frame (appropriate dynamic input)

5.2.4.1 *Method of analysis and material behaviour*

In earthquake engineering the effect of material behaviour on the choice of the method of analysis is a much greater issue than in non-seismic engineering. The problem can be divided into two categories depending on whether the material behaviour is brittle or ductile, i.e. whether it can be considered linear elastic or inelastic as described in Section 5.1. The normal analytical and design methods of dealing with these two states are summarized in Table 5.3. A brief discussion of the comparative merits of the eight methods of seismic analysis and design outlined in Table 5.3 is presented below.

Materials in the brittle category, such as masonry or porcelain, can be realistically analysed by methods (1), (2) and (3) in Table 5.3 because their linear elastic behaviour matches the analytical assumptions. The chief problems lie in choosing an adequate safety margin (load factor) within the elastic range, to cover the normal errors involved in assessing the loading, the geometric modelling and the ultimate strength. In masonry construction it is common practice to enhance the elastic load factor by developing as much inelastic deformation prior to collapse as possible with nominal ductility reinforcement. This increases the safety against strong shaking particularly of longer duration. In porcelain structures it is good practice to increase the earthquake protection by deliberately enhancing the damping. For further discussion of masonry and procelain see Sections 6.6 and 7.2 respectively.

Materials in the ductile category are more satisfactory for earthquake resistance than brittle materials because of their inelastic deformability, but are less convenient to analyse for the same reason. Of the methods used in analysing ductile materials (methods 4–8 in Table 5.3) only (7) and (8) attempt to model the stress–strain behaviour directly. Method (7) avoids the problems of non-linear analysis by keeping the structure elastic under full loading, as does method (3), and hence does not draw on the constructional economies afforded by inelastic deformation. It may be appropriate for efficient 'tube' structures, or multiple shear wall structures, or rich clients desiring great safety; otherwise method (7) is rare. Method (8), on the other hand, seeks the structural economies of inelastic behaviour, but is analytically expensive. At the present time hysteresis diagrams acceptable for use in this analysis are available only for steel. For steel structures which can be realistically reduced to regular plane frames, such as that referred to in Figures 5.4 and 5.7, method (8) may be economic from an analytical as well as from a construction point of view. It follows that most structures which will behave inelastically in (strong) earthquakes are designed by methods (4), (5) and (6) of Table 5.3 assuming linear elastic material behaviour in the analysis. Each of the three methods imply the application of an artificially reduced earthquake loading within the elastic capacity of the structure, and an approximate or arbitrary allowance for the inelastic deformations by ensuring certain ductility levels in highly stressed zones.

Table 5.3. Seismic analysis and design methods

Material behaviour	Method of analysis	Seismic loading	Design provisions
	Equivalent –static	Arbitrarily reduced	(1) Working stress or factored ultimate stress design, plus imposed nominal ductility
Linear elastic (brittle)	Linear dynamic	Arbitrarily reduced	(2) Working stress or factored ultimate stress design, plus imposed nominal ductility
		Full	(3) Ultimate stress design, plus imposed nominal ductility
	Equivalent –static	Arbitrarily reduced	(4) *Working stress or factored ultimate stress design, plus imposed arbitrary ductility
Inelastic (ductile)		Arbitrarily reduced	(5) *Working stress or factored ultimate stress design, plus imposed arbitrary ductility
	Linear dynamic	Arbitrarily reduced	(6) *Working stress or factored ultimate stress design, plus approximate analysis for ductility demands
		Full	(7) Structure intended to remain elastic, but nominal ductility imposed
	Inelastic dynamic	Full	(8) Ductility demands found from plastic hinge rotations

*Commonest methods in structural engineering.

This ductile design process, which attempts to do inelastic design by an 'equivalent-elastic' method, has great difficulties in;

(a)   allowing for inelastic deflection,
(b)   allowing for stiffness degradation,
(c)   determining the distribution of ductility demands, and
(d)   allowing for the duration of strong shaking.

These four factors are not independent and vary with the nature of the material, the structural form and the loading, as discussed below.

In endeavouring to define an equivalent-elastic loading two alternative methods of equating elastic and inelastic response are commonly considered (Figure 5.10).

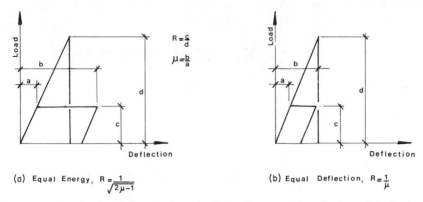

$$R = \frac{c}{d}$$

$$\mu = \frac{b}{a}$$

(a) Equal Energy, $R = \frac{1}{\sqrt{2\mu-1}}$ (b) Equal Deflection, $R = \frac{1}{\mu}$

Figure 5.10 Reduction factors R for seismic loading equating elastic and inelastic response in terms of (a) energy and (b) deflection

For simple elasto-plastic systems two possible reduction factors for the loading, $R = c/d$, are obtained in terms of the deflection ductility factor $\mu = b/a$ as shown. Until more research has been done, appropriate values of R and μ for a range of real structures will not be available. But it is often taken that values of μ in the range 4–6 are suitable for structures designed using the reduced loadings assumed in the equivalent-static loadings of codes of practice such as those used in California[24] and New Zealand.[25]

The above reduction factors and ductility factors, appropriate to much-studied earthquakes such as that of El Centro (1940), are fairly well established for regular buildings. But problems arise when considering other earthquake motions for equivalent-elastic design criteria. This is illustrated by comparing the El Centro earthquake with that of Parkfield (Figure 5.11).

It can be seen that the elastic response of a single degree-of-freedom system is much greater for the Parkfield than for the El Centro earthquake. Yet El Centro at magnitude 6·9 did much more damage than Parkfield at magnitude 5·6. The elastic response analysis has taken account of the higher peak acceleration of the Parkfield accelerogram (about 0·50 g) compared with El Centro's peak acceleration of 0·33 g, without being able to allow for the effect of the much greater duration of strong shaking which occurred at El Centro.

Clearly great care is necessary in selecting design earthquakes of different magnitude to ensure an appropriate relationship between elastic and inelastic response. A much greater reduction factor R for Parkfield would be required than for the El Centro earthquake.

The determination of the distribution of ductility demand throughout a multi-redundant structure using an equivalent-elastic analysis is also unreliable. The positions of maximum moment in a frame, determined elastically, will not necessarily indicate the order of plastic hinge formation.[17] However in ideally regular plane frames this approximation may be reasonable, and is often taken in practice.

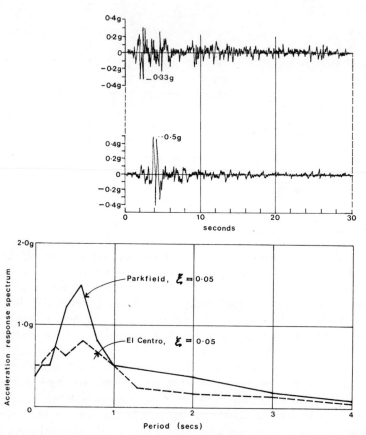

Figure 5.11 Accelerograms and elastic response spectra for El Centro and Parkfield earthquakes

Referring again to Table 5.3, it may be said that for regular steelwork structures design methods (4) and (5) are reasonable, despite the arbitrary relationship between reduction factor R and imposed ductility, because of the high ductility of steel. With less ductile materials such as reinforced and prestressed concrete, methods (4) and (5) are more suspect, although less so for small regular structures. Method (6) refers to ductile design procedures developed for reinforced concrete, where the distribution of ductility demand is determined with a collapse mechanism in mind.[26] The problems of ductility demand are further discussed in Chapter 6 in terms of various building materials.

5.3 EVALUATION OF CODES OF PRACTICE

5.3.1 Resumé of seismic countries and existing codes

There are over sixty countries in the world where earthquakes should be considered in design. Many of these countries as yet have no earthquake code, although the number lessens each year. Unfortunately, both the realism and thoroughness of the existing codes vary considerably, and such is the state of the art that every code could benefit by major improvements.

Table 5.4 lists most of the earthquake countries, identifies those known to have earthquake codes, and gives some comments thereon. It is not claimed that Table 5.4 is exhaustive, mainly because of the difficulties in defining earthquake countries. For example, how small a country, or state, or atoll should be listed? Omissions of this kind include many individual islands in the East and West Indies, which may be autonomous or governed by some remote state without an earthquake code, such as Great Britain. There also may be some argument as to whether earthquakes are a design consideration in all the countries listed. However, earthquakes do shake all countries, and are considered in the design of at least some structures in relatively unseismic areas not included in Table 5.4. For example in England nuclear reactors are designed against earthquakes. This illustrates the point that in any country it may be worth asking 'Should this structure be designed against earthquakes?'.

In many cases the value of base shear coefficient α determined from the code may be too low for the structure concerned, and all the variables concerned should be considered as discussed in earlier parts of this document. In some countries α will certainly be too low, and for reinforced concrete it is more likely to be low than for steelwork. In other countries, for the right building on a favourable site, for an earthquake return period of fifty years (say), the code value of α *may* be adequate without modification.

5.3.2 What earthquake codes should say

An earthquake code should give guidance on three main topics;

(i)   loading and risk;
(ii)   overall structural performance criteria;
(iii)   aseismic detailing.

At present, all codes are inadequate or misleading in one or more of these respects. The word 'risk' is deliberately put beside 'loading' to remind designers that loading is a probability problem, and that the risk of severe overloading due to earthquakes is harder to predict than for most other types of loading. A good earthquake code will discuss loading in terms of risk (probabilities and zoning), use of structure (consequences of collapse), type of structure, and soil conditions. The Spanish Code (1968) deals with these aspects of loading more logically than most other codes at the present time. Whether it derives loads which are in fact always satisfactory is another matter. Some further discussion of the equivalent static loading given in codes can be found in Section 5.2.

Criteria for *overall structural performance* should include recommendations on simplicity and symmetry in planning and constancy of structural sizes.

Table 5.4. Seismicity and codes of various countries

(1) Countries with earthquakes	(2) Codes in I.A.E.E. publication[27]	(3) Also has code	(4) Seismic risk rating	(5) Maximum base shear coefficient $\alpha = a/g$	(6) Comments
Afghanistan			high		
Albania			high		
Algeria	1955(F)*		medium	0·175	Code average
Argentina	post-1965(S)		high	0·120	Code average
Australia			low		
Austria	pre-1966(G)		low	0·005	Code meaningless, α too low
Bangla Desh			high		
Bolivia			medium		
Bulgaria	1964		medium	0·32 approx.	Code average
Burma			high		
Canada	1970		medium in parts	$\left(\begin{array}{c}0\cdot033\\ \text{shear}\end{array}\right)\left(\begin{array}{c}0\cdot017\\ \text{framed}\end{array}\right)$	Based on U.S.A. α too small. Based on U.S.A.
Caribbean Is.		1972 (draft)			
Chile	1972(S)		high	$\left(\begin{array}{c}0\cdot144\\ \text{shear}\end{array}\right)\left(\begin{array}{c}0\cdot096\\ \text{ductile}\end{array}\right)$	Code average
China			high in parts		
Colombia			high		
Congo Rep.			low		
Costa Rica			high		
Cuba	1964		low	$\left(\begin{array}{c}0\cdot10\\ \text{shear}\end{array}\right)\left(\begin{array}{c}0\cdot08\\ \text{framed}\end{array}\right)$	Code average
Cyprus			high		
Dominican Rep.			high		
Ecuador			high		

Country				Remarks
France	1967	high	$\left(\begin{smallmatrix}0\cdot24\\ \text{shear}\end{smallmatrix}\right)\left(\begin{smallmatrix}0\cdot12\\ \text{framed}\end{smallmatrix}\right)$ 0·22 approx.	Code average
Germany	1957(G)	medium in parts	0·10	Code average
Ghana		low		No detail, but α fairly safe
Gibraltar		low		
Greece	1959	high	0·16	Not very good
Guatemala		high		
Haiti		medium		
Honduras				
India	1970	high in parts	0·12	Code above average
Indonesia	1970	high in parts	0·15	Code below average
Iran	post-1955	high	0·10	'Code' not official
Iraq		high		
Israel	proposed	low	0·04	Code gives forces only
Italy	1962	high	0·10	Code poor
Jamaica	See Caribbean	medium		
Japan	1951–68	high	0·20	Code above average
Jordan		low		
Lebanon		low		
Libya		medium		
Malawi		low		
Mexico	1966	high	$\left(\begin{smallmatrix}0\cdot312\\ \text{shear}\end{smallmatrix}\right)\left(\begin{smallmatrix}0\cdot156\\ \text{framed}\end{smallmatrix}\right)$	Code average, α for Acapulco
Morocco		medium		
Mozambique		low		
Nepal		high		

114

Table 5.4—*continued*

(1) Countries with earthquakes	(2) Codes in I.A.E.E. publication[27]	(3) Also has code	(4) Seismic risk rating	(5) Maximum base shear coefficient $\alpha = a/g$	(6) Comments
New Guinea			high		
New Zealand	1965		high	0·16	Code average
Nicaragua			high		
Panama			high		
Peru	1968		high	$\begin{pmatrix}0\cdot16\\ \text{shear}\end{pmatrix}\begin{pmatrix}0\cdot08\\ \text{framed}\end{pmatrix}$	Code average, α values for $T^\dagger = 0\cdot1$
Phillippines	1972		high	$\begin{pmatrix}0\cdot16\\ \text{shear}\end{pmatrix}\begin{pmatrix}0\cdot08\\ \text{framed}\end{pmatrix}$	Code average, α values for $T^\dagger = 0\cdot1$
Portugal	1961		medium	0·15	Code below average
Rhodesia	pre-1968		low		
Rumania	1968(S)		medium	0·30 approx.	Code average
Spain			medium	0·40	Code above average
Syria			medium		
Taiwan			high		
Tanzania			low		
Tunisia			medium		
Turkey	1968		high	0·108	Code average
Uganda			low		
U.S.A.	1970		high in parts	$\begin{pmatrix}0\cdot133\\ \text{shear}\end{pmatrix}\begin{pmatrix}0\cdot067\\ \text{framed}\end{pmatrix}$	Code average
U.S.S.R.	1970		high in parts	0·250 (5 storeys)	Code above average

Venezuela	1967	medium		Code average
Yugoslavia	1964(Y)	high	$\left(\begin{array}{c}0\cdot110\\ \text{shear}\end{array}\right)\left(\begin{array}{c}0\cdot085\\ \text{framed}\end{array}\right)$ 0·30 approx.	Code average
Zambia		low		

* Superseded by 1967 French Code.

† T = fundamental period of vibration of the structure, in seconds.

Column (1) lists most countries where earthquakes may be a design consideration.

Column (2) lists those countries whose code is published[27]. The date of each individual code is given where known. Some of these codes are published in their original language only, and this is indicated by letters as follows:

(F) = in French; (G) = in German; (S) = Spanish; (Y) = in Yugoslavian

Column (3) indicates any other countries known by us to have codes. No doubt there will always be omissions in this column, as each year more countries are adopting codes. Also China may well have a code as yet unpublished in the West.

Column (4) gives a broad indication of the seismic risk in each country; the terms 'low', 'medium' and 'high' should not be taken too literally. It is interesting to observe that there are 20 or so countries with medium to high seismic risk which have not formally adopted an earthquake code.

Column (5) gives a rough comparison of the horizontal seismic forces assumed by the various national codes. Values are given for the base shear coefficient

$$\alpha = \frac{V}{G + Q}$$

where V is the seismic base shear, G is the dead load of the structure and Q is the live load on the structure. α has been calculated for steel or reinforced concrete buildings in each case, and is the highest value imposed by each code, i.e. it is for the highest risk zone, with the worst foundation conditions, and for the most valued public buildings (if such differentiations are made).

Numerical comparisons should not be made too literally as many other code features vary from country to country, e.g. seismic design stresses. Also a few codes, such as those of Bulgaria and Russia, use dynamic rather than static analysis, and only an approximate maximum value of α can be given.

Column (6) gives a broad indication of the value of each code to the designer. Some of the codes are very bad indeed, either because they give unsafe criteria or they give too little guidance. Even the best codes are at fault to some extent in either or both of these respects, and the conscientious designer may see fit to increase the seismic coefficients, or use larger gaps between adjacent buildings. In this column 'Code average' means that the code concerned should be used with caution, and some knowledgeable evaluation of the code should be obtained to identify its strengths and weaknesses.

Also it should give safe recommendations on minimum gaps between adjacent structures and limits on lateral drift during an earthquake.

Aseismic detailing is generally given explicitly or implicity within the codes for specific materials such as steel or reinforced concrete. Unfortunately these codes are often out of step with the general earthquake code itself, and may not give the strength of ductility implied by the levels of loading risk applied to the structure.

In conclusion, it is worth making the point again that even the best codes have their weaknesses and should not be used blindly. Some evaluation of each code should be obtained so that its recommendations can be strengthened, if necessary, for a given structure on a given site.

5.4 THEORY OF DYNAMICS AND SEISMIC RESPONSE

5.4.1 Introduction

Having given a general description in Section 5.2 of the main types of dynamic analysis used in earthquake engineering, a summary of the mathematical processes involved now follows.

As dynamic loading varies with time, the response of the structure also varies with time, and hence a full dynamic analysis involves determining the structural responses at each of a series of time intervals throughout the motion induced by the loading. Two different assumptions are generally used in specifying the deflected shape of the structure, the lumped mass approach and the generalized co-ordinate approach. In both cases the number of displacement components required to specify the position of all significant mass particles in the structure is called the number of degrees of freedom of the structure. Only the lumped mass system will be considered here. As it assumes that the mass of the structure is concentrated into a number of discrete parts, it is the simpler approach, and although consequently less accurate than the generalized co-ordinate approach it is satisfactory for most structural frames.

Fuller treatments of the following theory have been given by Clough and Penzien,[28] Biggs[29] and Newmark and Rosenblueth.[14]

5.4.2 Single-degree-of-freedom systems

To find the displacement history of a structure, it is necessary to solve the equations of motion of the system. There is one such equation of dynamic equilibrium for each degree of freedom. A common representation of a single-degree-of-freedom system is as shown in Figure 5.12.

In Figure 5.12 $F(t)$ is a force varying with time, k is the total spring constant of resisting elements, c is the damping coefficient (the damping force is usually taken as proportional to the velocity of the mass for ease of computation) and u is the displacement.

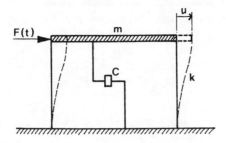

Figure 5.12 Idealized single-degree-of-freedom system

Generally, the equation of dynamic equilibrium is

$$F_I + F_D + F_S = F(t) \tag{5.1}$$

where the inertia force

$$F_I = m\ddot{u}$$

the damping force

$$F_D = c\dot{u}$$

the elastic force

$$F_S = ku$$

Thus

$$m\ddot{u} + c\dot{u} + ku = F(t) \tag{5.2}$$

For the case of earthquake excitation (Figure 5.13), the only external loading is in the form of an applied motion at ground level, $u_g(t)$, i.e. the total acceleration of the mass m is

$$\ddot{u}_t = \ddot{u} + \ddot{u}_g$$

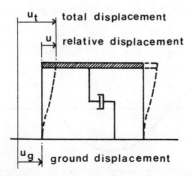

Figure 5.13 Single-degree-of-freedom system subjected to ground motion

118

Therefore

$$F_I = m\ddot{u}_t = m\ddot{u} + m\ddot{u}_g$$

and

$$F(t) = 0 = F_I + F_D + F_S$$

Therefore

$$m\ddot{u} + m\ddot{u}_g + c\dot{u} + ku = 0$$

or

$$m\ddot{u} + c\dot{u} + ku = F_{eff}(t) \tag{5.3}$$

where

$$F_{eff}(t) = -m\ddot{u}_g$$

is the effective load resulting from the ground motion, and is equivalent to the product of the mass of the structure and the ground acceleration.

5.4.2.1 Free vibrations (undamped)

In order to solve the equation of motion (equation 5.3), first consider the case of free vibration, $F(t) = 0$, with zero damping.
i.e.

$$m\ddot{u} + ku = 0$$

or

$$\ddot{u} + \omega^2 u = 0 \tag{5.4}$$

where $\omega = \sqrt{(k/m)}$ is the circular frequency of this free vibration system.

The solution of equation (5.4) is

$$u = A \sin \omega t + B \cos \omega t$$

Solving for A and B,

$$u = \dot{u}_0/\omega \sin \omega t + u_0 \cos \omega t \tag{5.5}$$

where u_0 and $\dot{u}_0$ are the initial displacement and velocity, respectively. The resulting simple harmonic motion is shown in Figure 5.14.

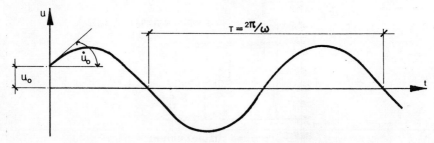

Figure 5.14 Undamped simple harmonic motion of single-degree-of-freedom system given initial displacement and velocity

The period of the above motion is

$$T = 2\pi/\omega$$

and the amplitude is

$$R = \sqrt{\left\{\left(\frac{\dot{u}_0}{\omega}\right)^2 + u_0^2\right\}}.$$

5.4.2.2 *Response to short impulse (undamped)*

An approximate solution for the response to a very short duration loading is easily derived from the free vibration results above. If the length of impulse $t_1 \ll T$, the period of vibration, it can be taken that $u_0 \approx 0$ and from impulse-momentum

$$m\Delta\dot{u} = \int F\,dt$$

therefore

$$\dot{u}_0 = \frac{\int F\,dt}{m}$$

Using these values of u_0 and $\dot{u}_0$ in equation (5.5)

$$u \approx \frac{\int F\,dt}{m\omega}\sin\omega t \tag{5.6}$$

This loading and response system is shown in Figure 5.15.

It is important to note that approximately the same response would be developed by any short duration impulses having equal values for $\int F\,dt$.

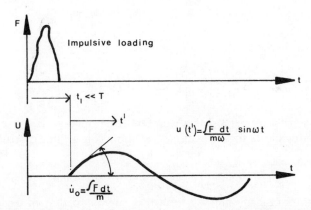

Figure 5.15 Response of single-degree-of-freedom system to impulsive undamped loading

120

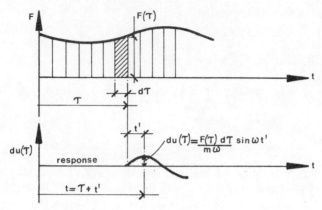

Figure 5.16 Response of single-degree-of-freedom system to arbitrary undamped loading

5.4.2.3 Response to arbitrary loading (undamped)

To find the response of a single-degree-of-freedom system to an arbitrary loading, the latter can be treated as a series of short impulses (Figure 5.16).

The displacement response due to any individual increment of loading ending at time τ and of duration $d\tau$, can be written down in the form of equation (5.6) as

$$du = \frac{F(\tau)}{m\omega} \sin \omega t' \, d\tau$$

or

$$du = \frac{F(\tau)}{m\omega} \sin \omega(t - \tau) \, d\tau$$

where

$$t' = t - \tau$$

The total response to the arbitrary loading is the sum of all the impulses of duration $d\tau$.

i.e.

$$u(t) = \int_0^t \frac{F(\tau)}{m\omega} \sin \omega(t - \tau) \, d\tau \tag{5.7}$$

This is an exact expression called the Duhamel integral. Because it depends upon the principle of superposition it is applicable to linear structures only.

5.4.2.4 Solution of equations of motion (damped)

The solution of the damped problem is similar to that of undamped systems, and only the key results will be given here.

5.4.2.5 Free vibrations (damped)

The equation of motion for a damped system may be written

$$\ddot{u} + 2\xi\omega\dot{u} + \omega^2 u = 0 \tag{5.8}$$

from which (for moderate damping)

$$u = e^{-\xi\omega t}\left(\frac{\dot{u}_0 + \xi\omega u_0}{\omega_D}\sin\omega_D t + u_0\cos\omega_D t\right) \tag{5.9}$$

where

$$\xi = \frac{c}{2m\omega}$$

is the damping ratio, and

$$\omega_D = \omega\sqrt{(1 - \xi^2)}$$

is the damped circular frequency.

5.4.2.6 Response to short impulse (damped)

The free vibration response is

$$u = \frac{\int F\,dt}{m\omega_D}e^{-\xi\omega t}\sin\omega_D t \tag{5.10}$$

5.4.2.7 Response to arbitrary loading (damped)

The damped form of the Duhamel integral becomes

$$u(t) = \int_0^t \frac{F(\tau)}{m\omega_D}e^{-\xi\omega(t-\tau)}\sin\omega_D(t - \tau)\,d\tau \tag{5.11}$$

To evaluate this response to an arbitrary loading history, such as would occur in an earthquake, many numerical integration processes are available, some of which are discussed by Clough and Penzien.[28]

5.4.2.8 Earthquake response

The response of damped single-degree-of-freedom structures to earthquake motion comes from the above as follows. Equation (5.11) can be rewritten in terms of the ground acceleration $\ddot{u}_g(\tau) = F(\tau)/m$, taking $\omega \approx \omega_D$, which is reasonable for small damping. Thus

$$u(t) = \frac{1}{\omega}\int_0^t \ddot{u}_g(\tau)e^{-\xi\omega(t-\tau)}\sin\omega(t - \tau)\,d\tau \tag{5.12}$$

Denoting the integral in the above equation by the response function

$$V(t) = \int_0^t \ddot{u}_g(\tau)\, e^{-\xi\omega(t-\tau)} \sin \omega(t-\tau)\, d\tau \tag{5.13}$$

The earthquake deflection response of a lumped mass system becomes

$$u(t) = \frac{1}{\omega} V(t) \tag{5.14}$$

The forces generated in the structure may best be found in terms of the effective acceleration

$$\ddot{u}_e(t) = \omega^2 u(t) \tag{5.15}$$

The effective earthquake force on the structure follows simply as

$$Q(t) = m\ddot{u}_e(t)$$
$$= m\omega^2\, u(t)$$

Therefore

$$Q(t) = m\omega V(t) \tag{5.16}$$

Thus the effective earthquake force (or base shear) is found in terms of the mass of the structure, its circular frequency, and the response function $V(t)$ expressed in equation (5.13). Equations (5.14) and (5.16) describe the earthquake response at any time t for a single-degree-of-freedom structure, and solutions to these equations depend upon the evaluation of equation (5.13) as discussed under equation (5.11).

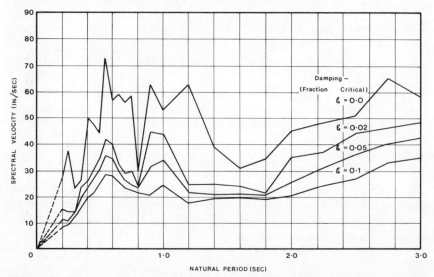

Figure 5.17 Velocity response spectra for El Centro earthquake of May 18, 1940 (N–S)

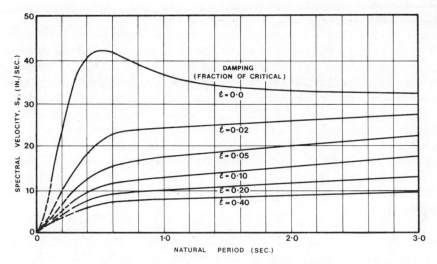

Figure 5.18 Averaged velocity response spectra, based on the spectral intensity of the 1940 El Centro earthquake

5.4.2.9 Response spectra

To obtain the entire history of forces and displacements during an earth-quake using the above equations is clearly a tedious and costly procedure. For many structures it will suffice to evaluate only the maximum responses. From equations (5.14) and (5.16) this means finding the maximum value of response function $V(t)$. This maximum value is called the spectral velocity S_v, or more accurately the spectral pseudo-velocity because it is not exactly the maximum velocity of a damped system. The spectral velocity is

$$S_v = \left\{ \int_0^t \ddot{u}_g(\tau) \, e^{-\xi\omega(t-\tau)} \sin \omega(t - \tau) \, d\tau \right\}_{max} \qquad (5.17)$$

From before it follows that the maximum displacement or spectral displacement

$$S_d = \frac{S_v}{\omega} \qquad (5.18)$$

and the spectral acceleration (or spectral pseudo-acceleration)

$$S_a = \omega S_v \qquad (5.19)$$

From these relationships, the maximum earthquake displacement response

$$u_{max} = S_d \qquad (5.20)$$

and the maximum effective earthquake force or base shear is

$$Q_{max} = m S_a \qquad (5.21)$$

124

If equation (5.17) is evaluated for single-degree-of-freedom structures of varying natural periods, a maximum velocity response curve (called a response
spectrum) can be plotted. A family of curves is usually calculated for any
given excitation, showing the effect of variation in the amount of damping
(Figure 5.17). The maximum responses of a single-degree-of-freedom structure
may be obtained directly from the spectra and equations (5.20) and (5.21).

The velocity spectrum of Figure 5.17 is for the ground motion of a specific
earthquake recorded at a specific site, and the sharp discontinuities in the
spectral curves indicate local resonances only. In any case the period of
vibration of a structure cannot be known with enough certainty to design
with either a peak spectral value or an adjacent trough. For general design
purposes an averaged spectrum as shown in Figure 5.18 will therefore be
more appropriate.

To obtain a realistic result from a response spectrum analysis, it is necessary to use a spectrum derived from an appropriate ground motion. The
subject of selection of response spectra is discussed in Sections 3.4 and 5.2.4.1.

5.4.3 Multi-degree-of-freedom systems

In the dynamic analysis of most structures it is necessary to assume that
the mass is distributed in more than one discrete lump. For most buildings
the mass is assumed to be concentrated at the floor levels, and to be subjected
to lateral displacements only. To illustrate the corresponding multi-degree-of-
freedom analysis, consider a three storey building (Figure 5.19). Each storey
mass represents one degree-of-freedom, each with an equation of dynamic
equilibrium,

$$F_{Ia} + F_{Da} + F_{Sa} = F_a(t)$$
$$F_{Ib} + F_{Db} + F_{Sb} = F_b(t)$$
$$F_{Ic} + F_{Dc} + F_{Sc} = F_c(t)$$

(5.22)

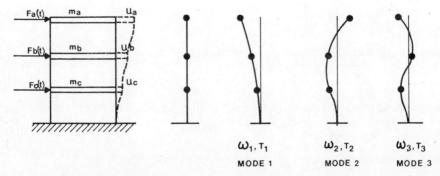

Figure 5.19 Multi-degree-of-freedom system subjected to dynamic loading

The inertia forces in equation (5.22) are simply

$$F_{Ia} = m_a \ddot{u}_a$$
$$F_{Ib} = m_b \ddot{u}_b \qquad (5.23)$$
$$F_{Ic} = m_c \ddot{u}_c$$

In matrix form

$$\begin{Bmatrix} F_{Ia} \\ F_{Ib} \\ F_{Ic} \end{Bmatrix} = \begin{bmatrix} m_a & 0 & 0 \\ 0 & m_b & 0 \\ 0 & 0 & m_c \end{bmatrix} \begin{Bmatrix} \ddot{u}_a \\ \ddot{u}_b \\ \ddot{u}_c \end{Bmatrix} \qquad (5.24)$$

Or more generally

$$\mathbf{F_I} = \mathbf{M\ddot{u}} \qquad (5.25)$$

where $\mathbf{F_I}$ is the inertia force vector, $\mathbf{M}$ is the mass matrix and $\mathbf{\ddot{u}}$ is the acceleration vector. It should be noted that the mass matrix is of diagonal form for a lumped sum system, giving no coupling between the masses. In more generalized shape co-ordinate systems, coupling generally exists between the co-ordinates, complicating the solution. This is a prime reason for using the lumped mass method.

The elastic forces in equation (5.22) depend on the displacements, and using stiffness influence coefficients they may be expressed

$$F_{Sa} = k_{aa}u_a + k_{ab}u_b + k_{ac}u_c$$
$$F_{Sb} = k_{ba}u_a + k_{bb}u_b + k_{bc}u_c \qquad (5.26)$$
$$F_{Sc} = k_{ca}u_a + k_{cb}u_b + k_{cc}u_c$$

In matrix form

$$\begin{Bmatrix} F_{Sa} \\ F_{Sb} \\ F_{Sc} \end{Bmatrix} = \begin{bmatrix} k_{aa} & k_{ab} & k_{ac} \\ k_{ba} & k_{bb} & k_{bc} \\ k_{ca} & k_{cb} & k_{cc} \end{bmatrix} \begin{Bmatrix} u_a \\ u_b \\ u_c \end{Bmatrix} \qquad (5.27)$$

Or more generally

$$\mathbf{F_S} = \mathbf{Ku} \qquad (5.28)$$

where $\mathbf{F_S}$ is the elastic force vector, $\mathbf{K}$ is the stiffness matrix, and $\mathbf{u}$ is the displacement vector. The stiffness matrix $\mathbf{K}$ generally exhibits coupling and will best be handled by a standard computerized matrix analysis.

By analogy with the expressions (5.26), (5.27) and (5.28), the damping forces in equation (5.22) may be expressed

$$\mathbf{F}_D = \mathbf{C}\dot{\mathbf{u}} \tag{5.29}$$

where $\mathbf{F}_D$ is the damping force vector, $\mathbf{C}$ is the damping matrix and $\dot{\mathbf{u}}$ is the velocity vector. In general it is not practicable to evaluate $\mathbf{C}$, and damping is usually expressed in terms of damping coefficients.

Using equations (5.25), (5.28) and (5.29), the equations of dynamic equilibrium (5.22) may be written generally as

$$\mathbf{F}_I + \mathbf{F}_D + \mathbf{F}_S = \mathbf{F}(t) \tag{5.30}$$

which is equivalent to

$$\mathbf{M}\ddot{\mathbf{u}} + \mathbf{C}\dot{\mathbf{u}} + \mathbf{K}\mathbf{u} = \mathbf{F}(t) \tag{5.31}$$

The matrix equation (5.31) for a multi-degree system is identical in form to the single-degree-of-freedom equation (5.2).

5.4.3.1 Vibration frequencies and mode shapes

As the dynamic response of a structure is dependent upon the frequency (or period T) and the displaced shape, the first step in the analysis of a multi-degree system is to find its free vibration frequencies and mode shapes. In free vibration there is no external force and damping is taken as zero. The equations of motion (5.31) become

$$\mathbf{M}\ddot{\mathbf{u}} + \mathbf{K}\mathbf{u} = 0 \tag{5.32}$$

But in free vibration the motion is simple harmonic

$$\mathbf{u} = \hat{\mathbf{u}}\sin \omega t$$

therefore

$$\ddot{\mathbf{u}} = -\omega^2 \hat{\mathbf{u}}\sin \omega t \tag{5.33}$$

where $\hat{\mathbf{u}}$ represents the amplitude of vibration.

Substituting in equation (5.32)

$$\mathbf{K}\hat{\mathbf{u}} - \omega^2\mathbf{M}\hat{\mathbf{u}} = 0 \tag{5.34}$$

Equation (5.34) is an eigenvalue equation and is readily solved for ω by standard computer programs. Its solution for a system having N degrees of freedom yields a vibration frequency ω_n and a mode shape vector ϕ_n for each of its N modes. ϕ_n represents the *relative* amplitudes of motion for each of the displacement components in mode n. It should be noted that equation (5.34) cannot be solved for absolute values of ϕ, as the amplitudes are arbitrary in free vibration.

Figure 5.19 shows the shapes of the three normal modes of a typical three storey building. The numerical value of the fundamental period

$$T_1 > T_2, \text{ and } T_2 > T_3.$$

An important simplification can be made in the equations of motion because of the fact that each mode has an independent equation of exactly equivalent form to that for a single-degree-of-freedom system. Because of the orthogonality properties of mode shapes, equation (5.31) can be written

$$\ddot{Y}_n + 2\xi_n\omega_n\,\dot{Y}_n + \omega_n^2\,Y_n = \frac{\phi_n^T\,\mathbf{F}(t)}{\phi_n^T\,\mathbf{M}\,\phi_n} \tag{5.35}$$

where Y_n is a generalized displacement in mode n, leading to the actual displacement (see equation 5.40) and ϕ_n^T is the row mode shape vector corresponding to the column vector ϕ_n.

5.4.3.2 Earthquake response analysis by mode superposition

The dynamic analysis of a multi-degree-of-freedom system can therefore be simplified to the solution of equation (5.34) for each mode, and the total response is then obtained by superposing the modal effects.

In terms of excitation by earthquake ground motion $\ddot{u}_g(t)$, equation (5.35) becomes

$$\ddot{Y}_n + 2\xi_n\omega_n\,\dot{Y}_n + \omega_n^2 Y_n = \frac{L_n}{\phi_n^T\,\mathbf{M}\,\phi_n}\,\ddot{u}_g(t) \tag{5.36}$$

where the earthquake participation factor

$$L_n = \phi_n^T\,\mathbf{M}\,\hat{\mathbf{I}} \tag{5.37}$$

in which $\hat{\mathbf{I}}$ is a unit column vector of dimension N.

The response of the nth mode at any time t demands the solution of equation (5.36) for Y_n. This may be done by evaluating the Duhamel integral (Section 5.4.2.3).

$$Y_n(t) = \frac{L_n}{\phi_n^T\mathbf{M}\phi_n}\cdot\frac{1}{\omega_n}\int_0^t \ddot{u}_g(\tau)\,e^{-\xi_n\omega_n(t-\tau)}\sin\omega_n(t-\tau)\,d\tau \tag{5.38}$$

The displacement of floor (or mass) i at time t is then obtained by superimposing the response of all modes evaluated at this time t

$$u_i = \sum_{n=1}^{N} \phi_{in}\,Y_n(t) \tag{5.39}$$

where ϕ_{in} is the relative amplitude of displacement of mass i in mode n. It should be noted that in structures with many degrees-of-freedom most of the vibrational energy is absorbed in the lower modes, and it is usually sufficiently accurate to superimpose the effects of only the first few modes.

The earthquake forces in the structure may then be expressed in terms of the effective accelerations

$$\ddot{Y}_{n_{\text{eff}}}(t) = \omega_n^2 \, Y_n(t) \qquad (5.40)$$

from which the acceleration at any floor i is

$$\ddot{u}_{in_{\text{eff}}}(t) = \omega_n^2 \, \phi_{in} Y_n(t) \qquad (5.41)$$

and the earthquake force at any floor i at time t is

$$q_{in}(t) = m_i \omega_n^2 \phi_{in} Y_n(t) \qquad (5.42)$$

Superimposing all the modal contributions, the earthquake forces in the total structure may be expressed in matrix form as

$$\mathbf{q}(t) = \mathbf{M}\boldsymbol{\phi}\omega^2 \mathbf{Y}_n(t) \qquad (5.43)$$

where ϕ is the square matrix of relative amplitude distributions in each mode and ω^2 is the diagonal matrix of ω^2 for each of the n modes.

From equations (5.39) and (5.43) the entire history of displacement and force response can be defined for any multi-degree-of-freedom system, having first determined the modal response amplitudes of equation (5.39).

5.4.3.3 Response spectrum analysis for multi-degree systems

As with single-degree-of-freedom structures considerable simplification of the analysis is achieved if only the maximum response to each mode is considered rather than the whole response history. If the maximum value $Y_{n\,\text{max}}$ of the Duhamel equation (5.38) is calculated, the distribution of maximum displacements in that mode is

$$\mathbf{u}_{n\text{max}} = \boldsymbol{\phi}_n \mathbf{Y}_{n\text{max}} = \boldsymbol{\phi}_n \frac{L_n}{\boldsymbol{\phi}_n^T \mathbf{M}\boldsymbol{\phi}_n} \frac{S_{vn}}{\omega_n} \qquad (5.44)$$

and the distribution of maximum earthquake forces in that mode is

$$\mathbf{q}_{n\text{max}} = \mathbf{M}\boldsymbol{\phi}_n \omega_n^2 Y_{n\text{max}} = \mathbf{M}\boldsymbol{\phi}_n \frac{L_n}{\boldsymbol{\phi}_n^T \mathbf{M}\boldsymbol{\phi}_n} S_{an} \qquad (5.45)$$

In equation (5.44) and (5.45), S_{vn} and S_{an} are the spectral velocity and spectral acceleration for mode n, and are as defined in Section 5.4.2.9.

Equations (5.44) and (5.45) enable the maximum response in each mode to be determined. As the modal maxima do not necessarily occur at the same time, nor necessarily have the same sign, they cannot be combined to give the precise total maximum response. The best that can be done in a response spectrum analysis is to combine the modal responses on a probability basis. Various approximate formula for superposition are used, the most common being the root-sum-square procedure. As an example the maximum deflection at the top of a three-storey structure (three masses) would be

$$u_{a\,max} \approx \sqrt{(u_{a\,1\,max}^2 + u_{a\,2\,max}^2 + u_{a\,3\,max}^2)} \qquad (5.46)$$

This approximation is usually, but not necessarily, conservative.

Considerable savings in computation are made by the further approximation of using the responses of only the first few modes in this equation. Usually the first three to six modes are all that need to be included, as most of the energy of vibration is absorbed in these modes.

Most analyses utilizing response spectra, take the spectral velocities S_{vn} for all modes from a single-degree-of-freedom spectrum. This approximation is reasonable for uniform and regular structures, but for irregular structures with larger changes of stiffness more general forms of analysis are advisable. Apart from the obvious possibility of using a full modal analysis or direct integration technique, it is possible to create response spectra for systems of more than one degree-of-freedom. Penzien[30] has used two degree-of-freedom response spectra developed for analysing buildings having large set-backs of the facade at high level.

5.4.4 Non-linear and inelastic earthquake response

The importance of and difficulties involved in carrying out realistic inelastic analyses has been discussed in Sections 5.1 and 5.2. A very brief summary of the processes involved follows. Further discussions of non-linear analysis may be found in the literature.[14,28,31]

The mode superposition techniques discussed above are necessarily limited to the study of linear material behaviour only. To analyse the effects of non-linear inelastic response during strong-motion earthquakes, a step-by-step procedure is necessary as outlined below.

5.4.4.1 Step-by-step integration

A number of step-by-step integration procedures are possible. Generally the response history is divided into very short time increments, during each of which the structure is assumed to be linearly elastic. Between each interval the properties of the structure are modified to match the current state of deformation. Therefore the non-linear response is obtained as a sequence of linear responses of successively differing systems. One method of step-by-step integration is now described. In each time increment the following computations are made.

(1)　The stiffness of the structure for that increment is computed, based on the state of displacement existing at the beginning of the increment.

(2)　Changes of displacement are computed assuming the accelerations to vary linearly during the interval.

(3)　These changes of displacement are added to the displacement state of the beginning of the interval to give the displacements at the end of the interval.

(4)　Stresses appropriate to the total displacements are computed.

In the above procedure the equations of motion must be integrated in their original form during each time increment. For this purpose equation (5.31) may be written

$$\mathbf{M}\Delta\ddot{\mathbf{u}} + \mathbf{C}\Delta\dot{\mathbf{u}} + \mathbf{K}(t)\Delta\mathbf{u} = \Delta\mathbf{F}(t) \tag{5.47}$$

where $\mathbf{K}(t)$ is the stiffness matrix for the time increment beginning at time t, and $\Delta\mathbf{u}$ is the change in displacement during the interval. The determination of $\mathbf{K}$ for each increment is the most demanding part of the analysis, as all the individual member stiffnesses must be found each time for their current state of deformation.

5.5 DYNAMICS OF SOILS AND SOIL-STRUCTURE SYSTEMS AND FOUNDATION DESIGN

5.5.1 Introduction

Due to the lack of reliable data over the wide range of soil and soil-structure problems encountered in earthquake engineering, codes of practice offer little (and conflicting) guidance on aseismic foundation design. In this section an attempt has been made to coordinate the well informed design office techniques currently used in this subject. Because of the importance of subsoil conditions and the soil-structure relationship on the seismic behaviour of structures, the main aspects of aseismic foundation design are discussed in this section under the following headings;

5.5.2 dynamic behaviour of soils;
5.5.3 dynamics of soil-structure systems;
5.5.4 foundation design.

Although a selection of the best available design data is presented, lack of detailed knowledge often permits only broad generalizations to be made, and it is stressed that sound engineering judgement rather than elaborate analysis will lead to successful aseismic foundation design. As a supplement to the specific references quoted in the following discussion, the reader may find it useful to read general texts on dynamics of soils such as that of Richart et al.[32]

5.5.2 Dynamic behaviour of soils

Soil behaviour under dynamic loading depends on the strain magnitude, the strain rate, and on the number of cycles of loading. Some soils increase in strength under rapid cyclic loading, while saturated sands or sensitive clays may lose strength with vibration. The behaviour of soils in earthquakes will be discussed under the following three sub-headings;

(1)   settlement of dry sands;
(2)   liquefaction of saturated cohesionless soils;
(3)   dynamic design parameters of soils (shear modulus and damping).

5.5.2.1 Settlement of dry sands

It is well known that loose sands can be compacted by vibration. In earthquakes such compaction causes settlements which may have serious effects on all types of construction. It is therefore important to be able to assess the degree of vulnerability to compaction of a given sand deposit. Unfortunately this is difficult to do with accuracy, but it appears that sands with relative density less than 60 percent or with standard penetration resistance less than fifteen are susceptible to significant settlement. The amount of compaction achieved by any given earthquake will obviously depend on the magnitude and duration of shaking as well as on relative density, as demonstrated by the laboratory test results plotted in Figure 5.20.

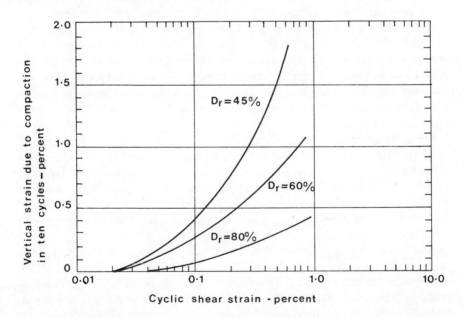

Figure 5.20 Effect of relative density on settlement in ten cycles (after Silver and Seed[33])

Attempts have been made to predict the settlement of sands during earthquakes and a simple method[14] is presented below. It should be noted that this ignores the effect of important factors such as confining pressure and number of cycles, but no fully satisfactory method of settlement prediction as yet exists.

There is a critical void ratio e_{cr} above which a granular deposit will compact when vibrated. If the void ratio of the stratum is $e > e_{cr}$ the maximum amount of settlement possible can be shown to be

$$\Delta H = \frac{e_{cr} - e}{1 - e} H \qquad (5.48)$$

where H is the depth of the stratum.

The critical void ratio can be obtained from

$$e_{cr} = e_{min} + (e_{max} - e_{min}) \exp\left[-0.75a/g\right] \qquad (5.49)$$

where

e_{min} = minimum possible void ratio as determined by testing;
e_{max} = maximum possible void ratio;
a = amplitude of applied acceleration;
g = acceleration due to gravity.

5.5.2.2 Liquefaction of saturated cohesionless soils

Under earthquake loading some soils may compact, increasing the pore water pressure and causing a loss in shear strength. This phenomenon is generally referred to as liquefaction. Gravel or clay soils are not susceptible to liquefaction. Dense sands are less likely to liquefy than loose sands, while hydraulically deposited sands are particularly vulnerable due to their uniformity. Liquefaction can occur at some depth causing an upward flow of water. Although this flow may not cause liquefaction in the upper layers it is possible that the hydrodynamic pressure may reduce the allowable bearing pressures at the surface.

Extensive liquefaction at Niigata (Japan) during the 1964 earthquake[34] led to increased attempts to quantify liquefaction potential,[35,36] and some of the results of these studies are presented here. As yet no generally accepted unified criterion has been developed for liquefaction potential, although attempts have been made to relate it individually to relative density (Table 5.5), to standard penetration resistance (Figure 5.21), and to particle size distribution (Figure 5.22). Unfortunately these separate criteria may give conflicting results for a given site and further analysis may be required, by a method such as that suggested by Shannon et al.[35] in their Appendix B.

If liquefaction is likely to be a hazard, the use of deep foundations or piling may be necessary in order to avoid unacceptable settlement or foundation failure during an earthquake. In most cases specialist advice on liquefaction should be taken.

Table 5.5. Liquefaction potential related to relative density D_r, of soil (after Seed and Idriss[36])

Maximum ground surface acceleration	Liquefaction very likely	Liquefaction depends on soil type and earthquake magnitude	Liquefaction very unlikely
0·10 g	$D_r < 33\%$	$33\% < D_r < 54\%$	$D_r > 54\%$
0·15 g	$D_r < 48\%$	$48\% < D_r < 73\%$	$D_r > 73\%$
0·20 g	$D_r < 60\%$	$60\% < D_r < 85\%$	$D_r > 85\%$
0·25 g	$D_r < 70\%$	$70\% < D_r < 92\%$	$D_r > 92\%$

5.5.2.3 Dynamic design parameters of soils

The information given in this section will help to provide the basic parametric data for dynamic response analyses of soil or soil-structure systems. Emphasis is placed on the discussion of the key dynamic design parameters, namely shear modulus and damping, and ways of estimating suitable values are suggested. For completeness, typical values are also given for soil

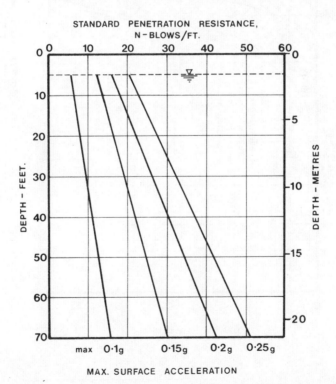

Figure 5.21 Standard penetration resistance values above which liquefaction is unlikely to occur under any conditions (after Seed and Idriss[36])

134

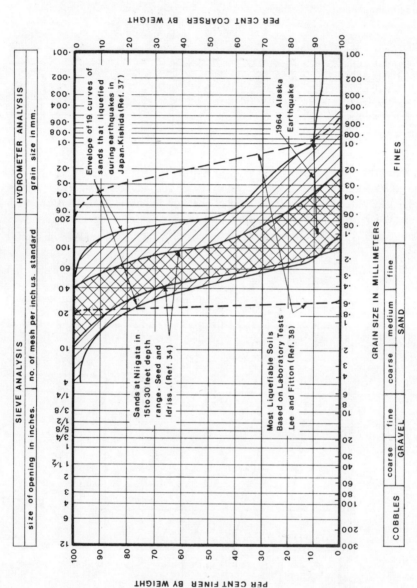

Figure 5.22 Liquefaction potential related to particle size (after Shannon *et al.*[35])

densities, modulus of elasticity and Poisson's ratio. A more detailed discussion of most of these parameters may be found elsewhere.[35] In order to obtain appropriate values of these parameters for a given site, suitable field and laboratory tests as discussed in Section 3.3 may be necessary.

5.5.2.3(i) Shear modulus

For small strains the shear modulus of a soil can be taken as the mean slope of the stress-strain curve. At large strains the stress-strain curve becomes markedly non-linear so that the shear modulus is far from constant but is dependent on the magnitude of the shear strain (Figure 5.23).

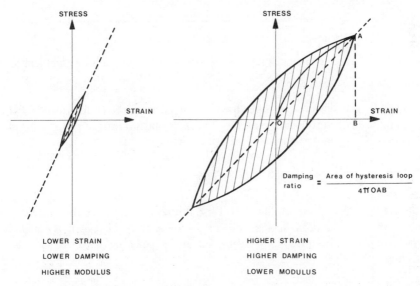

Figure 5.23 Illustration defining the effect of shear strain on damping and shear modulus of soils (after Seed and Idriss[39])

There are various field and laboratory methods available for finding the shear modulus G of soils. Field tests concentrate on finding the shear wave velocity v_s and calculating the shear modulus from the relationship

$$G = \rho v_s^2 \qquad (5.50)$$

where ρ is the mass density of the soil. Laboratory methods generally measure G more directly from stress–strain tests. It is clear from Figure 5.23 that the level of strain at which G is measured must be known. Average relationships of shear modulus to strain are shown for clay and sand in Figure 5.24.

Shear strains developed during earthquakes may increase from about 10^{-3} percent in small earthquakes to 10^{-1} percent for large motions, and the maximum strain in each cycle will be different. Whitman[41] suggests that

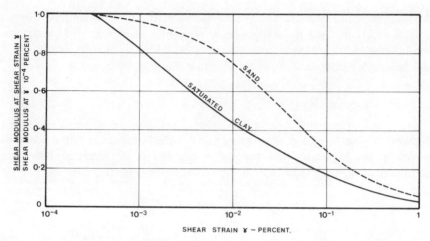

Figure 5.24 Average relationships of shear modulus to shear strain for sand and saturated clays (After Seed and Idriss[40])

for earthquake design purposes a value of two thirds G measured at the maximum strain developed may be used. Alternatively an appropriate value of G can be calculated from the relationship

$$G = \frac{E}{2(1 + v)} \tag{5.51}$$

where E is Young's modulus and v is Poisson's ratio. In the absence of any more specific data, low strain values of E may be taken from Table 5.7. Values of Poisson's ratio from Table 5.8 may be used in the above formula.

5.5.2.3(ii) Damping

The second key dynamic parameter for soils is damping. Two fundamentally different damping phenomena are associated with soils, namely material damping and radiation damping.

(a) *Material damping.* Material damping (or internal damping) in a soil occurs when any vibration wave passed through the soil. It can be thought of as a measure of the loss of vibration energy resulting primarily from hysteresis in the soil. Damping is conveniently expressed as a fraction of critical damping, in which form it is referred to as the damping ratio. A physical definition of the damping ratio for soils is shown graphically in Figure 5.23. Published data on damping ratios are sparse, and consist only of values deduced from tests on small samples, or theoretical estimates. It should be appreciated that to date no *in situ* determinations of material damping have been made, and that damping ratios can only be used in analyses in a comparative sense. Accepting a philosophy that dynamic soils analysis will be warranted

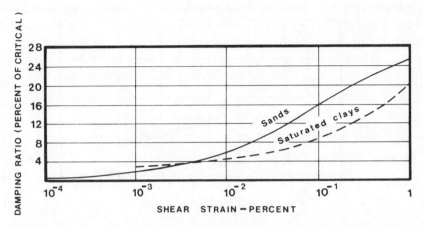

Figure 5.25 Average relationship of internal damping to shear strain for sands and saturated clays (After Seed and Idriss[40])

for some projects, at least for its qualitative information, a means of choosing values of material damping is required. Some material damping values are therefore given in Figure 5.25. These represent average values of laboratory test results on sands and saturated clays as presented elsewhere.[40] In the absence of any other information it may be reasonable to take the damping of gravels as for sand.

(b) *Radiation damping.* In considering the vibration of foundations radiation damping is present as well as material damping. Radiation damping is a measure of the energy loss from the structure through radiation of waves away from the footing, i.e. it is a purely geometrical effect. Like material damping, it is very difficult to measure in the field. The theory for the elastic half-space has been used to provide estimates for the magnitude of radiation damping. Whitman and Richart[42] have calculated values of radiation damping for circular footings for machines by this method and their results are reproduced in Figure 5.26.

As with the values for material damping, the limitations of the values in Figure 5.26 must be emphasized. Because they are only theoretical values for a particular type of footing, they should be applied with circumspection. In the analysis of foundations of buildings the usefulness of Figure 5.26 may be for qualitative rather than quantitative assessments, but the following generalizations may be helpful. For horizontal and vertical translations, radiation damping may be quite large (>10 percent of critical), while for rocking or twisting it is quite small (about 2 percent of critical) and may be ignored in most practical design problems.

A further limitation of the half-space theory is that it takes no account of the reflective boundaries provided by harder soil layers or by bedrock at some distance vertically or horizontally from the structure. Any such reflection of radiating waves will naturally reduce the beneficial radiation damping

138

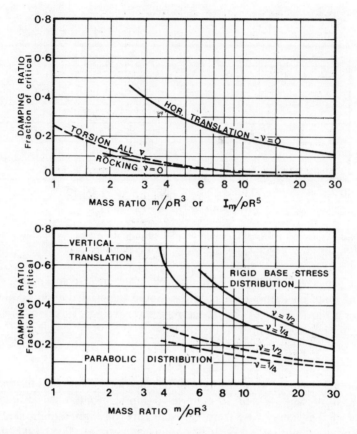

Figure 5.26 Values of equivalent damping ratio for radiation damping, of machines, derived from the theory of circular footings on elastic half-space (after Whitman and Richart[42])

effect. Various aspects of radiation damping are discussed in Section 5.5.3 below.

In Figure 5.26 m is the mass of the foundation block plus machinery, I_m is the mass moment of inertia of the foundation block plus machinery, R is the radius (or equivalent radius) of the soil contact area at the foundation base, ρ is the mass density of the soil and v is Poisson's ratio for the soil. For rectangular bases the equivalent radius is given by the following;

for translation

$$R = \left(\frac{BL}{\pi}\right)^{1/2}$$

for rocking

$$R = \left(\frac{BL^3}{3\pi}\right)^{1/4}$$

for twisting

$$R = \left\{ \frac{BL(B^2 + L^2)}{6\pi} \right\}^{1/4}$$

5.5.2.3(iii) Other basic soil properties

Typical values of soil density (ρ), modulus of elasticity (E), and Poisson's ratio (v) are given below in Tables 5.6, 5.7 and 5.8 respectively.

Table 5.6. Typical mass densities of basic soil types

Soil type	Mass density ρ (Mg/m³)*			
	Poorly graded soil		Well graded soil	
	Range	Typical value	Range	Typical value
Loose sand	1·70–1·90	1·75	1·75–2·00	1·85
Dense sand	1·90–2·10	2·00	2·00–2·20	2·10
Soft clay	1·60–1·90	1·75	1·60–1·90	1·75
Stiff clay	1·90–2·25	2·07	1·90–2·25	2·07
Silty soils	1·60–2·00	1·75	1·60–2·00	1·75
Gravelly soils	1·90–2·25	2·07	2·00–2·30	2·15

*Values are representative of moist sands and gravels and saturated silts and clays.

Table 5.7. Typical modulus of elasticity values for soils and rocks

Soil type	E (MN/m²)	E/c_u
Soft clay	up to 15	300
Firm, stiff clay	10 to 50	300
Very stiff, hard clay	25 to 200	300
Silty sand	7 to 70	
Loose sand	15 to 50	
Dense sand	50 to 120	
Dense sand and gravel	90 to 200	
Sandstone	up to 50,000	400
Chalk	5,000 to 20,000	2000
Limestone	25,000 to 100,000	600
Basalt	15,000 to 100,000	600

Note the values of E vary greatly for each soil type depending on the chemical and physical condition of the soil in question. Hence the above wide ranges of E value provide only vague guidance prior to test results being available. The ratio E/c_u may be helpful, if the undrained shear strength c_u is known, although the value of this ratio also varies for a given soil type. See elsewhere for information on E values for Clays,[43] Sands[44] and Rocks.[45]

Table 5.8. Typical values of Poisson's ratio for soils

Soil type	Poisson's ratio v
Sand	0·35
Saturated clay	0·50
Most other soils	0·40

5.5.3 Seismic behaviour of soil-structure systems

The importance of the nature of the sub-soil for the seismic response of structures has been demonstrated in many earthquakes, but a reasonable understanding of the factors involved has only recently begun to emerge. For example it seems clear from studies of recent earthquakes that the relationship between the periods of vibration of structures and the period of the supporting soil is profoundly important regarding the seismic response of the structure. In the case of the 1970 earthquake at Gediz, Turkey, part of a factory was demolished in a town 135 km from the epicentre while no other buildings in the town were damaged. Subsequent investigations revealed that the fundamental period of vibration of the factory was approximately equal to that of the underlying soil. Further evidence of the importance of periods of vibration was derived from the medium sized earthquake of Caracas in 1967 which completely destroyed four buildings and caused extensive damage to many others. The pattern of structural damage has been directly related to the depth of soft alluvium overlying the bedrock.[46] Extensive damage to medium-rise buildings (5–9 storeys) was reported in areas where depth to bedrock was less than 100 m while in areas where the alluvium thickness exceeded 150 m the damage was greater in taller buildings (over 14 storeys). The depth of alluvium is of course directly related to the periods of vibration of the soil. Considering shear waves travelling vertically through a soil layer of depth H, the periods of horizontal vibration of the soil are given by

$$T_n = \frac{4H}{(2n - 1)v_s}$$
(5.52)

where n is an integer, 1,2,3..., and v_s is the velocity of the shear wave.

In order to evaluate the seismic response of a structure at a given site, the dynamic properties of the combined soil-structure system must be understood. The nature of the sub-soil may influence the response of the structure in three ways.

(i)   The phenomenon of *soil amplification* may occur, in which the seismic excitation at bedrock is modified during transmission through the overlying soils to the foundation. This may cause attenuation or amplification effects.

(ii)   The fixed base dynamic properties of the structure may be signifi-
        cantly modified by the presence of soils overlying bedrock. This
        will include changes in the mode shapes and periods of vibration.
(iii)   A significant part of the vibrational energy of the flexibly supported
        structure may be dissipated by material damping and radiation
        damping in the supporting medium.

Items (ii) and (iii) above are investigated under the general title of *soil-structure interaction* which may be defined as the interdependent response relationship between a structure and its supporting soil. The behaviour of the structure is dependent in part upon the nature of the supporting soil and similarly the behaviour of the stratum is modified by the presence of the structure.

It follows that *soil amplification* (item (i) above) will also be influenced by the presence of the structure, as the effect of soil-structure interaction is to produce a difference between the motion at the base of the structure and the free-field motion which would have occurred at the same point in the absence of the structure. In practice however, this refinement in determining the soil amplification is seldom taken into account, the free-field motion generally being that which is applied to the soil-structure model as discussed in the following section. Because of the difficulties involved in making dynamic analytical models of soil systems, it has been common practice to ignore soil-structure interaction effects simply treating structures as if rigidly based regardless of the soil conditions. However intensive study in recent years has produced considerable advances in our knowledge of soil-structure interaction effects and also in the analytical techniques available, as discussed below.

5.5.3.1 Dynamic analysis of soil-structure systems

Comprehensive dynamic analysis of soil-structure systems is the most demanding analytical task in earthquake engineering. The cost, complexity, and validity of such exercises are major considerations. There are two main problems to be overcome. Firstly the large computational effort which is generally required for the foundation analysis makes the choice of foundation model very important; four main methods of modelling the foundation are discussed in the next section. Secondly there are great uncertainties in defining a design ground motion which not only represents the nature of earthquake shaking appropriate for the site, but also represents a suitable level of risk.

Ideally the earthquake motion should be applied at bedrock to the complete soil-structure system. This is not a very realistic method at the present time because much less is known about bedrock motion than surface motion and there is a great scatter in possible results for the soil amplification effect defined above. At present the most realistic methods of analysis seem to be those which apply the *free-field* motion to the base of the structure, the

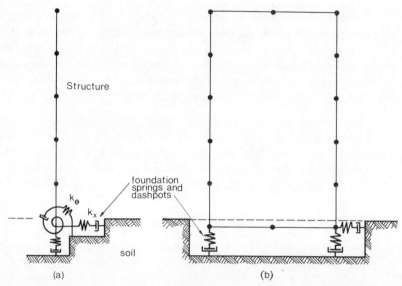

Figure 5.27 Soil-structure analytical models representing the soil flexibility simply by springs at the base of the structure

free field motion being that which would occur at the surface in the absence of the structure. This may be done most simply using simple springs at the base of the structure (Figure 5.27), as described in Section 5.5.3.2(i), or using a sub-structuring technique in which the foundation dynamic character- istics are predetermined and superposition of soil and structure response is carried out. The latter technique has been described by Penzien and Tseng[47] using half-space modelling of the soil, and by Vaish and Chopra[48] who illustrate their presentation with finite element modelling. These two types of soil model are discussed in the next section.

It should be noted that where the dynamic behaviour is expressed in fre- quency-dependent terms, the problem must be analysed in the *frequency domain* not the *time domain*. For this purpose acceleration-time records must be transformed into acceleration-frequency terms using Fourier transform methods before application to the system. An inverse transformation is required to obtain the response time record. These techniques are described in the above two papers.[47,48]

For projects in which soil-structure interaction effects are likely to be important, the choice of analytical method requires careful consideration. The reader will find useful extra guidance in the state-of-the-art reports by Kausel and Roesset[49] and by Seed *et al.*[50]

5.5.3.2 Soil models for dynamic analysis

A realistic dynamic model of the soil requires the representation of soil stiffness, material damping and radiation damping, allowing for strain-depen- dence (non-linearity) and variation of soil properties in three dimensions.

While various analytical techniques exist for handling different aspects of the above soil behaviour they all suffer from varying combinations of expensiveness or inaccuracy. Therefore there is some difficulty for any given project in choosing an analytical model for the soil which will permit an appropriate level of understanding of the soil-structure system.

The methods of modelling the soil may be divided into four categories of varying complexity;

(i)   equivalent static springs and viscous damping at base level only;

(ii)   shear beam analogy using continua or lumped masses and springs distributed vertically through the soil profile;

(iii)   elastic or viscoelastic half-space;

(iv)   finite elements.

A brief discussion of each of the above modelling methods follows below.

5.5.3.2(i) *Springs at base level*

The simplest method of modelling the soil is to use springs at base level to represent the horizontal, rocking, vertical and torsional stiffness of the soil (Figure 5.27). In the system shown in Figure 5.27(b), the stiffness of the individual vertical springs must be chosen to sum to either the required rocking stiffness or the required total vertical stiffness, as it is unusual to achieve both conditions simultaneously. This generally does not matter in analyses in which vertical and horizontal excitations are not applied simultaneously.

A convenient method for determining the overall foundation spring stiffnesses is to use the zero-frequency (static) stiffnesses derived from elastic half-space theory as given in Table 5.9. It should be noted that the values in Table 5.9 are for a homogeneous elastic half-space, but may be factored to give some equivalence to layered soils or to allow for a given degree of non-linearity in the soil behaviour.

The spring stiffnesses are dependent on the shear modulus which in turn varies with the level of shear strain. Hence for linear elastic calculations, spring stiffnesses should be calculated corresponding to a value of shear strain which is less than the maximum expected shear strain. For instance, if the spring stiffness at low strain is k_0, then a value of k equal to $0.67 k_0$ may be used in the analysis. Alternatively a series of comparative analyses may be done using a range of values of k, particularly if *in situ* tests have not been made, in this case it may be appropriate to select values of k from the following ranges:

for translation

$$0.5 k_0 \leqslant k \leqslant k_0$$

for rocking

$$0.33 k_0 \leqslant k \leqslant k_0$$

Table 5.9. Spring stiffnesses k for rigid base resting on elastic half-space

Motion	Circular footings	Rectangular footings
Vertical	$\dfrac{4GR}{1-v}$	$\dfrac{G}{(1-v)}\beta_z\sqrt{(BL)}$
Horizontal	$\dfrac{32(1-v)GR}{7-8v}$	$2G(1+v)\beta_x\sqrt{(BL)}$
Rocking	$\dfrac{8GR^3}{3(1-v)}$	$\dfrac{G\beta_\phi BL^2}{1-v}$
Torsion	$\dfrac{16GR^3}{3}$	

G is the shear modulus for the soil where $G = E/\{2(1+v)\}$, v is Poisson's ratio for the soil, R is the radius of the footing, B, L, are the plan dimensions of rectangular pads, and $\beta_x, \beta_z, \beta_\phi$ are coefficients given in Figure 5.28.

In some computer programs spring supports may not be available and members with appropriate area and stiffness can be used instead.

When using the very simple foundation stiffness modelling of Figure 5.27 difficulties arise in accurately representing the effects of material damping and radiation damping in the foundation, as the total amount of equivalent viscous damping for the foundation (Section 5.5.2.3ii) is likely to exceed considerably that for the superstructure. A conservative compromise between the structural and soil damping values will generally be necessary.

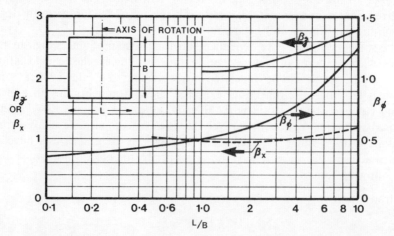

Figure 5.28 Coefficients β_x, β_z and β_ϕ for rectangular footings (after Whitman and Richart[42])

Also the damping in the soil in different modes of vibration varies considerably. As most currently available dynamic analysis computer programs are written for equal damping in all modes, some intermediate value of damping has to be chosen which hopefully will lead to the most realistic result. The value of damping used should not vary too greatly from that of the mode in which most of the vibrational work is done. Hence a trial mode shape analysis may have to be done to determine which modes predominate. Use of too high or too low a value of damping will lead to unconservative or conservative results respectively.[51]

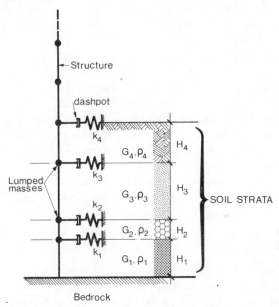

Figure 5.29 Soil-structure analytical model representing the soil vertical profile by lumped parameter system of springs and dashpots

5.5.3.2(ii) *Lumped masses and springs*

The shear beam approach with lumped masses and springs may be used to model the soil layers overlying bedrock (Figure 5.29) although difficulties arise in choosing appropriate stiffness and damping values for the soil. Nonlinearity may be allowed for using iterative linear analyses such as those used in soil amplification studies (Section 3.4.3.2i), or by non-linear foundation springs as in the research on piled bridge foundations reported by Penzien.[52] The continuum method is described in item (ii) of Section 3.4.3.1.

5.5.3.2(iii) *Elastic or viscoelastic half-space*

Modelling the foundation as a homogeneous linear elastic or viscoelastic half-space in which the stiffness and damping are treated as frequency-dependent provides a very useful means of allowing for the radiation damping effect. Luco and Westmann[53] and Veletsos and Wei[54] have evaluated the complex expressions for dynamic stiffness (impedance) k^*, of rigid circular footings on a homogeneous elastic half-space which are of the form

$$k^*(\omega) = k(\omega) + i\omega c(\omega) \tag{5.53}$$

in which the real part of the expression can be considered as a stiffness and the imaginary part represents the radiation damping. The zero-frequency value of the stiffness $k(\omega = 0)$ is as given in Table 5.9.

The viscoelastic formulation of foundation impedance is an improvement on the elastic case because material damping may be incorporated by including a constant term in the imaginary part of the expression, that is $c(\omega)$ is replaced by a term of the form $c_1 + c(\omega)$. This has been done by Veletsos and Nair[55] who considered material damping to be linearly hysteretic.

Non-linear soil behaviour cannot be explicitly modelled in frequency domain solutions used for the above formulations, but the viscoelastic hysteretic model can be thought of as representing a limited degree of non-linearity.

The half-space model also permits the study of layering, such as by Luco,[56] but further work is required on the basically closed-form solution provided by this technique before its results can be properly compared with the finite element method.

Various attempts have also been made to study the behaviour of foundations embedded into a half-space, such as those of Bielak,[57] Luco et al.,[58] and Novak and Beredugo.[59]

5.5.3.2(iv) Finite elements

The use of finite elements for modelling the foundation of a soil-structure system is the most comprehensive (if most expensive) method available. Like the half-space model it permits radiation damping and three-dimensionality, but has the major advantage of easily allowing changes of soil stiffness both vertically and horizontally to be explicitly formulated. Embedment of footings is also readily dealt with. Although a full three-dimensional model is generally too expensive, three dimensions should be simulated. This can be achieved either by an equivalent two-dimensional model, or for structures with cylindrical symmetry an analysis in cylindrical coordinates can be used.[60]

In order to simulate radiation of energy through the boundaries of the element model three main methods are available.

(a) *Elementary boundaries* that do not absorb energy and rely on the distance to the boundary to minimize the effect of reflected waves.

(b) *Viscous boundaries* which attempt to absorb the radiating waves, modelling the far field by a series of dashpots and springs, as used by Lysmer and Kuhlemeyer.[61] The accuracy of this method is not very good for thin surface layers or for horizontal excitation, although an improved version has been developed by Ang and Newmark.[62]

(c)   *Consistent boundaries* are the best absorptive boundaries at present available, reproducing the far field in a way consistent with the finite element expansion used to model the core region. This method was developed by Lysmer and Waas[63] and generalized by Kausel.[60] The latter method among other things allows the lateral boundary to be placed directly at the side of the foundation, with a considerable reduction in the number of degrees of freedom.

Non-linearity of soil behaviour can be modelled with non-linear finite elements, but the necessary *time-domain* analysis, is very expensive. Alternatively non-linearity could theoretically be simulated in repetitive linear model analyses with adjustment of modulus and damping in each cycle as a function of strain level. In *frequency-domain* solutions (for example, when using consistent boundaries) non-linearity can be approximately simulated again using an iterative approach.

As in the half-space solutions, material damping may be accounted for by using a viscoelastic finite element model, as used by Kausel.[60,64]

5.5.3.3 *Results of soil-structure interaction studies*

In recent years there have been intensive theoretical investigations of the dynamics of soil-structure systems using soil modelling techniques as described above. Although many of the conclusions of these studies are still tentative, requiring experimental verification, some of the results are physically or intuitively sound. A brief summary of the more important conclusions is therefore included here.

5.5.3.3(i)

Perhaps the leading question to be answered about soil-structure interaction is: 'For what soil conditions will the rigid base assumption lead to significant errors in the response calculations?'. Veletsos and Meek[65] have suggested that consideration of soil-structure interaction is only warranted for values of the ratio

$$\frac{v_\mathrm{s}}{fh} < 20 \tag{5.54}$$

where v_s is the shear wave velocity in the soil, f is the fixed-base frequency of the single degree-of-freedom structure and h is its height. Substituting $f \approx 30/h$ for framed buildings, and $f \approx 45/h$ for shear wall buildings in the above equation implies that soil-structure interaction effects may be important for framed buildings when $v_\mathrm{s} \leqslant 600$ m/s, or for shear wall buildings when $v_\mathrm{s} \leqslant 900$ m/s. It is of interest to note that shear wave velocities of this order are used to define rock-like material for site response purposes in the Californian code[66] as noted in Section 3.3.2.2 above.

It is of interest that equation (5.54) correctly predicts that soil-structure interaction is important for the concrete gravity oil platforms studied by

Watt *et al.*[67] Radiation damping effects were found to reduce the base shear of a platform on 'very hard' ground (v_s = 480 m/s) by about 50 percent (the relevant value of v_s/fh was 6·6), despite the fact that the foundation was effectively rigid regarding its effect on the mode shapes and periods.

5.5.3.3(ii)

The periods of vibration of a given structure increase with decreasing stiffness of the sub-soil. This logical phenomenon has been widely remarked, such as by Veletsos and Meek[65] and Watt *et al.*[67] The latter found this effect to be very marked for a large offshore oil platform, where the fundamental period was 2·95 s for the rigid foundation condition and 5·9 s when allowance was made for a sub-stratum of 'firm' overconsolidated clay (Figure 5.30a).

5.5.3.3(iii)

The mode shapes of a given structure change as some function of the soil stiffness. Coupled with this effect there may be a corresponding change in the predominant mode; it can be seen from Figure 5.30(b) that for a given oil platform structure the dominant mode changed from third to first with decreasing soil stiffness.

5.5.3.3(iv)

According to Veletsos and Nair[55] hysteretic action in the soil should be considered in soil-structure systems. They found that for realistic systems hysteretic action has the effect of increasing the overall damping of the system, and reducing the deformation of the structure.

5.5.3.3(v)

Radiation damping in the foundation generally leads to a reduction in response of a structure. For large concrete gravity platforms this reduction may be as much as 50 percent as shown for the overturning moments in Figure 5.30(a).

5.5.3.3(vi)

Because of the complexity and expense of rigorously computing the effects of radiation damping in the foundation, an equivalent viscously damped response spectrum technique would be desirable. For estimating an equivalent viscous damping for a soil-structure system, the foundation damping (radiation plus hysteretic) is not directly additive to the structural damping. Veletsos and Nair[55] have suggested an approximate method for doing this, in which the overall damping for tall slender interacting structures may be less than that applicable to fixed-base structures. This applies to structures

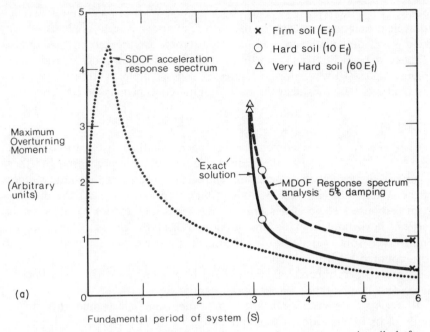

Figure 5.30(a) Variation of peak seismic base moment of a concrete gravity oil platform, resulting from three different foundation stiffnesses

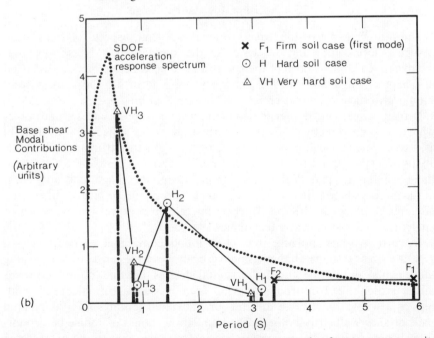

Figure 5.30(b) Base shear contributions of the significant modes, for a concrete gravity oil platform with three different foundation stiffnesses

founded on either elastic or viscoelastic media. Watt et al.[67] found that the equivalent damping concept for use with response spectrum analyses of concrete gravity oil platforms was not satisfactory, as the amount of damping required depended strongly on the part of the structure under consideration (Figure 5.30a). However for structures with more uniform stiffness and mass distributions the equivalent viscous damping concept may be reliable.

5.5.3.3(vii)

The effects of embedment have been studied by various workers.[57,58,59,64,68,69] To date the findings are not very comprehensive but there is general agreement that increasing embedment increases the static stiffness of the system, decreases the periods of vibration and decreases the displacement responses. These effects are evident in all four modes of vibration i.e. vertical, horizontal, rocking and torsion. Where backfill is softer than the undisturbed soil, the effects of embedment are obviously reduced. This can be accounted for in the theory.

In cases where theoretical results have been compared with experimental, agreement is qualitative rather than quantitative.

It is clear that deep embedment has important effects and that further study of its implications is warranted.

5.5.4 Aseismic foundation design

The following discussion covers only the seismic aspects of foundation design. Before completing the design of the foundations it is assumed that the dynamic characteristics of the sub-soil have been determined as discussed in Chapter 3 and Section 5.5.2, and a suitable form for the sub-structure should also have been chosen as suggested in Section 4.5.

It then remains to design the foundations for appropriate seismic forces which arise (a) directly from the deformation of the adjacent soil, and (b) as a result of the earthquake forces acting in the superstructure. While our ability to estimate the seismic forces from (b) above is now quite advanced, there remains a great deal of uncertainty about the magnitude and effect of the forces induced directly by the ground. This is true despite the increasing attempts to elucidate the soil-structure interaction problem by sophisticated analytical and experimental techniques.

In current design practice it is often found convenient to consider two separate stress systems: (i) the seismic vertical stresses (e.g. due to overturning movements) and (ii) the seismic horizontal stresses (e.g. due to the base shear on the structure). Overturning moments are not usually a problem for buildings as a whole, but can be difficult for individual footings such as column pads or shearwall strip footings. The foundation should of course be proportioned so as to keep the maximum bearing pressures due to the overturning moments and gravity loads within the allowable seismic value for the soil

Table 5.10. Allowable bearing pressures on soils under public buildings in New Zealand[71]

Soil types		Allowable bearing pressures (kN/m²) Long term loads		Total loads (including seismic loads)		Standard penetration blow count (N)	Apparent cohesion c_u (kN/m²)
Soft or broken rock		960		1440			
Gravel	Dense	285–570		285–570			
	Medium	96–285		72–215			
		Well graded	Uniform	Well graded	Uniform		
Sand*	Dense	240–525	120–265	240–525	120–265		
	Medium	96–240	48–120	72–180	40–90		
Clay†	Very stiff	190–380		285–570			
	Medium stiff	48–190		72–285		4–15	25–100
	Soft	0–48		0–48		0–4	0–25
Peat, silts made ground		To be determined after investigation					

*Reduce bearing pressures by half below the water table.
†Alternatively: Allow 1·2 times c_u for round and square footings, and 1·0 times c_u for length/width ratios of more than 4·0. Interpolate for intermediate values.

concerned. But unfortunately there is little agreement on what constitutes safe seismic bearing pressures on sedimentary soils. Most earthquake codes do not discuss the effect of soil type on bearing pressures. It appears that most soils are capable of sustaining higher short-term loads than long-term loads, with the exception of some sensitive clays which lose strength under dynamic loading.

The bearing pressures used in the design of public buildings in New Zealand quoted in Table 5.10 may be helpful; here the bearing pressures are reduced by 25 percent for medium gravel and medium sand, and increased by 50 percent for rock and very stiff or medium stiff clay. The values in Table 5.10 may in some cases be over-conservative, and well informed geotechnical advice should in any case be taken for the actual soil conditions for each project.

Further guidance on some aspects of aseismic foundation design has been given by Zeevaert.[70]

The horizontal interaction stresses between the soil and the foundation are arguably more problematical than the vertical stresses, as comparatively little is known about allowable seismic passive pressures and the effect of seismic active pressure in different foundation situations. Indeed it is customary to assume even more arbitrary distributions for horizontal stress between foundations and soil than for vertical stress. The main problems (peculiar to earthquakes) of foundation design as presently understood occur in transferring the base shear of the structure to the ground, and in maintaining structural integrity of the foundation during differential soil deformations. Some design guidance on these problems now follows under the headings of;

(i)   shallow foundations;
(ii)   deep foundations;
(iii) piled foundations.

5.5.4.1 Shallow foundations

The horizontal seismic shear force at the base of the structure must be transferred through the sub-structure to the soil. With shallow foundations it is normal to assume that most of the resistance to lateral load is provided by friction between the soil and the base of the members resisting horizontal load. Other footings and slabs in contact with the ground may also be assumed to provide shear resistance if they are suitably connected to the main resisting elements. The total available resistance to lateral movement of the structure may be taken to be equal to the product of the dead load carried by the elements considered and the coefficient of sliding friction between the soil and the sub-structure. Typical values of friction angles for foundations are given in Table 5.11.

In some cases further horizontal resistance will arise from the passive soil pressures developed against subsurface elements. If this resistance is taken

Table 5.11. Typical friction angles and adhesion values for bases without keys[72]

Interface materials	Friction angle (degrees)	Adhesion (kN/m²)
Mass concrete on the following foundation material:		
Clean sound rock	35–45	
Clean gravel, gravel–sand mixtures, coarse sand	29–31	
Clean fine to medium sand, silty medium to coarse sand, silty or clayey gravel	24–29	
Clean fine sand, silty or clayey fine to medium sand	19–24	
Fine sandy silt, non-plastic silt	17–19	
Very stiff and hard residual or preconsolidated clay	22–26	
Medium stiff and stiff clay and silty clay	17–19	
Formed concrete on the following foundation material:		
Clean gravel, gravel–sand mixtures, well graded rock fill with spalls	22–26	
Clean sand, silty sand–gravel mixture, single size hard rock fill	17–22	
Silty sand, gravel or sand mix with silt or clay	17	
Fine sandy silt, non-plastic silt	14	
Soft clay and clayey silt		10–35
Stiff and hard clay and clayey silt		35–60

into account it is usual to restrict the calculated total restraint by reducing either the frictional force or the passive resistance force by fifty percent. In order to ensure that the passive restraint can be developed, appropriate measures must be taken on site, such as adequate compacting of backfill against sides of footings.

Shallow foundations are often of a form that is highly vulnerable to damage from differential horizontal and vertical ground movements during earthquakes. It is therefore good practice even in quite low structures, especially those founded on soft soils, to provide ties between column pads. In the absence of a more realistic method an arbitrary design criterion for such ties is to make them capable of carrying compression and tension loads equal to ten percent of the maximum vertical load in adjacent columns (Section 6.2.7). However, it may be possible to resist some or all of these horizontal forces by passive action of the soil, particularly for light buildings. The designer may also have a choice between providing the tie action at the bottom floor level (in tie beams or in the slab), or at some other position in relation to the foundations.

5.5.4.2 Deep foundations

Unfortunately at the present time an authoritative aseismic design rationale for deep foundations does not exist, as discussed by Barnes.[73] Designers must rely mainly on normal structural and geotechnical static design techniques, supplemented where appropriate by consideration of known seismic phenomena such as seismically enhanced soil pressures. The natural stiffness and strength of the box shape of deep foundations should be utilized to advantage in distributing the seismic forces from the soil and the superstructure throughout the foundation with an adequate safety factor.

Although less susceptible to damage from ground motions than isolated pad footings, deep box foundations nevertheless require proper design to withstand strong earthquakes. This was exemplified by the virtual destruction of underground water tanks in the 1971 San Fernando earthquake.[74] This failure demonstrated the importance of internal walls to provide an egg-crate type of stiffening; it also showed the valuable contribution that concrete keys could have provided to the strength of construction joints which moved 0.67 m in shear despite the presence of steel reinforcement normal to the joint.

5.5.4.3 Piled foundations

The proper aseismic design of piled foundations will include consideration of the vertical and horizontal stresses and the structural integrity of the foundation. Vertical seismic loads in individual piles may vary greatly depending on their position in relation to the rest of the pile group and to the superstructure. Some piles, particularly those at the edges or corners of pile systems, may have to carry large tensile as well as compression forces during earthquakes. In any case care will be needed to ensure that the strata contiguous to and below the piles have sufficient adhesive, shear and bearing strength in seismic conditions.

Lack of structural integrity has caused failure of piled foundations in earthquakes, such as that of San Fernando, 1971. Sufficient continuity reinforcement must be provided between the piles and the pile cap, and the piles themselves must obviously be able to develop the required tensile, compression and bending strength. Where plastic hinges are likely to form in the tops or bottoms of reinforced concrete piles, suitable confinement reinforcement must be provided, as for columns (Sections 6.2.7 and 6.2.10).

The most difficult aspect of the aseismic design of piles is that of lateral strength, as little is known of the stress deformations involved in soil-pile interaction during earthquakes. Some light has been thrown on the likely behaviour of long piles in deep sensitive clay in a sophisticated non-linear analysis of a bridge described by Penzien.[52] In this case it was found that if subjected to an earthquake like that of El Centro (1940), the piles would have been deformed to their yield curvatures.

The lateral resistance of end bearing piles may be determined by similar methods to those for friction piles, or as discussed by Zeevaert,[70] with allowance being made for moment fixity at the bottoms of the piles.

Of the various design methods for the lateral strength of piles, that for friction piles originated by Broms[75] has been adopted in this text. Broms assumed that the ultimate lateral resistance of a pile is developed when the soil yields and plastic hinges develop in the pile. This approach forms the basis of the design procedures outlined below.

Lateral resistance of friction piles

It is assumed that there are three different modes of pile failure, depending on the pile length L.

Short pile (Figure 5.31a). When the pile is short, failure occurs as the soil yields along the full length of the pile. The limiting case is reached when a hinge develops in the top of the pile at the same time as the soil yields. This occurs when $L = L_1$, and for $L < L_1$ the pile is classed as short. L_1 can be evaluated in terms of the soil properties and the moment capacity of the pile.

Long pile (Figure 5.31c). If the pile is sufficiently long, failure occurs when two hinges develop in the pile. This occurs if the pile is firmly enough embedded in the bearing stratum, otherwise the soil will fail and the pile will rotate (Figure 5.31b). A limiting length L_2 can be determined such that when $L > L_2$ the pile is classed as long.

Intermediate length pile. If the pile length is between L_1 and L_2 the pile is classed as intermediate.

The design procedure is as follows.

(a) Decide pile lengths on the basis of vertical load bearing considerations.
(b) For the given soil conditions and pile dimensions establish L_1 and L_2.
(c) Establish the type of lateral failure by comparing L to L_1 and L_2.
(d) Having classed the pile, calculate the ultimate lateral resistance V_u.
(e) Check that the lateral deflection of the piles is tolerable.

When the pile is of intermediate length it may be easiest to calculate V_u at L_1 and L_2 and interpolate for V_u at L.

156

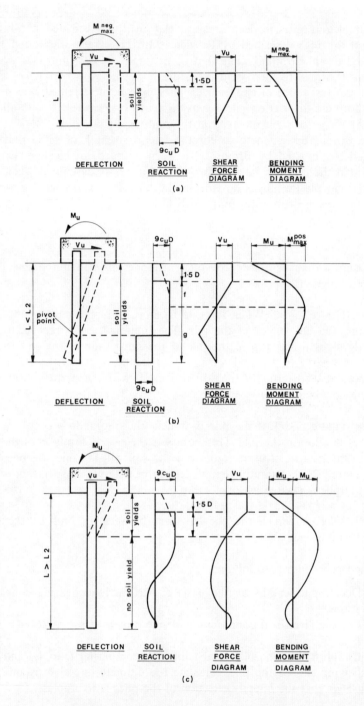

Figure 5.31 Lateral resistance of fixed head piles in cohesive soil. (a) Short pile, (b) intermediate pile, (c) long pile (developed from Broms[75])

157

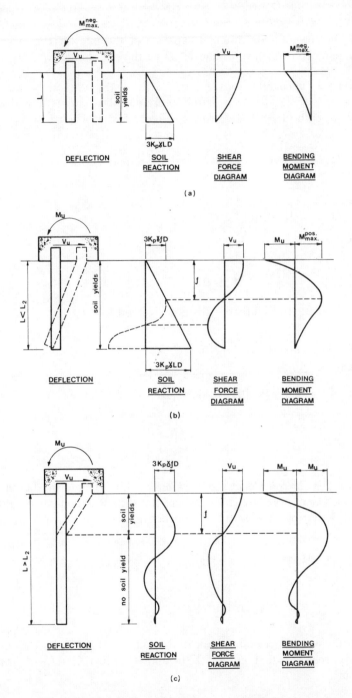

Figure 5.32 Lateral resistance of fixed head piles in cohesionless soil. (a) Short pile, (b) intermediate pile, (c) long pile (developed from Broms[75])

Evaluation of limit lengths L_1 *and* L_2

Let M_u be the ultimate moment capacity of the pile for the given axial load and reinforcement layout, and let D be the pile diameter.

(i) *Friction piles in cohesive soils (Figure 5.31)*. Let the ultimate bearing capacity of the soil be $9c_u$, where c_u is the undrained shear strength, and assume $c_u = 0$ to a depth of $1.5D$. Referring to Figure 5.31(a) when $L = L_1$

$$M_u = 9c_u D(L_1 - 1.5D)\left\{\frac{(L_1 - 1.5D)}{2} + 1.5D\right\}$$

i.e.

$$M_u = 4.5c_u D(L_1^2 - 2.25D^2)$$

Therefore

$$L_1 = \left\{2.25D^2 + \frac{M_u}{4.5c_u D}\right\}^{1/2} \tag{5.55}$$

Similarly, referring to Figures 5.31(b) and (c), when $L = L_2$

$$L_2 = f + g + 1.5D \tag{5.56}$$

where f can be found from

$$2.25c_u Df^2 + 6.75c_u D^2 f - M_u = 0$$

and

$$g = \left[\frac{M_u}{2.25c_u D}\right]^{1/2}$$

(ii) *Friction piles in cohesionless soils (Figure 5.32)*. Assume the lateral earth pressure coefficient at failure is $3K_p$, where K_p is the Rankine passive pressure coefficient. Referring to Figure 5.32(a), when $L = L_1$

$$M_u = 3K_p \gamma \frac{DL_1^2}{2} \cdot \frac{2L_1}{3}$$

$$L_1 = \left[\frac{M_u}{K_p \gamma D}\right]^{1/3} \tag{5.57}$$

where γ is the weight per unit volume of the soil.

Similarly, referring to Figures 5.32(b) and (c), when $L = L_2$, L_2 may be found from

$$M_u = V_u L_2 - \frac{K_p \gamma L_2^2 D}{2} \tag{5.58}$$

using

$$V_{\text{u}} = 1{\cdot}5K_{\text{p}}\gamma Df^2 \tag{5.59}$$

and

$$f = \left[\frac{2M_{\text{u}}}{K_{\text{p}}D}\right]^{1/3}$$

5.6 EARTH-RETAINING STRUCTURES

5.6.1 Introduction

As the non-seismic design of earth-retaining structures is well discussed elsewhere,[76,77,78,79] little other than seismic considerations are dealt with here. The principal types of structure covered in this section are retaining walls and basement walls.

The magnitude of the seismic earth pressures acting on an earth-retaining structure in part depends on the relative stiffness of the structure and the associated soil mass. Two main categories of soil-structure interaction are usually defined, namely;

(i)   flexible structures which move away from the soil sufficiently to minimize the soil pressures, such as slender free-standing retaining walls;
(ii) rigid structures such as basement walls or tied-back retaining walls.

In case (i) above, active pressures will occur, and the amount of movement required to produce the active state is of the order indicated in Table 5.12.

Table 5.12. Movement of wall required to produce active state

Soil	Wall movement/height
Cohensionless, dense	0·001
Cohensionless, loose	0·001–0·002
Firm clay	0·01–0·02
Loose clay	0·02–0·05

The amount of wall movement which will occur during earthquakes depends mainly on the foundation fixity and the wall flexibility. Unless a more exact analysis is made, the following earth pressure coefficients may be used.

(a) *Flexible.* Walls founded on non-rock materials or cantilever walls higher than 5 m, assume active soil state, and use K_{AE}.
(b) *Intermediate.* Cantilever walls less than 5 m high founded on rock or piles, use $0{\cdot}5\,(K_0 + K_{\text{AE}}) + \Delta K_{\text{AE}}$.
(c) *Rigid.* Counterfort or gravity wall founded on rock or piles; at-rest soil state, use $K_0 + \Delta K_{\text{AE}}$.

K_{AE} is the coefficient of total active earthquake earth pressure, ΔK_{AE} is the coefficient of the increment of active earth pressure due to an earthquake, and K_0 is the coefficient of earth pressure at rest. These coefficients are further discussed in the following section.

5.6.2 Seismic earth pressures

In the present state of knowledge, the recommended method of obtaining seismic earth forces is that using equivalent-static coefficients.[80,81] Only for exceptional structures would dynamic analyses using finite elements seem warranted.

In the equivalent-static method, a horizontal earthquake force equal to the weight of the soil wedge multiplied by a seismic coefficient is assumed to act at the centre of gravity of the soil mass. This earthquake force is additional to the static forces on the wall.

Relatively few failures of retaining walls have been reported in earth-quakes.[81] It is considered that many retaining walls do not need to be designed for seismic forces, as those walls adequately designed for static earth pressures will be able to withstand moderate earthquakes without distress.

Table 5.13. Seismic coefficients for earth-retaining structures[79]

Importance category	Seismic coefficient $\alpha = (a/g)$		
	Zone* A	Zone* B	Zone* C
1	0·24	0·18	0·12
2	0·17	0·13	0·09

*New Zealand Seismic Zones.[25] Zone A represents moderately high seismic risk on a world scale, the maximum design earthquake being of magnitude 7·8 approximately.

The New Zealand Ministry of Works[79] has suggested the following importance categories for use in determining seismic earth pressure coefficients for retaining walls, as set out in Table 5.13.

Importance category 1
Major retaining walls supporting important structures or services, and where failure would have disastrous consequences, such as cutting vital services or causing serious loss of life.
Importance category 2
Free-standing structures at least 6 m high, not in locations as in category 1, but where replacement would be difficult or costly and other consequences would be serious.
Importance category 3
For less important structures than above, no specific provision for earthquake loading need be made.

The seismic coefficients corresponding to the above importance categories, as recommended for the three seismic zones of New Zealand, are given in Table 5.13. These coefficients apply to stress levels and load factors described in Section 5.6.5.

5.6.2.1 Active seismic pressures in dry cohesionless soils

The most commonly used solution[79,80,82] is that derived by Mononobe and Okabe[83,84] based on Coulomb's theory. The effect of an earthquake is represented by a static horizontal force equal to the weight of the wedge of soil multiplied by the seismic coefficient α. Referring to Figure 5.33, the Mononobe–Okabe equations are as follows:

The total force on a wall due to the static and earthquake active earth pressures due to *dry* cohensionless soils is

$$P_{AE} = \tfrac{1}{2}K_{AE}\gamma_d H^2 \tag{5.60}$$

where

$$K_{AE} = \frac{\cos^2(\phi' - \beta - \theta)}{\cos\theta\cos^2\beta\cos(\delta + \beta + \theta)\left\{1 + \sqrt{\left[\dfrac{\sin(\phi' + \delta)\sin(\phi' - \omega - \theta)}{\cos(\delta + \beta + \theta)\cos(\beta - \omega)}\right]}\right\}^2} \tag{5.61}$$

and

$$\cot(\alpha_{AE} - \omega) = -\tan(\phi' + \delta + \beta - \omega) + \sec(\phi' + \delta + \beta - \omega)$$
$$\times \sqrt{\left\{\frac{\cos(\beta + \delta + \theta)\sin(\phi' + \delta)}{\cos(\beta - \omega)\sin(\phi' - \theta - \omega)}\right\}} \tag{5.62}$$

where

  α = the seismic coefficient (Table 5.13);
  α_{AE} = slope angle of failure plane in an earthquake (Figure 5.33);
  β = the angle of the back face of the wall to the vertical;
  γ_d = the unit weight of the dry soil;
  δ = the angle of wall friction;
  θ = $\tan^{-1}\alpha$;
  ϕ' = the effective angle of shearing resistance;
  ω = the slope angle of the backfill.

If the earthquake earth pressure is calculated on a vertical plane through the rear of the heel of the wall, $\beta = 0$ and $\delta = \omega$.

To find the point of application of P_{AE}, the total active force is divided into its two components P_A (from static loading) and the earthquake increment, $\Delta P_{AE} = P_{AE} - P_A$. The forces P_A and ΔP_{AE} are applied at the positions

162

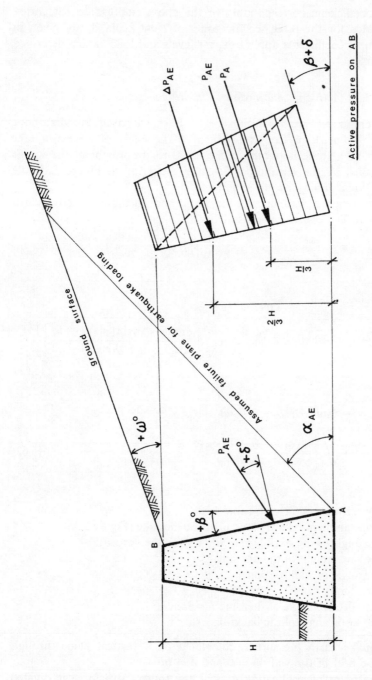

Figure 5.33 Active pressures due to dry cohesionless soil on a retaining wall during an earthquake, for use with Mononobe–Okabe equations (Coulomb conditions)

shown in Figure 5.33, from which the position of P_{AE} and the pressure distribution may be found.

In the conditions assumed in Coulomb's theory where the shearing resistance is mobilized between the back of the wall and the soil, the earthquake earth pressure is calculated directly (Figure 5.33). For walls cast against formwork, the wall friction δ may be taken as $\frac{2}{3}\phi'$. The static active force P_A (Figure 5.33) may be found from the Coulomb equation

$$P_A = \tfrac{1}{2}K_A\gamma_d H^2 \tag{5.63}$$

where

$$K_A = \frac{\cos^2(\phi' - \beta)}{\cos^2\beta\cos(\delta + \beta)\left[1 + \sqrt{\left\{\dfrac{\sin(\phi' + \delta)\sin(\phi' - \omega)}{\cos(\phi' + \beta)\cos(\omega - \beta)}\right\}}\right]^2} \tag{5.64}$$

In the conditions assumed in Rankine's theory (Figure 5.34), the earth pressure is calculated on a vertical plane and is taken as acting parallel to the ground surface. The Mononobe–Okabe equations (5.60) to (5.62) are evaluated for Rankine conditions in Figures 5.35 to 5.38 for various values of α, β, ω and ϕ'. The position of the total force P_{AE} derived from Figures 5.35 to 5.38 may be found from its components P_A and ΔP_{AE} ($= P_{AE} - P_A$) as shown on Figure 5.34. In this case P_A should be derived from Rankine's equations

$$P_A = \tfrac{1}{2}K_A\gamma_d H^2 \tag{5.63}$$

where

$$K_A = \cos\omega\,\frac{\cos\omega - (\cos^2\omega - \cos^2\phi')^{1/2}}{\cos\omega + (\cos^2\omega - \cos^2\phi')^{1/2}} \tag{5.65}$$

5.6.2.2 Active seismic pressures in cohesionless soils containing water

For cohesionless soils containing water the above solution using the Mononobe–Okabe equations is not realistic, and attempts to use them by applying factors to the densities and using the apparent angle of internal friction ϕ_u[79] may be grossly unconservative. In the case of loose saturated sands liquefaction leads to virtually zero values of ϕ_u and the Mononobe–Okabe equations would be used to solve the wrong problem.

The undrained situation is not only undesirable physically but also difficult to analyse, hence it is recommended that good drainage should be provided to obviate the problem. Such drainage should be effective to well below the potential failure zone behind the wall, and also in front of the wall if cohesionless soils exist there in order that the required passive resistance is available.

164

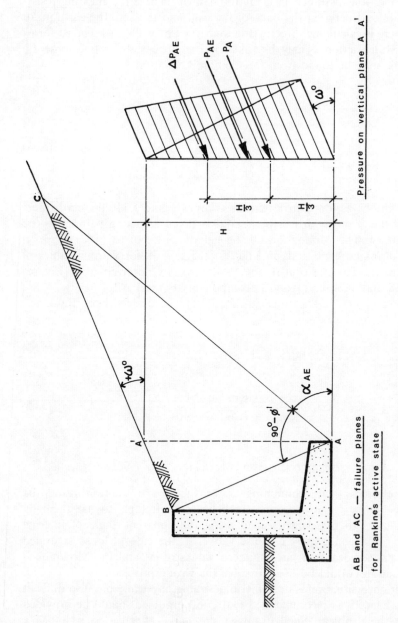

AB and AC — failure planes
for Rankine's active state

Pressure on vertical plane A A'

Figure 5.34 Active pressures due to dry cohesionless soil on a retaining wall during an earthquake (Rankine conditions)

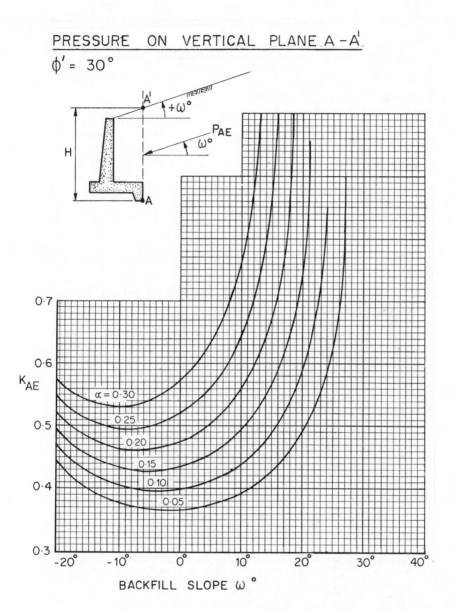

Figure 5.35 Active earthquake earth pressure coefficients K_{AE} for dry cohesionless soil with uniform sloping backfill and $\phi' = 30°$ (New Zealand Ministry of Works[79])

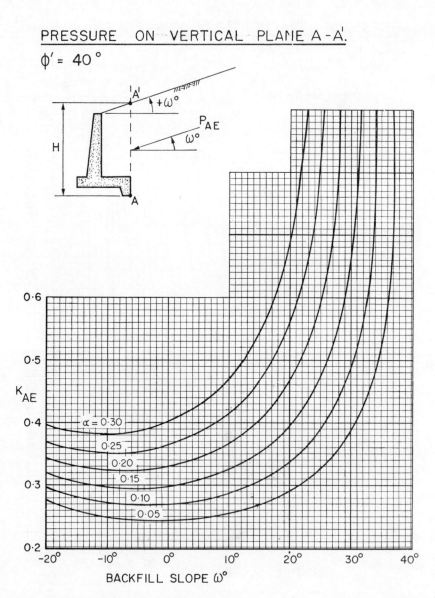

Figure 5.36 Active earthquake earth pressure coefficients K_{AE} for dry cohesionless soil with uniform sloping backfill and $\phi' = 40^\circ$ (New Zealand Ministry of Works[79])

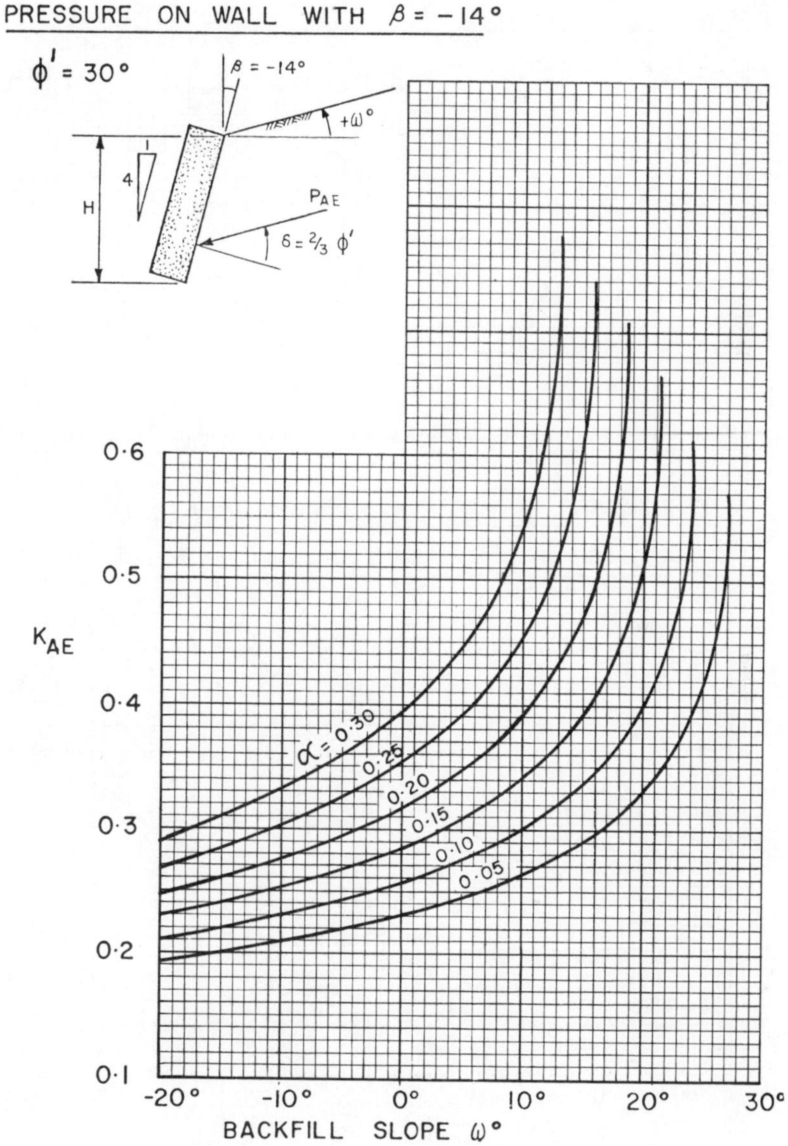

PRESSURE ON WALL WITH $\beta = -14°$

Figure 5.37 Active earthquake earth pressure coefficients K_{AE} for dry cohesionless soil with uniform sloping backfill and $\phi' = 30°$ (New Zealand Ministry of Works[79])

168

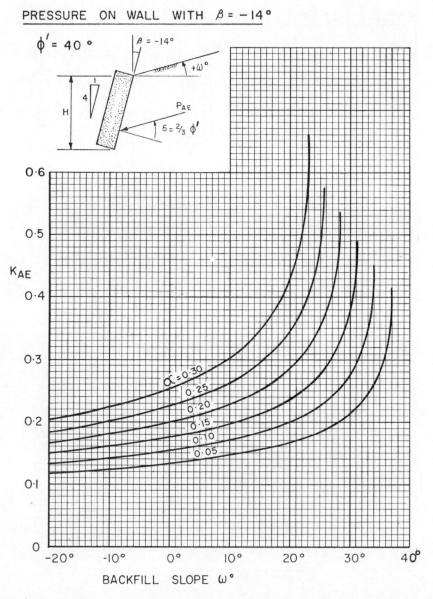

PRESSURE ON WALL WITH $\beta = -14°$

$\phi' = 40°$

$\beta = -14°$

$+\omega°$

H

4

1

P_{AE}

$\delta = \frac{2}{3}\phi'$

K_{AE}

0·6

0·5

0·4

0·3

0·2

0·1

0

$\alpha = 0·30$

0·25

0·20

0·15

0·10

0·05

-20°   -10°   0°   10°   20°   30°   40°

BACKFILL SLOPE $\omega°$

Figure 5.38 Active earthquake earth pressure coefficients K_{AE} for dry cohesionless soil with uniform sloping backfill and $\phi' = 40°$ (New Zealand Ministry of Works[79])

5.6.2.3 Active seismic pressures in cohesive soils or with irregular ground surface

The trial wedge method (Figure 5.39) offers the easiest derivation of seismic earth pressure when the material is cohesive or the surface of the ground is irregular. This figure is drawn for Rankine conditions, and where the ground surface is very irregular the direction of P_{AE} may be taken as approximately parallel to a line drawn between points A and C. For Coulomb conditions the principles of the trial wedge method are similar, and the direction of P_{AE} will be at an angle δ to the surface on which the pressure is calculated, similar to Figure 5.33.

Note that in seismic conditions tension cracks may be ignored on the assumption that this introduces relatively small errors compared with others involved in the analysis. For saturated soils the appropriate density will have to be taken in determining W on Figure 5.39.

5.6.2.4 Completely Rigid Walls

Where soil is retained by a rigid wall, pressures greater than active develop. In this situation the static and earthquake earth pressures may be taken as approximately

$$P_E = \tfrac{1}{2}\gamma H^2(K_0 + \Delta K_{AE}) \tag{5.66}$$

where γ is the total unit weight of the soil, and K_0 is the coefficient of at-rest earth pressure.

As with the active pressure case discussed above, this equation should not be applied to saturated sands. For a vertical wall and horizontal ground surface, and for all normally consolidated materials, K_0 may be taken as

$$K_0 = 1 - \sin\phi' \tag{5.67}$$

where ϕ' is the effective angle of shearing resistance. For other wall angles and ground slopes, K_0 may be assumed to vary proportionally to K_A. The at-rest earth pressure force

$$P_0 = \tfrac{1}{2}K_0\gamma H^2 \tag{5.68}$$

may be assumed to act at a height $H/3$ above the base of the wall.

For gravity retaining walls the at-rest force should be taken as acting normal to the back of the wall, while for cantilever and counterfort walls it should be calculated on the vertical plane through the rear of the heel and taken as acting parallel to the ground surface.

The symbols γ, H and ΔK_{AE} in equation (5.66) have the meanings given variously in the preceding sections of this chapter.

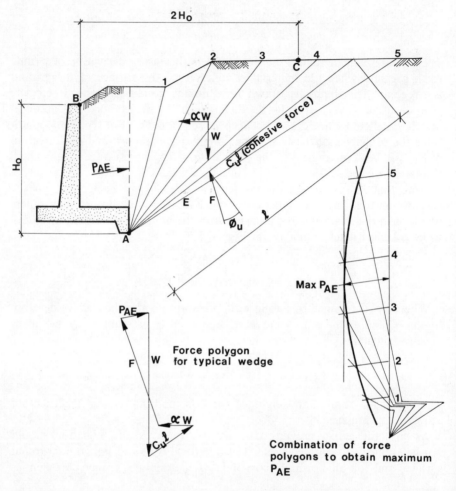

170

Figure 5.39 Trial wedge method for earthquake loading in Rankine conditions for cohesive soil or irregular ground surface

5.6.3 Stability of retaining walls

Factors of safety in terms of moments or forces aiding stability and causing instability of free-standing retaining walls should be calculated for the following modes of failure;

(i)   sliding of the wall outwards from the retained soil;
(ii)   overturning of the wall about its toe;
(iii) foundation bearing failure;
(iv)   slip circle failure in the surrounding soil.

The New Zealand Ministry of Works[79] recommends that the factors of safety in the above cases are respectively

(i) $FS_E \geqslant 1\cdot2$ $FS_s \geqslant 1\cdot5$
(ii) $FS_E \geqslant 1\cdot5$ $FS_s \geqslant 2\cdot0$
(iii) $FS_E \geqslant 2\cdot0$ $FS_s \geqslant 3\cdot0$
(iv) $FS_E \geqslant 1\cdot3$ $FS_s \geqslant 1\cdot5$

where FS_E is the factor of safety in an earthquake, and FS_s is the factor of safety in static conditions.

5.6.4 Section design of retaining walls

(i) When using the equivalent-static earthquake loadings as suggested in Section 5.6.2, and ultimate strength design methods, the section capacity[79] U should be

$$U \geqslant 1\cdot35(DL + 1\cdot35EP + W) \tag{5.69}$$

or

$$U \geqslant 1\cdot08[kDL + 1\cdot25(EQ + W)] \tag{5.70}$$

where

DL = dead load of the structural element;
EP = static earth pressure acting on the element (plus surcharge);
EQ = earthquake earth pressure acting on the element;
W = hydrostatic water pressure;
k = 1·2 or 0·8 whichever is more severe, to allow for vertical acceleration.

(ii) Again referring to New Zealand practice, for working stress design when all load factors are taken as unity, normal permissible static stresses may be increased by one third when applying the equivalent-static earthquake loadings.

5.7 SHEAR WALLS

5.7.1 Introduction

Great structural advantage may be taken from reinforced concrete shear walls in aseismic construction, provided they are properly designed and detailed for strength and ductility. Enough research information is now available to warrant discussion of this important structural element in a section of its own.

Favourably positioned shear walls can be very efficient in resisting horizontal wind and earthquake loads. The considerable stiffness of shear walls not only reduces the deflection demands on other parts of the structure, such as beam-column joints, but may also help to ensure development of all available plastic hinge positions throughout the structure prior to failure. A valuable bonus of shear wall stiffness is the protection afforded to non-structural components in earthquakes due to the small storey drift compared with beam and column frames. Further discussion of the advantages of stiff and flexible construction can be found in Section 4.2.

A notable example of the growing confidence being accorded *in situ* shear wall construction is the forty-four-storey Parque Central apartment buildings in Caracas,[85] built after the 1967 Caracas earthquake.

The loading applied to individual shear walls for design purposes may be obtained by analytical techniques discussed earlier in this chapter and elsewhere. It should be noted that simpler methods of analysis, particularly equivalent-static seismic analysis, may give markedly inaccurate force distributions especially in upper storeys due to the interaction of shear walls with rigid-jointed frames. This interaction may have undesirable effects, resulting in unusually high ductility demands throughout the whole of the structure.

Most shear walls are fairly lightly loaded vertically and behave essentially as cantilevers, i.e. as vertical beams fixed at the base. A discussion of the basic design criteria for shear walls follows under the headings of tall and squat cantilevers. Coupled shear walls are then discussed in a special case of cantilever shear walls. Irregular arrangements of openings in shear walls require individual consideration and generally defy analysis by conventional means. Such structures may invite disaster by concentrating energy absorption in a few zones which are unable to develop the ductility necessary for survival.

5.7.2 Cantilever shear walls

A single cantilever shear wall can be expected to behave as an ordinary flexural member if its height to depth ratio H/h is greater than about $2 \cdot 0$. In fact the ACI code[86] only distinguishes tall from squat shear walls when the height to depth ratio is $0 \cdot 5$ or less. Some distinctions between the two types of shear walls are made in the following two sections.

Having obtained the design ultimate axial force N_u, moment M_u and shear force V_u for a given wall it will usually be appropriate first to check the wall size and reinforcement for bending strength. This should be followed by a check that its ductility is adequate, and then that the wall's shear strength is somewhat greater than its bending strength. While considering the shears it should be ensured that the safe maximum applied shear stress is never exceeded and that the construction joints are adequately reinforced. These considerations are discussed more fully below.

5.7.2.1 Bending strength of cantilever shear walls

Rectangular walls. When rectangular walls are designed for small bending moments, the designer may be tempted to use a uniform distribution of vertical steel as for walls in non-seismic areas, but it may be shown from first principles that with this steel arrangement the ductility reduces as the total steel content increases.

When the flexural steel demand is larger, it will be better to place much of the flexural steel near the extreme fibres, while retaining a minimum of 0·25 percent vertical steel in the remainder of the wall (see also Section 5.7.2.4). Apart from efficient bending resistance this steel arrangement will considerably enhance the rotational ductility.

In rectangular shear walls in which the reinforcement is concentrated at the extremities, the bending strength may be calculated from first principles following accepted codes of practice, or use may be made of column design charts which are frequently available. As design charts for *uniformly* reinforced members are not so readily available, their bending strength is discussed below.

The ultimate bending strength of a uniformly reinforced rectangular shear wall (with $H/h > 1·0$) has been derived from assumptions in the ACI Code[86] by Cardenas *et al.*[87] as follows:

$$M_u = 0·5 A_s f_y h \left(1 + \frac{N_u}{A_s f_y} \right) \left(1 - \frac{c}{h} \right) \tag{5.71}$$

where

$$\frac{c}{h} = \frac{\alpha + \beta}{2\beta + 0·85\beta_1};$$

$$\alpha = \frac{1·2 A_s f_y}{b h f_{cu}};$$

$$\beta = \frac{1·2 N_u}{b h f_{cu}};$$

M_u = design resisting moment (ultimate) (N mm);
A_s = total area of vertical reinforcement (mm^2);
f_y = yield strength of vertical reinforcement (N/mm^2);
h = horizontal length of shear wall (mm);
c = distance from extreme compression fibre to neutral axis (mm);
b = thickness of shear wall (mm);
N_u = design axial load (ultimate), positive if compressive (N);
f_{cu} = characteristic cube compressive strength of concrete (N/mm^2);
β_1 = 0·58 for strength f_{cu} up to 32·5 N/mm^2 and reduced continuously at a rate of 0·05 for each 8 N/mm^2 of strength in excess of 32·5 N/mm^2.

Alternatively the bending strength of uniformly reinforced rectangular walls can be predicted from non-linear beam theory as discussed by Salse and Fintel,[88] who derived the axial load-moment interaction curves shown in Figure 5.40.

174

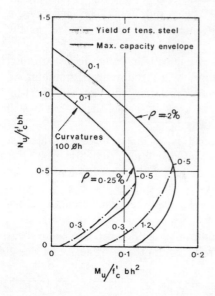

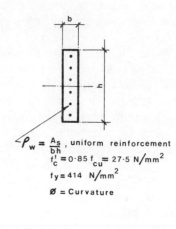

Figure 5.40 Axial load-moment interaction curves for rectangular uniformly reinforced shear walls (after Salse and Fintel[88])

Flanged shear walls. Flanged shear walls are desirable for their high bending resistance and ductility, and arise in the form of I-sections or as channel sections which may be coupled together as lift shafts. As for rectangular shear walls, the derivation of interaction curves for axial load and bending of flanged shear walls is relatively easy working from first principles and with the aid of a small computer. Figure 5.47 shows the interaction curves for a channel section with bending about the minor axis.

Behaviour effects of different reinforcement arrangements can be seen in Figure 5.41 which shows axial load-moment interaction curves for I-sections or channel sections derived from non-linear beam theory. The curves are general for all values of b and h, and the web reinforcement is 0·25 percent in all cases except curve (1). It should be noted that curves (1) and (3) both represent sections containing 3 percent of steel in the flanges, the considerably enhanced strength of curve (1) being largely due to the assumptions of *high concrete confinement* in the flanges in this case.

Squat shear walls. In squat shear walls, i.e. those with a height to depth ratio $H/h \leqslant 1·0$, it is not possible to separate the considerations of flexure and shear from each other, and the preceding methods of calculating bending resistance are not really applicable. As the question of shear strength assumes pre-eminence, squat shear wall design is considered in Section 5.7.2.3.

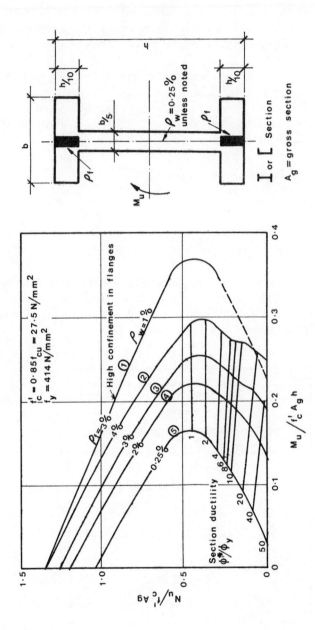

Figure 5.41 Axial load-moment interaction curves for I-section or [-section reinforced concrete shear walls (after Salse and Fintel[88])

176

5.7.2.2. Ductility of cantilever walls

The general problem of ductility is an involved one which is discussed elsewhere (Sections 5.1.2, 6.2.4), but suffice it here to say that adequate ductility under seismic loadings implies inelastic cyclic deformations without appreciable loss of strength.

As mentioned above, shear walls will exhibit greater ductility in bending if much of the reinforcement is concentrated near the extreme fibres, and consequently flanged sections are more ductile than rectangular walls. A comparison of the ductility of rectangular and I (or [ ) sections is given in Figure 5.42 where it was taken that

$$\text{available section ductility} = \frac{\phi^*}{\phi_y} \qquad (5.72)$$

where

$\phi^* = $ curvature at maximum moment;

$\phi_y = $ curvature at initiation of tension steel yield.

The ductility calculation was based on monotonic loading only, and hence Figure 5.42 serves better for comparative purposes than quantitative; the true ductility under reversible loading may be less than that shown depending on the reinforcement quantities and disposition.

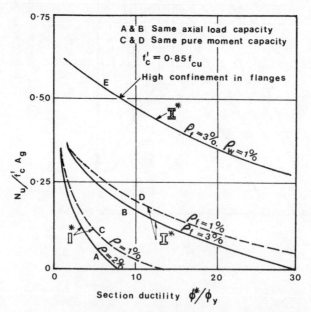

Figure 5.42 Ductility as affected by cross-sectional shape, steel distribution and concrete confinement (after Salse and Fintel[88])

From Figure 5.42 it can be seen that both increasing steel percentages and increasing axial loads will decrease ductility. By comparing curve A with B, and curve C with D, it can be seen that the section ductility for I shapes is three to four times greater than that for uniformly reinforced rectangular sections. By comparing curve E with the remainder in Figure 5.42 the great effect on ductility of concrete confinement in the flanges can be seen.

In design situations it may be convenient to refer to an interaction diagram as shown on Figure 5.41 which incorporates ductility factors, thus allowing suitable strength and ductility to be chosen simultaneously.

Squat shear walls, i.e. those with height to depth ratio $H/h \lesssim 1.0$, are not amenable to the above ductility calculations. Their relatively large flexural capacity may be associated with large enough shear forces to destroy the wall in a brittle manner. Paulay[89] recommends that if a ductile (flexural) failure mechanism is desired for squat shear walls, then the nominal shear stresses associated with the flexural capacity of the wall must be moderate, say $v_u \leqslant 0.45\sqrt{f_{cu}}$. This requirement will limit the flexural steel content in the wall.

5.7.2.3 Shear strength of cantilever shear walls

The shear strength of all but squat rectangular beams can be assessed as for ordinary beams, and the approach of the ACI Code[86,87,90] may be adopted in most circumstances. American shear design practice is based on the assumption that the shear capacity of concrete beams is made up of two independent parts, that carried by the concrete and that carried by the web reinforcement. Except as discussed below, the shear stress carried by the concrete in a rectangular wall may be taken as the lesser of

$$v_c = 0.25\sqrt{f_{cu}} + \frac{N_u}{4bh} \tag{5.73}$$

or

$$v_c = 0.46\sqrt{f_{cu}} + \frac{h[0.095\sqrt{f_{cu}} + 0.015(N_u/bh)]}{[(M_u/V_u) - (h/2)]} \tag{5.74}$$

but the shear strength of the concrete need not be taken as less than

$$v_c = 0.15\sqrt{f_{cu}} \tag{5.75}$$

where

M_u, N_u and V_u are the maximum ultimate applied loads acting on the section (N mm and N);

b and h are the thickness and length of section (mm);

f_{cu} is the characteristic cube crushing strength of the concrete (N/mm^2).

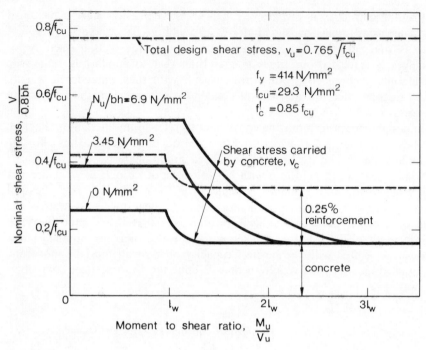

Figure 5.43 Shear carried by rectangular shear walls according to ACI Code (after Cardenas et al.[87] Reproduced by permission of the American Concrete Institute)

The contribution of the reinforcement based on the truss analogy is expressed in terms of the nominal shear stress v_s, and the horizontal reinforcement ratio ρ_h, as

$$v_s = \rho_h f_y \tag{5.76}$$

Figure 5.43 shows the minimum shear strength of a rectangular shear wall, identifying the contributions of the concrete and the minimum recommended steel content of 0·25 percent, as well as showing the beneficial effect of axial compression N_u. The curves have been plotted for f_{cu} of 29·3 N/mm^2 and f_y of 414 N/mm^2 and indicate a minimum shear strength of tall shear walls of the order of $0·31\sqrt{f_{cu}}$ and for low walls (small M_u/V_u) of about $0·41\sqrt{f_{cu}}$. Note that an overall shear stress of $0·765\sqrt{f_{cu}}$ should never be exceeded.

Notwithstanding the above minimum strengths, however, it is recommended[89] that where yielding of the flexural steel can occur (i.e. near the base of a wall) the shear contribution of the cracked concrete should be ignored, and horizontal stirrup reinforcement should be provided for the total shear. Such reinforcement should be designed for a stress slightly less than yield under ultimate load. Assuming a 45° potential diagonal crack, this reinforcement will have to be provided over a height equal to the depth of the wall section.

Squat shear walls, i.e. those with height to depth ratio $H/h \leqslant 1.0$ require different design and detailing treatment from taller walls. In order to obtain a ductile (i.e. flexural) failure mechanism in squat walls, the nominal shear stresses associated with the flexural capacity of the wall should be moderate $(v_u \leqslant 0.45\sqrt{f_{cu}})$ thus limiting the flexural steel content in the wall. Also, because the flexural failure mechanism implies large cracks, no concrete shear stress should be allowed, and *all* the shear force should be resisted by web reinforcement. The vertical and horizontal steel will need to be equal, the principal role of the vertical steel being to resist the overturning moment.

5.7.2.4 Horizontal construction joints in shear walls

Earthquake damage in shear walls has often occurred at horizontal construction joints in the form of sliding movements. In order to reliably prevent sliding due to reversible seismic shear forces, a rough concrete interface combined with sufficient vertical reinforcement must be provided; allowance may be made for the beneficial effect of gravity loading. Load-reversal tests described by Paulay[89] indicate that the reinforcing content required is

$$\rho_v \geqslant \frac{1.25v_u - (N_u/A_s)}{f_y} \geqslant 0.0025 \qquad (5.77)$$

The above equation is plotted in Figure 5.44, where it can be seen that with the minimum vertical steel content of 0.25 percent, moderate shear stress can be resisted without any assistance from vertical loads.

In computing the value of N_u for use in the above equation, care should be taken to deduct a suitable upwards seismic axial force from the gravity loading. In strong ground motion areas an upwards acceleration of 20–30 percent of gravity may be appropriate.

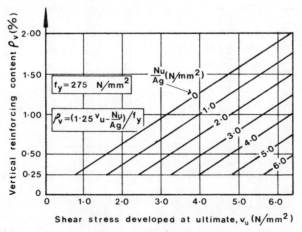

Figure 5.44 Steel requirements across horizontal construction joints in shear walls (after Paulay[89])

It is important that the steel content ρ_v calculated above should be provided at all portions of a joint, as concentrations of reinforcement will provide the necessary clamping action in their locality only. It should also be noted that the method of joint preparation does not matter as long as a coarse textured surface is produced to which freshly placed concrete can bond.

5.7.3 Coupled shear walls

5.7.3.1 Design approach

It is common practice nowadays to utilize the inherent lateral resistance of adjacent shear walls by coupling them together with beams at successive floor levels. Vertical access shafts punctured by door openings, as shown in Figure 5.45, form the classical example of this type of member. The analysis of coupled shear walls requires consideration of axial deformations of the walls and shear distortions of the coupling beams. The laminar or continuum technique which replaces the discrete beams by an equivalent set of continuous laminae, provides a convenient method of analysis.[91,92,93,94]

Ideally the designer would like the coupled walls to act as a box or I-unit as if the openings did not exist; such a structure would be much stronger than the two constituent channel units acting independently. This implies the development of high shears in the coupling beams acting as a web, and the existence of large longitudinal forces in each wall unit. The failure of the coupling beams in coupled shear walls exposed to strong earthquakes indicates insufficient ductility of the beams. This has been due partly to inadequate detailing of the beams and partly to the use of elastic analysis which

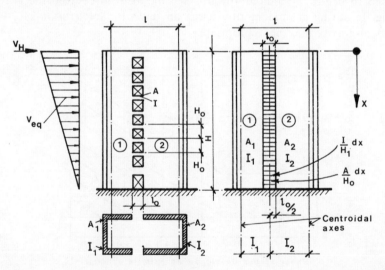

Figure 5.45 A typical coupled shear wall structure and its mathematical laminar model (after Paulay[89])

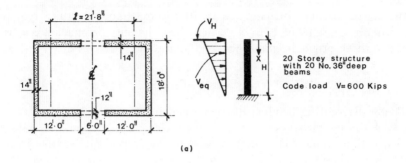

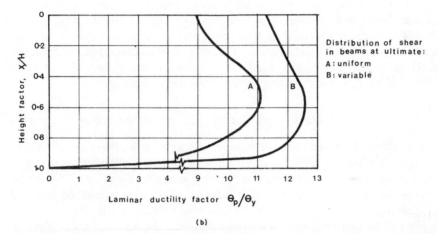

Figure 5.46 Example core wall structure, dimensions, loads and calculated ductility demand in coupling beams (after Park and Paulay[96])

will not adequately predict the distribution of ductility demand throughout the member.

Paulay has suggested a deterministic elasto-plastic analysis based on a laminar model,[95] in which a desirable sequence of plastic hinge formations in the coupling beams is dictated. The plastic hinges in the walls, which are also major gravity load carrying units, should be the last ones to form. As the two walls are subjected to axial forces generated by the lateral load, their ductility will be restricted, so that any delay in the formation of their hinges should be beneficial in terms of reduced ductility demand.

Using the above-mentioned deterministic design approach, Paulay[89] analysed the core of a twenty-storey building with the dimensions and loads shown in Figure 5.46(a).

The ductilities required in the coupling beams are shown in Figure 5.46(b) in terms of maximum and yield rotations θ_p and θ_y. It can be seen that the maximum ductility was about $\theta_p/\theta_y = 11$, if all the beams were of equal strength (curve A); while when the beams were given strengths proportional to the elastic strength demand (curve B) the maximum ductility required was virtually the same at 12·6. Note that these ductility demands are a revision[96] to those originally published by Paulay.[89]

5.7.3.2 The strength of coupled walls

Having derived the bending moments and forces acting on the wall elements of the coupled system, it will be necessary to design the walls to withstand those forces. The bending moment pattern will be similar to that of simple cantilever walls. In addition, because of the coupling system there will be considerable axial forces which may produce net tensions in the walls.

It is evident that the design considerations are as for cantilever walls discussed above. In the design of a high-rise structure with many similar horizontal sections to consider it may be worth producing a family of axial load-moment interaction diagrams (Figure 5.47) with the aid of a small computer program. It is to be noted that similar diagrams for different ratios of biaxial bending may be necessary for the same section.

5.7.3.3 The strength and ductility of coupling beams

The classical failure mode of coupling beams in earthquakes is that of diagonal tension. To avoid this brittle type of failure the shear strength of the beams must always exceed the bending strength regardless of the design loads. Thus an upper limit on the flexural steel content may be imposed, particularly when such beams are deep relative to their spans.

Firstly the currently accepted upper limit for the nominal shear force (in Newtons) in a beam is given by

$$V = 0.77 \, \phi_v \, bd \, \sqrt{f_{cu}} \tag{5.78}$$

where ϕ_v is the undercapacity factor for shear, and secondly the shearing force corresponding to the flexural capacity of a typical beam (Figures 5.45 and 5.48) is

$$V_u = \frac{2M_u}{l_0} = \frac{2(d - d_1) \, A_s f_y}{l_0 \phi_m} \tag{5.79}$$

where

l_0 = the length of the coupling beam;
ϕ_m = the undercapacity factor in flexure.

Equating equations (5.78) and (5.79) and taking $\phi_m = 0.9$ and $\phi_v = 0.85$, the maximum flexural steel content for conventional beams is

$$\frac{A_s}{bd} \leqslant \frac{0.39 \, l_0 \sqrt{f_{cu}}}{(d - d_1) f_y} \tag{5.80}$$

where f_{cu} and f_y are measured in N/mm².

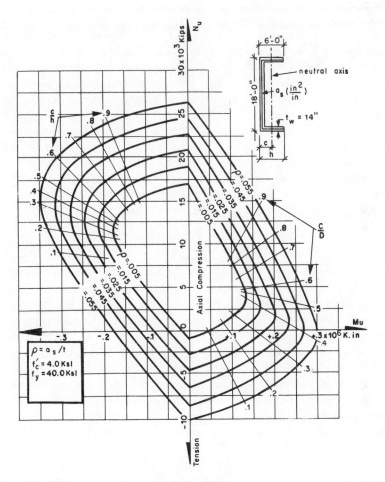

Figure 5.47 Axial force-moment interaction diagram for channel shaped wall section (after Paulay[89])

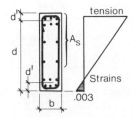

Figure 5.48 Section through conventionally reinforced coupling beam

184

Unfortunately the high ductility in pure bending of beams with equal top and bottom reinforcement is not exhibited in beams with a span/depth ratio of less than two. According to Paulay[89] this is because the shearing forces and diagonal cracking causes tension to develop in both the top and bottom reinforcement over the whole span. Theory and research[97,98] have demonstrated that the compression reinforcement in deep coupling beams does not behave as such; and therefore as the concrete is not relieved in compression by the reinforcement, the ductility is reduced. Further, conventional web reinforcement is unhelpful in carrying the shear across the diagonally cracked compression zones of the beams, and direct sliding shear failures can occur. For both the above reasons Paulay suggests that equation (5.80) should be modified to limit the shear capacity of conventionally reinforced coupling beams by restricting the flexural steel content to

$$\rho_{max} = \frac{0 \cdot 25 l_0 \sqrt{f_{cu}}}{(d - d_1)f_y} \tag{5.81}$$

Where coupling beams may experience high seismic stresses, diagonal reinforcement of the type shown in the Reinforced Concrete Detail 6.6 provides far greater seismic resistance than conventional steel arrangements, as the comparison of ductilities in Figure 5.49 shows.

Conventionally reinforced deep coupling beams having a ductility ratio of 4–5 (Figure 5.49) clearly would be unsatisfactory for the structure examined in Figures 5.45 and 5.46, whereas the diagonal reinforcement arrangement easily provides the required ductility ratio of about 12. Thus in moderate or strong ground motion areas, diagonal reinforcement of deep coupling beams is seen to be required. The importance of restraining the diagonals

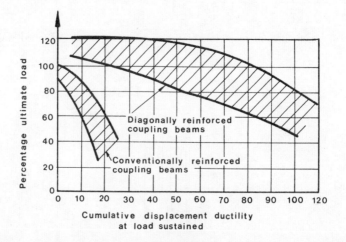

Figure 5.49 Comparison of ductilities of diagonally and conventionally reinforced deep coupling beam (after Paulay[89])

against buckling in compression must be realized, however, and careful detailing to suit this and still allow the proper placement of concrete will be necessary.

5.8 INTERACTION OF FRAMES AND INFILL PANELS

5.8.1 Introduction

Walls are often created in buildings by infilling parts of the frame with stiff construction such as bricks or concrete blocks. Unless adequately separated from the frame (Section 8.2), the structural interaction of the frame and infill panels must be allowed for in the design. This interaction has a considerable effect on the overall seismic response of the structure and on the response of the individual members. Many instances of earthquake damage to both the frame members and infill panels have been recorded.[99,100,101] The currently available analytical techniques used in studying frame/panel interaction are briefly discussed below.

5.8.2 The effect of infill panels on overall seismic response

The principal effects of infill panels on the overall seismic response of structural frames are;

(i)   to increase the stiffness and hence increase the base shear response in most earthquakes;

(ii)   to increase the overall energy absorption capacity of the building;

(iii)   to alter the shear distribution throughout the structure.

The more flexible the basic structural frame, the greater will be the above-mentioned effects. As infill is often made of brittle and relatively weak materials, in strong earthquakes the response of such a structure will be strongly influenced by the damage sustained by the infill and its stiffness-degradation characteristics.

In order to fully simulate the earthquake response of an infilled frame, a complex non-linear time-dependent finite-element dynamic analysis would be necessary. At the present time no *practical* computer program capable of such an analysis has been published, and such are the problems involved in modelling the structural behaviour of normal masonry infill[102-107] that only rudimentary dynamic analyses are likely to be warranted for some time to come.

For many structures a response spectrum analysis in which the infill panels are simulated by simple finite elements, will be very revealing. Figure 5.50 shows the results of such an analysis of a multi-storey hotel building, in which all of the bedroom floors (4th to 20th) have alternate partitions in brickwork. Curve A shows the horizontal earthquake shear distribution up the shear core ignoring the brickwork, while curve B shows the shears when an approximate allowance for the brick walls is made.

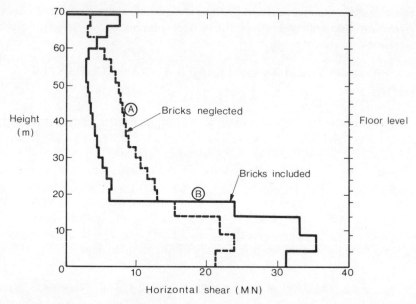

Figure 5.50 Horizontal seismic shear diagram for lift core of 20-storey hotel building showing effect of brick partitions above 4th floor

Allowing for the brick reduces the fundamental period of the structure from 1·96 s to 1·2 s, and correspondingly increases the base shear on the shear core from 21·0 MN to 31·0 MN. The effect on the distribution of shear is particularly dramatic; it can be seen how the brick walls carry a large portion of the shear until they terminate at the fourth floor level; below this level the shear walls of the core must of course take the total load (see also Section 4.2.4).

In carrying out this simple type of dynamic analysis, difficulty may be experienced in selecting a suitable value of shear modulus G for the infill material. Not only is the G value notoriously variable for bricks, but the infill material may not even have been chosen at the time of the analysis. Either a single representative value may have to be assumed, or it may be desirable to take a lower and a higher likely value of G in two separate analyses for purposes of comparison.

Further examples of the effect of infill on mode shapes and periods of vibration of structural frames are briefly reported by Lamar and Fortoul.[104] In their examples the period of the first mode is generally reduced by a factor of three or four when comparing the 'infill-included' with the 'infill-excluded' cases. The comparable mode shapes also vary considerably.

Interesting use of the dynamic properties of infill panels has been made in Japan by Muto.[108] Tall steel-framed buildings have been fitted with precast concrete panels, which not only stiffen the otherwise highly flexible frames against excessive sway under lateral loading, but also considerably improve the energy-absorption and ductility characteristics of the structures in strong earthquakes. The precast panels are specially designed and detailed with slits and reinforcement to have prescribed elastic and post-elastic deformation behaviour, and are attached to the beams of the steel frame at discrete points. These provisions, coupled with considerable experimental performance data, make the Muto slitted shear panels readily amenable to rational analysis in distinct contrast to ordinary masonry infill.

5.8.3 The effect of infill panels on member forces

As mentioned above there is no practical means of predicting accurately the seismic interaction between infill panels and structural frames. Only fairly crude assessments can be made of the stresses in the panels and in the adjacent members. However from a straightforward finite-element response spectrum analysis some basic design information may be derived. While to take such data as definitive seismic design criteria would be misleading, sensible use of the computed forces in design would nevertheless be much better than ignoring the presence of the infill as has often been the case in the past. By carrying out comparative analyses with and without the infill panels, at least a qualitative idea of the effect of the infill can be obtained.

(i) *Infill panels.* The shear stresses computed in the infill panels should give a reasonable indication whether or not the infill will survive the design earthquake. Despite being very approximately determined, the shear stress level will also help in determining what reinforcement to use in the panel and whether to tie the panel to the frame (Section 6.6.5).

(ii) *Frame members.* The design of the beams and columns abutting the infill is generally the least satisfactory aspect of this form of aseismic construction. Because of the approximations in the analytical model, the stresses in the frame members are ill-defined. Failures tend to occur at the tops and bottoms of columns, due to shears arising from interaction with the compression diagonal which exists in the infill panel during the earthquake.[101] Unfortunately no comprehensive design criteria for this problem have yet been established, and further research examining the frame rather than the panel stresses is required. If the analysis indicates the failure of the infill panels, the frame should be analysed with any failed panels deleted, so that appropriate frame stresses may be taken into consideration. The fail-safe structure will not necessarily be less highly stressed in individual frame members than the original undamaged structure, although the resulting more flexible structure will generally have a reduced overall response.

REFERENCES

1. Blake, R. E. 'Basic vibration theory', in *Shock and Vibration Handbook* (Ed. C. M. Harris and C. E. Crede) Vol. I, McGraw-Hill, New York, 1961.

188

2. California Institute of Technology. 'Analyses of strong motion earthquake accelerograms', *Vol. III—Response Spectra*, Earthquake Engineering Research Laboratory, California Institute of Technology, Pasadena. Part A, EERL 72–80.
3. Albert C. Martin and Associates. 'Inelastic dynamic analysis, Bank of New Zealand Headquarters Building, Wellington', *Report prepared for Brickell, Moss, Rankine and Hill, Wellington*, A. C. Martin and Associates, Los Angeles, March, 1973.
4. Sinha, B. P., Gerstle, K. H., and Tulin, L. G. 'Stress-strain relationships for concrete under cyclic loading', *Journal American Concrete Institute*, **61**, No. 2, 195–211 (Feb. 1964).
5. Park, R., Kent, D. C., and Sampson, R. A. 'Reinforced concrete members with cyclic loading', *Jnl. Structural Division, ASCE*, **98**, ST7, 1341–1360 (July, 1972).
6. Blakeley, R. W. G. 'Prestressed concrete seismic design', *Bulln. New Zealand Society for Earthquake Engineering*, **6**, No. 1, 2–21 (March, 1973).
7. Williams, D., and Scrivener, J. E. 'Response of reinforced masonry shear walls to static and dynamic cyclic loading', *Proc. 5th World Conference on Earthquake Engineering, Rome*, **2**, 1491–1494 (1973).
8. Park, R. 'Theorisation of structural behaviour with a view to defining resistance and ultimate deformability', *Bulln. New Zealand Society for Earthquake Engineering*, **6**, No. 2, 52–70 (June, 1973).
9. Lord, J., Meyer, C., Hoerner, J. B., and Zayed, M. 'Inelastic dynamic analysis of a 60 storey building', *Proc. 5th World Conference on Earthquake Engineering, Rome*, **1**, 1353–1356 (1973).
10. Jennings, P. C., Matthiesen, R. B., and Hoerner, J. B. 'Forced vibrations of a tall steel-frame building', *Intl. Jnl. Earthquake Engineering and Structural Dynamics*, **1**, No. 2, 107–132 (1972).
11. Blume, J. A. 'The motion and damping of buildings relative to seismic response spectra', *Bull. Seismological Society of America*, **60**, No. 1, 231–259 (Feb. 1970).
12. Despeyroux, J. 'Applications of precast prestressed concrete in seismic-zone structures', *Report presented at the FIP Symposium on Seismic Structures, Tbilisi, 1972*, London, 1972.
13. Reay, A. M., and Shepherd, R. 'Steady state vibration tests of a six storey reinforced concrete building', *Bull. New Zealand Society for Earthquake Engineering*, **4**, No. 1, 94–107 (March, 1971).
14. Newmark, N. M., and Rosenblueth, E. *Fundamentals of Earthquake Engineering*, Prentice Hall, Englewood Cliffs, New Jersey, 1971.
15. Takanashi, K. 'Inelastic lateral buckling of steel beams subjected to repeated and reversed loadings', *Proc. 5th World Conference on Earthquake Engineering, Rome*, **1**, 795–798 (1973).
16. Nigam, N. G., and Housner, G. W. 'Elastic and inelastic response of framed structures during earthquakes', *Proc. 4th World Conference on Earthquake Engineering, Santiago*, **2**, A4, 89–104 (1969).
17. Clough, R. W., Benuska, K. L., and T. Y. Lin and Assoc. 'FHA study of seismic design criteria for high-rise buildings', *U.S. Dept. of Housing and Urban Development, F.H.A.*, HUD TS–3, 168–170 (August 1966).
18. Housner, G. W. 'Design spectrum', in *Earthquake Engineering* (Ed. R. L. Wiegel), Prentice-Hall, Englewood Cliffs, New Jersey, 1970, Chap. 4, pp. 73–106.
19. Clough, R. W. 'Earthquake response of structures', in *Earthquake Engineering* (Ed. R. L. Wiegel), Prentice-Hall, Englewood Cliffs, New Jersey, 1970, Chap. 12, pp. 307–334.
20. Shepherd, R. 'Determination of seismic design loads in a framed structure', *New Zealand Engineering*, **22**, No. 2, 56–61 (Feb. 1967).
21. Newmark, N. M., 'Current trends in the seismic analysis and design of high-rise structures', in *Earthquake Engineering* (Ed. R. L. Wiegel), Prentice-Hall, 1970, Chap. 16, pp. 403–424.

22. Newmark, N. M., and Hall, W. J. 'A rational approach to seismic design standards for structures', *Proc. 5th World Conference on Earthquake Engineering, Rome*, **2**, 2266–2275 (1973).
23. Shepherd, R., and McConnel, R. E. 'An application of spectral techniques to inelastic seismic response', *Bulln. New Zealand Society for Earthquake Engineering*, **3**, No. 4, 173–178 (Dec. 1970).
24. Seismology Committee, S.E.A.O.C. *Recommended lateral force requirements and commentary*, Structural Engineers Association of California, 1973.
25. New Zealand Standards Institute. *Basic design loads, earthquake provisions*, New Zealand Model Building Bylaw (NZSS 1900), 1965, Chap. 8.
26. Park, R. 'Ductility of reinforced concrete frames under seismic loading', *New Zealand Engineering*, **23**, No. 11, 427–435 (Nov. 1968).
27. I.A.E.E. *Earthquake resistant regulations—a word list*, International Association for Earthquake Engineering, Tokyo, 1973.
28. Clough, R. W., and Penzien, J. *Dynamic of structures*, McGraw-Hill, New York, 1975.
29. Biggs, J. M. *Introduction to Structural Dynamics*, McGraw-Hill, New York, 1964.
30. Penzien, J. 'Earthquake response of irregularly shaped buildings', *Proc. 4th World Conference on Earthquake Engineering, Santiago*, **2**, A3, 75–90 (1969).
31. Avinash Singhal, 'Inelastic earthquake responses of multi-storey buildings', *The Structural Engineer*, No. 9, **49**, 397–412 (Sept. 1971).
32. Richart, F. E., Hall, J. R., and Woods, R. D. *Vibrations of Soils and Foundations*, Prentice-Hall, Englewood Cliffs, New Jersey, 1970.
33. Silver, M. L., and Seed, H. B. 'The behaviour of sands under seismic loading conditions', *Report No. EERC 69—16*, Earthquake Engineering Research Centre, University of California, Berkeley, Dec. 1969.
34. Seed, H. B. and Idriss I. M. 'Analysis of soil liquefaction; Niigata earthquake', *Jnl. Soil Mechanics and Foundations Division*, ASCE, **93**, No. SM3, 83–108 (May, 1967).
35. Shannon and Wilson, Agbabian-Jacobson Associates, Seattle and Los Angeles. 'Soil behaviour under earthquake loading conditions', *State of art evaluation for U.S. Atomic Energy Commission*, Jan. 1972).
36. Seed, H. B., and Idriss, I. M. 'Simplified procedure for evaluating soil liquefaction potential', *Jnl. Soil Mechanics and Foundation Division*, ASCE, **97**, SM9, 1249–1273 (Sept. 1971).
37. Kishida, H. 'A note on liquefaction of hydraulic fill during the Tokachi-Oki earthquake', *Second Seminar on Soil Behaviour and Ground Response during Earthquakes*, University of California, Berkeley, Aug. 1969.
38. Lee, K. L. and Fitton, J. A. 'Factors affecting the cyclic loading strength of soil', *Special Technical Publication No. 450, Symposium on Vibration Effects of Earthquakes on Soils and Foundations*, ASTM, 1968, pp. 71–95.
39. Seed, H. B., and Idriss, I. M. 'Influence of soil conditions on ground motions during earthquakes', *Jnl. Soil Mechanics and Foundations Division*, ASCE, **95**, No. SM1, 99–137 (Jan. 1969).
40. Seed, H. B., and Idriss, I. M. 'Soil moduli and damping factors for dynamic response analyses', *Report No. EERC 70—10, Earthquake Engineering Research Center*, University of California, Berkeley, Dec. 1970.
41. Whitman, R. V. 'Soil-structure interaction', and 'Evaluation of soil properties for site evaluation and dynamic analysis of nuclear plants', *Seismic Design for Nuclear Power Plants* (Ed. R. J. Hansen), M.I.T. Press, 1970, pp. 245–305.
42. Whitman, R. V., and Richart, F. E. 'Design procedures for dynamically loaded foundations', *Jnl. Soil Mechanics and Foundations Division*, ASCE, **93**, No. SM6, 169–191 (Nov. 1967).

190

43. D'Appolonia, D. J., Poulos, H. G., and Ladd, C. C. 'Initial settlement of structures on clay', *Jnl. Soil Mechanics and Foundations Division, ASCE*, **97,** No. SM10, 1359–1377 (Oct. 1971).
44. Lambe, T. W., and Whitman, R. V. *Soil Mechanics*, Wiley, New York, 1969, p. 155.
45. Deere, D. U. 'Geological considerations', in *Rock Mechanics* (Ed. K. G. Stagg and O. C. Zienkiewicz), Wiley, New York, 1968, pp. 1–20.
46. Seed, H. B., Whitman, R. V., Dezfulian, H., Dobry, R., and Idriss, I. M. 'Soil conditions and building damage in 1967 Caracas earthquake', *Jnl. Soil Mechanics and Foundations Division, ASCE*, **98,** No. SM8, 787–806 (Aug. 1972).
47. Penzien, J., and Tseng, W. S. 'Seismic analysis of gravity platforms including soil-structure interaction effects', *Proc. Offshore Technology Conference, Houston, Texas,* Paper No. 2674 (1976).
48. Vaish, A. K., and Chopra, A. K. 'Earthquake finite element analysis of structure-foundation systems', *Jnl. Engineering Mechanics Division*, **100,** No. EM6, 1101–1116 (Dec. 1974).
49. Kausel, E., and Roesset, J. M. *Soil structure interaction problems for nuclear containment structures*, presented at August 1974 ASCE Power Division Speciality Conference, Boulder, Colorado.
50. Seed, H. B., Lysmer, J., and Hwang, R. 'Soil-structure interaction analyses for seismic response', *Jnl. Geotechnical Engineering Division, ASCE*, **101,** No. GT5, 439–457 (May, 1975).
51. Roesset, J. M., Whitman, R. V., and Dobry, R. 'Modal analysis for structure with foundation interaction', *Jnl. Structural Division, ASCE*, **99,** No. ST3, 399–416 (Mar. 1973).
52. Penzien, J. 'Soil-pile foundation interaction', in *Earthquake Engineering* (Ed. R. L. Wiegel), Prentice-Hall, Englewood Cliffs, New Jersey, 1970, pp. 349–381.
53. Luco, J. E., and Westmann, R. A. 'Dynamic response of circular footings', *Jnl. Engineering Mechanics Division, ASCE*, **97,** No. EM5, 1381–1395 (Oct. 1971).
54. Veletsos, A. S., and Wei, Y. T. 'Lateral and rocking vibrations of footings', *Jnl. Soil Mechanics and Foundations Division, ASCE*, **97,** No. SM9, Sept. 1971, 1227–1248 (Sept. 1971).
55. Veletsos, A. S., and Nair, V. V. D. 'Seismic interaction of structures on hysteretic foundations', *Jnl. Structural Division, ASCE*, **101,** No. ST1, 109–129 (Jan. 1975).
56. Luco, J. E. 'Impedance functions for a rigid foundation on a layered medium', *Nuclear Engineering and Design*, **31,** No. 2, 204–217 (1974).
57. Bielak, J. 'Dynamic behaviour of structures with embedded foundations', *Earthquake Engineering and Structural Dynamics*, **3,** No. 3, 259–274 (Jan–March, 1975).
58. Luco, J. E., Wong, H. L., and Trifunac, M. D. 'A note on the dynamic response of rigid embedded foundations', *Earthquake Engineering and Structural Dynamics*, **4,** No. 2, 119–127 (Oct–Dec. 1975).
59. Novak, M., and Beredugo, Y. O. 'Vertical vibration of embedded footings', *Jnl. Soil Mechanics and Foundations Division*, **98,** No. SM12, 1291–1310 (Dec. 1972).
60. Kausel, E. 'Forced vibrations of circular foundations on layered media', *MIT Research Report R74—11*, Massachusetts Institute of Technology, Jan. 1974.
61. Lysmer, J., and Kuhlemeyer, R. L. 'Finite dynamic model for infinite media', *Jnl. Engineering Mechanics Division, ASCE*, **95,** No. EM4, 859–877 (Aug. 1969).
62. Ang, A. H., and Newmark, N. M. 'Development of a transmitting boundary for numerical wave motion calculations', *Report to Defence Atomic Support Agency*, Contract DASA-01-0040, Washington, D.C., 1971.
63. Lysmer, J., and Waas, G. 'Shear waves in plane infinite structures', *Jnl. Engineering Mechanics Division, ASCE*, **98,** EM1, 85–105 (Feb. 1972).
64. Kausel, E., and Roesset, J. M. 'Dynamic stiffness of circular foundations', *Jnl. Engineering Mechanics Division, ASCE*, **101,** No. EM6, 771–785 (Dec. 1975).

65. Veletsos, A. S., and Meek, J. W. 'Dynamic behaviour of building-foundation systems', *Earthquake Engineering and Structural Dynamics*, **3,** No. 2, 121–138 (Oct–Dec. 1974).

66. Seismology Committee, SEAOC. *Recommended lateral force requirements and commentary*, Structural Engineers Association of California, 1974.

67. Watt, B. J., Boaz, I. B., and Dowrick, D. J. 'Response of concrete gravity platforms to earthquake excitations', *Proc. Offshore Technology Conference, Houston, Texas*, Paper No. 2673 (1976).

68. Novak, M. 'Effect of soil on structural response to wind and earthquake', *Earthquake Engineering and Structural Dynamics*, **3,** No. 1, 79–96 (July–Sept. 1974).

69. Novak, M., and Sachs, K. 'Torsional and coupled vibrations of embedded footings', *Earthquake Engineering and Structural Dynamics*, **2,** No. 1, 11–33 (July–Sept. 1973).

70. Zeevaert, L. *Foundation engineering for difficult sub-soil conditions*, Van Nostrand Reinhold, New York, 1972.

71. New Zealand Ministry of Works. 'Design of Public Buildings', *Code of Practice PW 81/10/1*, 1970.

72. U.S. Department of Navy. *Design manual—Soil mechanics, foundations and earth structures*, Navfac DM–7, 1971.

73. Barnes, S. B. 'Some special problems in the design of deep foundations', *Proc. 4th World Conference on Earthquake Engineering, Chile*, **III,** A—6, 29–36 (1969).

74. Wyllie, L. A., McClure, F. E., and Degenkolb, H. J. 'Performance of underground structures at the Joseph Jensen filtration plant', *Proc. 5th World Conference on Earthquake Engineering, Rome*, **1,** 66–75 (1973).

75. Broms, B. B. 'Design of laterally loaded piles', *Jnl. Soil Mechanics and Foundations Division, ASCE*, **91,** No. SM3, 79–98 (May, 1965).

76. Huntington, W. C. *Earth pressures and retaining walls*, Wiley, New York, 1957.

77. Scott, R. F. *Principles of Soil Mechanics*, Addison-Wesley, 1963.

78. Terzaghi, K., and Peck, R. B. *Soil Mechanics in Engineering Practice*, Wiley, New York, 1967.

79. New Zealand Ministry of Works. *Retaining wall design notes*, Issue C, July, 1973.

80. Japan Society of Civil Engineers. *Earthquake resistant design for civil engineering structures, earth structures and foundations in Japan*, 1973.

81. Seed, H. B., and Whitman, R. V. 'Design of earth retaining structures for dynamic loads', *ASCE specialty conference on lateral stresses in ground and the design of earth retaining structures, New York*, 103–148 (1970).

82. Indian Standards Institution, New Delhi. Criteria for earthquake resistance design of structures, *Indian Standard*, IS: 1893, 1966.

83. Mononobe, N. 'Earthquake-proof construction of masonry dams', *Proc. World Engineering Conference*, **9,** 275 (1929).

84. Okabe, S. 'General theory of earth pressure', *Jnl. Japanese Society of Civil Engineers*, **12,** No. 1 (1926).

85. Paparoni, M., Ferry Borges, J., and Whitman, R. V. 'Seismic studies of Parque Central buildings', *Proc. 5th World Conference on Earthquake Engineering, Rome*, **2,** 1991–2000 (1973).

86. American Concrete Institute. Building code requirements for reinforced concrete (ACI 318–71), *ACI Standard* 318–71, 1971.

87. Cardenas, A. E., Hanson, J. M., Corley, W. G. and Hognestad, E. 'Design provisions for shear walls'. *ACI Jnl.*, **70,** No. 3, 221–230 (March, 1973).

88. Salse, E. A. B., and Fintel, M. 'Strength, stiffness and ductility properties of slender shear walls', *Proc. 5th World Conference on Earthquake Engineering, Rome*, **1,** 919–928 (1973).

89. Paulay, T. 'Some aspects of shear wall design', *Bulln. New Zealand Society for Earthquake Engineering*, **5,** No. 3, 89–105 (Sept. 1972).

90. Corley, W. G. and Hanson, J. M. 'Design of earthquake-resistant structural walls', *Proc. 5th World Conference on Earthquake Engineering*, Rome, **1**, 933–936 (1973).
91. Beck, H. 'Contribution to the analysis of shear walls subject to lateral loads', *ACI Jnl.* **59**, 1055–1070 (Aug. 1962).
92. Rosman, R. 'Approximate analysis of shear walls subject to lateral loads', *ACI Jnl.*, **61**, 717–733 (June, 1964).
93. Burns, R. J. 'An approximate method of analysing coupled shear walls subject to triangular loading', *Proc. 3rd World Conference on Earthquake Engineering*, New Zealand, **III**, IV–123 to IV–140 (1965).
94. Paulay, T. 'A discussion of the analysis of coupled shear walls', by Coull, A., and Choudbury, J. R. *ACI Jnl.*, **65**, 237–239 (Mar. 1968).
95. Paulay, T. 'An elasto-plastic analysis of coupled shear walls', *ACI Jnl.*, **67**, No. 11, 915–922 (Nov. 1970).
96. Park, R., and Paulay, T. *Reinforced Concrete Structures*, Wiley, New York, 1975.
97. Paulay, T. 'Reinforced concrete shear walls', *New Zealand Engineering*, **24**, No. 10, 315–321 (Oct. 1969).
98. Paulay, T. 'Simulated seismic loading of spandrel beams', *Jnl. Structural Division*, *ASCE*, **97**, No. ST9, 2407–2419 (Sept. 1971).
99. Esteva, L., Rascón, O. A., and Gutiérrez, A. 'Lessons from some recent earthquakes in Latin America', *Proc. 4th World Conference on Earthquake Engineering*, Chile, **III**, J2, 65, 66 and 73 (1969).
100. Meehan, J. F. 'Performance of school buildings in the Peru earthquake of May 31, 1970', *Bulln. Seismological Society of America*, **61**, No. 3, June, 1971, 591–608.
101. Stratta, J. L., and Feldman, J. 'Interaction of infill walls and concrete frames during earthquakes', *Bulln. Seismological Society of America*, **61**, No. 3, 609–612 (June, 1971).
102. Mallick, D. V., and Severn, R. J. 'Dynamic characteristics of infilled frames', *Proc. Institution of Civil Engineers*, **39**, 261–287 (1968).
103. Stafford Smith, B., and Carter, C. 'A method of analysis for infilled frames', *Proc. Institution of Civil Engineers*, **44**, 31–48 (1969).
104. Lamar, S., and Fortoul, C. 'Brick masonry effect in vibration of frames', *Proc. 4th World Conference on Earthquake Engineering*, Chile, **II**, A3, 91–98 (1969).
105. Moss, P. J., and Carr, A. J. 'Aspects of the analysis of frame-panel interaction', *Bulln. New Zealand Society for Earthquake Engineering*, **4**, No. 1, 126–144 (March, 1971).
106. Dawson, R. V., and Ward, M. A. 'Dynamic response of framed structures with infill walls', *Proc. 5th World Conference on Earthquake Engineering*, Rome, **2**, 1507–1516 (1973).
107. Wen, R. K., and Natarajan, P. S. 'Inelastic seismic behaviour of frame-wall systems', *Proc. 5th World Conference on Earthquake Engineering*, Rome, **1**, 1343–1352 (1973).
108. Muto, K. 'Earthquake resistant design of 36-storied Kasumigaseki building', *Proc. 4th World Conference on Earthquake Engineering*, Chile, **III**, J4, 15–33 (1969).

Chapter 6

Structural design and detailing for earthquake resistance

6.1 INTRODUCTION

The choice of structural materials has been discussed in Section 4.3, where some comparisons were made between the performance of different materials in earthquakes. In Sections 5.1 and 5.2 the seismic response of various materials was introduced and the means of analysis discussed. The questions of seismic response and analysis have been expanded upon as necessary for the individual materials discussed in this chapter. Design and detailing problems for different stress conditions are then considered in terms of earthquake behaviour.

In the space of one book it is not possible to treat all matters of seismic design fully, but reference has been made to leading publications which were considered to represent the state of the art at the time of writing.

6.2 *IN SITU* REINFORCED CONCRETE DESIGN AND DETAIL

6.2.1 Introduction

There is more information available about the seismic performance of reinforced concrete than any other material. No doubt this is because of its widespread use and because of the difficulties involved in ensuring its adequate ductility. Well designed and well constructed reinforced concrete is suitable for most structures in earthquake areas, but achieving both these prerequisites is problematical even in areas of advanced technology.

Reinforced concrete is generally desirable because of its availability and economy, and its stiffness can be used to advantage to minimize seismic deformations and hence reduce the damage to non-structure. Difficulties arise due to reinforcement congestion when trying to achieve high ductilities in framed structures, and at the time of writing the problem of detailing beam-column joints to withstand strong cyclic loading had not been resolved. It should be recalled that no amount of good detailing will enable an ill-conceived structural form to survive a strong earthquake. The principles for determining good seismic-resistant structural form are discussed in Chapter 4.

The following notes provide an introduction to earthquake resistance of reinforced concrete, which has been discussed in depth by Park and Paulay.[1]

6.2.2 Seismic response of reinforced concrete

The seismic response of structural materials has been discussed generally in Section 5.1, where some stress-strain diagrams were presented. The hysteresis loops of Figure 5.5(d) indicate that considerable ductility without strength loss can be achieved in doubly reinforced beams having adequate confinement reinforcing. This is in distinct contrast to the loss of strength and stiffness degradation exhibited by plain unconfined concrete under repeated loading as shown in Figure 5.5(c). Because the hysteretic behaviour of reinforced concrete is so dependent on the amount and distribution of the longitudinal and transverse steel, mathematical models of hysteresis curves are still largely a matter for research as reviewed by Park.[2] While most testing has been done on rectangular beams, Krawinkler and Popov[3] have studied the hysteretic behaviour of T-beams demonstrating the importance of reinforcement in the flange or slab portion.

Even in well reinforced concrete members the root cause of failure under earthquake loading is usually concrete cracking. Degradation occurs in the cracked zone under cyclic loading. Cracks do not close up properly when the tensile stress drops because of permanent elongation of reinforcement in the crack, and aggregate interlock is destroyed in a few cycles. In hinge and joint zones reversed diagonal cracking breaks down the concrete between the cracks completely, and *sliding shear* failure occurs (Figure 6.1).

6.2.3 Principles of earthquake-resistant design

In reinforced concrete structures the essential features of earthquake resistance are embodied in ensuring the following;

(i)   beams should fail before columns;
(ii)   failure should be in flexure rather than shear;
(iii) premature failure of joints between members should be prevented;
(iv)   ductile rather than brittle failure should be obtained.

The above four criteria are now discussed in turn.

(i) Some codes[4] have specific strength factors which try to ensure that beams fail before columns, but this situation can be further facilitated by the use of mild steel for the longitudinal reinforcement of beams, and higher strength steels for columns. The less predictable but generally greater strength increase due to strain hardening of high strength steels can be used to advantage this way.

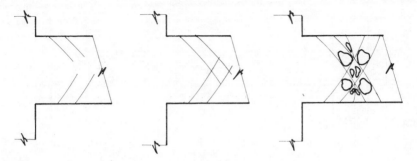

Figure 6.1 Progressive failure of reinforced concrete hinge zone under seismic loading

(ii) To prevent shear failure occurring before bending failure, it is good practice to design so that the flexural steel in a member yields while the shear reinforcement is working at a stress less than yield (say 90 percent). In beams a conservative approach to safety in shear is to make the shear strength equal to the maximum shear demands which can be made on the beam in terms of its bending capacity.

Referring to Figure 6.2, the shear strength of the beam should correspond to

$$V_{max} = \frac{M_{u1} - M_{u2}}{l} + V_g \qquad (6.1)$$

where V_g is the dead load shear force and

$$M_u = A_s f_{su} z$$

where A_s is all the steel in the tension zone, Figure 6.2(b), f_{su} is the maximum steel strength after strain hardening, say the 95 percentile for the steel samples, Figure 6.2(c), and z is the lever arm.

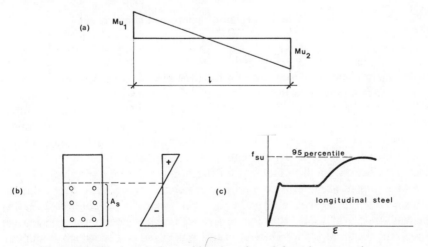

Figure 6.2 Shear strength considerations for reinforced concrete beams

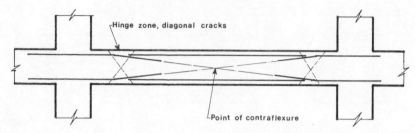

Figure 6.3 Prevention of sliding shear in plastic hinge zones using sloping main steel to carry shear after heavy concrete cracking

The possibility of *sliding shear failure* occurring in a hinge zone has been discussed above (Figure 6.1). Recent experiments by Paulay at Canterbury University have shown that the shear can be carried through the broken concrete zone by sloping the main bars through the hinge zone towards a point of centraflexure at the centre of the beam (Figure 6.3).

(iii) Joints between members such as beam-column joints are susceptible to failure earlier than the adjacent members due to destruction of the joint zone, in a manner similar to that shown in Figure 6.1. This is particularly true for external columns as reported by Park and Paulay.[5] The latter work considered the plane frame condition, and the three-dimensional joint with beams in two orthogonal directions has not yet been adequately researched. For example, Megget[6] reports that the expected favourable restraint is obtained by including stud beams laterally to a planar bending test. However such lateral beams may also be expected to be much less effective in restraining the beam-column panel zone when they in turn are subjected to strong cyclic bending. Hence Megget's test results are unconservative for the general joint condition.

6.2.4 Ductile design of reinforced concrete

Ductility has been discussed in terms of inelastic behaviour in Section 5.1.3, and the problems of assessing the ductility demands of a structure have been considered in Section 5.2.4.1. At the present time the most practical way of determining the ductility demand for a reinforced concrete structure is by method (6) of Table 5.3. Simplified methods of determining the hinge rotations in reinforced concrete frames have been suggested by Hollings[7] and Park.[8] These techniques involve the assumption of hinge mechanisms for the frame such as shown in Figure 6.4, and the imposition of an arbitrary lateral deflection ductility factor μ on the frame.

As mentioned above it is preferable that beams should fail before columns. Considering ten storeys above the column hinges of a column sidesway mechanism, Park[8] found that for an overall frame deflection ductility factor $\mu = 4$, the required section ductility ratio was $\phi_u/\phi_y = 122$, which is impossibly high. ϕ_u and ϕ_y are the hinge curvatures at ultimate and first yield

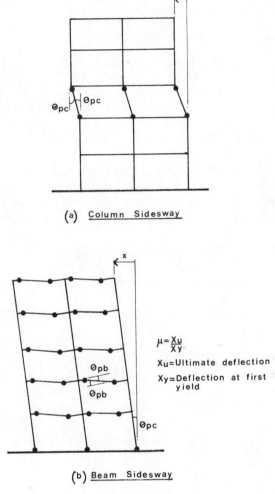

(a) Column Sidesway

$\mu = \dfrac{X_u}{X_y}$

X_u=Ultimate deflection

X_y=Deflection at first yield

(b) Beam Sidesway

Figure 6.4 Alternative plastic hinge mechanisms for a typical multi-storey frame (after Park.[8] Reproduced from *New Zealand Engineering* by permission of the N.Z. Institution of Engineers)

respectively. On the other hand for a beam sidesway mechanism the required section ductility was found to be less than 20.

Simple mechanisms such as shown in Figure 6.4 are only guesses at the actual rotational demands due to an earthquake, but they are likely to produce better aseismic structures than if no rotation assessment is made. An alternative method is to assess hinge rotations on the basis of the maximum deflection diagram produced by a linear elastic analysis. This is again an approximation because the pattern of plastic hinges will not necessarily correspond to the pattern of peak elastic stresses (Section 5.2.4.1).

Having made an estimate of the ductility demands in the structure, the members should be detailed to have the appropriate section ductility as discussed below.

6.2.5 Available ductility for reinforced concrete members

The available section ductility of a concrete member is most conveniently expressed as the ratio of its curvature at ultimate moment ϕ_u to its curvature at first yield ϕ_y. The expression ϕ_u/ϕ_y may be evaluated from first principles, the answers varying with the geometry of the section, the reinforcement arrangement, the loading and the stress-strain relationships of the steel and the concrete. Various idealizations of the stress-strain relationships give similar values for ductility, and the following methods of determining the available ductility should be satisfactory for most design purposes. It should be noted that the ductility of shear walls is discussed elsewhere (Section 5.7).

6.2.5.1 Single reinforced sections

Consider conditions at first yield and ultimate moment as shown in Figure 6.5.

Assuming an under-reinforced section, first yield will occur in the steel, and the curvature

$$\phi_y = \frac{\epsilon_{sy}}{(1 - k)d} = \frac{f_y}{E_s(1 - k)d} \tag{6.2}$$

where

$$k = \sqrt{\{(\rho n)^2 + 2\rho n\}} - \rho n \tag{6.3}$$

Strictly this formula for k is true for linear elastic concrete behaviour only, i.e. for

$$f_{ce} = \frac{2\rho f_y}{k} \leqslant 0 \cdot 6 f_{cu}$$

and for higher concrete stresses the true non-linear concrete stress block should be used. However according to Blume et al.[9] equation (6.3) provides a reasonable estimate of k even if the computed concrete stress is as high as $0 \cdot 85 f_{cu}$. Referring again to Figure 6.7 it can be shown that the ultimate curvature is

$$\phi_u = \frac{\epsilon_{cu}}{c} = \frac{\beta_1 \epsilon_{cu}}{a} \tag{6.4}$$

where

$$a = \frac{A_s f_y}{0 \cdot 72 f_{cu} b} \tag{6.5}$$

199

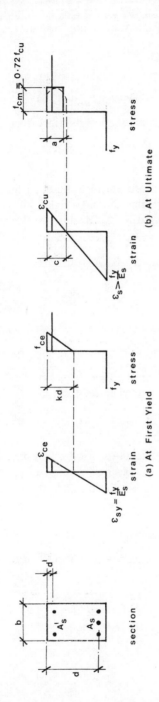

(a) At First Yield

(b) At Ultimate

Figure 6.5 Reinforced concrete section in flexure

200

and β_1, which describes the depth of the equivalent rectangular stress block, may be taken as in Clause 10.2.7 of the A.C.I. Code[4] to be

$$\beta_1 = 0.85$$

for $f_{cu} \leqslant 32.5$ N/mm², otherwise

$$\beta_1 = 0.026 f_{cu} - 0.0063(f_{cu} - 32.5) \tag{6.6}$$

From the above derivation the available section ductility may be written as

$$\frac{\phi_u}{\phi_y} = \frac{\epsilon_{cu} d(1 - k)E_s}{cf_y} \tag{6.7}$$

The ultimate concrete strain ϵ_{cu} is given various values in different codes for different purposes. For estimating the ductility available from reinforced concrete in a strong earthquake, a value of 0.004 may be taken as representing the limit of useful concrete strain.

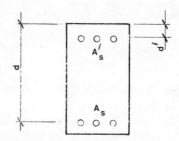

Figure 6.6 Doubly reinforced section

6.2.5.2. Doubly reinforced sections

The ductility of doubly reinforced sections (Figure 6.6) may be determined from the curvature in the same way as for singly reinforced sections above.

Once again the expression for available section ductility is

$$\frac{\phi_u}{\phi_y} = \frac{\epsilon_{cu} d(1 - k)E_s}{cf_y} \tag{6.7}$$

but to allow for the effect of compression steel ratio ρ', the expressions for c and k become

$$c = \frac{a}{\beta_1}$$

i.e.

$$c = \frac{(\rho - \rho')f_y d}{0.72 f_{cu}\beta_1} \tag{6.8}$$

and

$$k = \sqrt{\{(\rho + \rho')^2 n^2 + 2[\rho + (\rho'd/d)]n\}} - (\rho + \rho')n \tag{6.9}$$

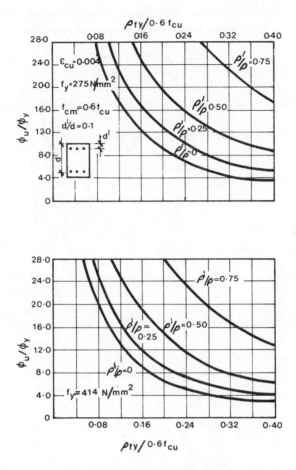

Figure 6.7 Variation of ϕ_u/ϕ_y for singly and doubly reinforced unconfined concrete (after Blume *et al.*[9])

The above equations assume that the compression steel is yielding, but if this is not so, the *actual* value of the steel stress should be substituted for f_y. And as k has been found assuming linear elastic concrete behaviour, the qualifications mentioned for singly reinforced members also apply.

Computations of ductility are clearly best done with the aid of a small computer, and may be presented graphically as in Figure 6.7.

In Figure 6.7 the ultimate compressive force in the concrete was taken as $0.6f_{cu}bc$ acting at a distance of $0.4c$ from the extreme compressive fibre. It can be seen from Figure 6.7 that;

(a) ductility reduces with increasing tension steel ρ;
(b) ductility increases with increasing compression steel ρ';
(c) ductility reduces with increasing yield stress f_y.

6.2.5.3 The effect of confinement on ductility

That the ductility and strength of concrete is greatly enhanced by confining the compression zone with closely spaced lateral steel has been demonstrated by various workers.[10,11,12] In order to quantify the ductility of confined concrete, a number of stress-strain curves for confined concrete have been derived from research.[9,13,14,15] Unfortunately the difference in effectiveness of different shapes and arrangements of the lateral confinement steel has not yet been established. It is known that rectangular all-enclosing links are moderately effective in small columns, but are of little use in large columns. Although this can be to some extent remedied by the use of inter-mediate lateral ties anchored to the all-enclosing links, the most efficient confinement method for large columns has yet to be ascertained.

While it is well known that circular binders are greatly superior to rectangular ones, for lack of definitive research data it is commonly assumed that rectangular binders are half as effective as circular.

The procedure for calculating the section ductility ϕ_u/ϕ_y is the same as that for unconfined concrete described above, the only difference being in determining an appropriate value of ultimate concrete strain ϵ_{cu} for use in equation (6.7). As all of the previously mentioned sources of stress-strain curves for confined concrete[9,13,14,15] have their limitations, it may be convenient to choose that of Corley[15] as it involves the least computational effort. Corley has recommended that a lower bound for the maximum concrete strain for concrete confined with rectangular links is

$$\epsilon_{cu} = 0.003 + 0.02\,(b/l_c) + \left(\frac{\rho_v f_{yv}}{138}\right)^2 \qquad (6.10)$$

where b/l_c is the ratio of the beam width to the distance from the critical section to the point of contraflexure, ρ_v is the ratio of volume of confining steel (including the compression steel) to volume of concrete confined, and f_{yv} is the yield stress of the confining steel (N/mm^2).

Because of the high strains at ultimate curvature, the increased tensile force due to strain hardening should be taken into account, or the calculated ultimate curvature may be too large and the estimated ductility will be unconservative. Spalling of the concrete in compression is ignored in Corley's method.

Example 6.1—Section ductility of reinforced concrete beam

Consider the beam shown in Figure 6.8. The confining steel consists of 12 mm diameter mild steel bars ($f_y = f_{yv} = 275\,N/mm^2$) at 75 mm centres, and the concrete strength is $f_{cu} = 25\,N/mm^2$. Estimate the section ductility ϕ_u/ϕ_y.

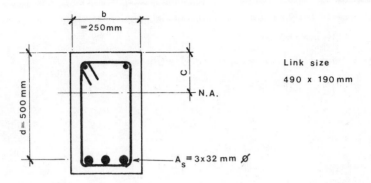

Figure 6.8 Reinforced concrete beam for ductility calculation example

To find the curvature at first yield, first estimate the depth of the neutral axis using equation (6.3), the section being effectively singly reinforced. As the modular ratio $n = 9$, and $\rho = 0.0193$

$$\rho n = 0.174$$

and

$$k = \sqrt{\{(\rho n)^2 + 2\rho n\}} - \rho n$$

$$= 0.441$$

Although this implies a computed maximum concrete stress greater than $0.85 f_{cu}$, the triangular stress block gives a reasonable approximation. Using equation (6.2) the yield curvature is found to be

$$\phi_y = \frac{f_y}{E_s(1 - k)d} = \frac{275}{2 \times 10^5 \times (1 - 0.441)500}$$

i.e.

$$\phi_y = 4.92 \times 10^{-6} \, \text{radian/mm}$$

To find the ultimate curvature for the confined section first determine the ultimate concrete strain from equation (6.10). Assume that for this beam

$$\frac{b}{l_c} = \frac{1}{8}$$

and

$$\rho_v = \frac{113 \times 2(490 + 190)}{490 \times 190 \times 75}$$

and

$$\epsilon_{cu} = 0.003 + 0.02(b/l_c) + \left(\frac{\rho_v f_{yv}}{138}\right)^2 \tag{6.10}$$

$$= 0.003 + \frac{0.02}{8} + \left(\frac{0.022 + 275}{138}\right)^2$$

i.e.

$$\epsilon_{cu} = 0.00742$$

Next find the depth of the neutral axis at ultimate from

$$c = \frac{a}{\beta_1}$$

$$= \frac{A_s f_y}{\beta_1 \times 0.72 f_{cu} b}$$

$$= \frac{2412 \times 275}{0.85 \times 0.72 \times 25 \times 250}$$

i.e.

$$c = 173 \, \text{mm}$$

Hence the ultimate curvature is

$$\phi_u = \frac{\epsilon_{cu}}{c}$$

$$= \frac{0.00742}{173}$$

i.e.

$$\phi_u = 4.29 \times 10^{-5} \, \text{radian/mm}$$

The available curvature ductility for the confined section can now be found,

$$\frac{\phi_u}{\phi_y} = \frac{4.29 \times 10^{-5}}{4.92 \times 10^{-6}} = 8.7$$

It is of interest to observe that the ultimate concrete strain $\epsilon_{cu} = 0.00742$, computed in the above example, is about twice the figure of 0.004 normally assumed for unconfined concrete. Hence the available section ductility has been roughly doubled by the use of confinement steel. This can be checked by reference to the curves of Figure 6.7 which gives values of ϕ_u/ϕ_y for unconfined flexural members. Now for the example beam

$$\frac{\rho f_y}{0.6 f_{cu}} = \frac{0.193 \times 275}{0.6 \times 25} = 0.354$$

and as the beam is singly reinforced, $\rho'/\rho = 0$. Hence from Figure 6.7 it can be seen that for the unconfined section

$$\phi_u/\phi_y \approx 4.25$$

which is about half the figure of $\phi_u/\phi_y = 8.7$ determined above.

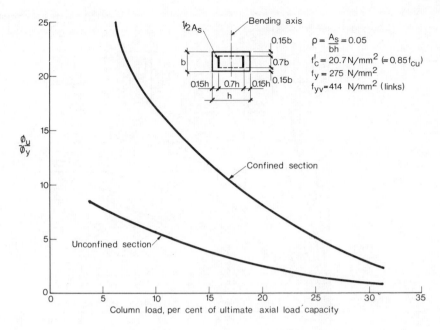

Figure 6.9 ϕ_u/ϕ_y for columns with confined or unconfined concrete (after Blume *et al.*[9])

6.2.5.4 Ductility of reinforced concrete members with flexure and axial load

Axial load unfavourably affects the ductility of flexural members, as can be seen from Figure 6.9. Indeed it has been shown[16] that only with axial loads less than the balanced load does ductile failure occur.

It is evident from Figure 6.9 that for practical levels of axial load, columns must be provided with confining reinforcement. Insufficient research has as yet been done for generalized methods of ductility calculation to be available. However, for rectangular columns with closely spaced links, and in which the longitudinal steel is mainly concentrated in two opposite faces, the ratio ϕ_u/ϕ_y may be estimated from Figure 6.10.

In Figure 6.10, A_s is the area of tension reinforcement and

$$\beta_h = \frac{1 \cdot 2 A_h f_{yh}}{s h_h f_{cu}} \tag{6.11}$$

where A_h is the cross-sectional area of the links, f_{yh} is the yield stress of the link reinforcement, s is the spacing of the link reinforcement, and h_h is the longer dimension of the rectangle of concrete enclosed by the links.

The value ϕ_u/ϕ_y for a particular section is obtained by following a path parallel to the arrowed zigzag on the diagram.

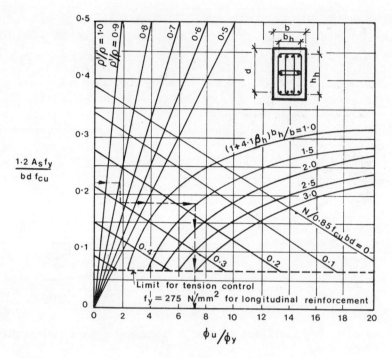

Figure 6.10 ϕ_u/ϕ_y for columns of confined concrete (after Blume *et al.*[9])

6.2.6 *In situ* concrete detailing—general requirements

The following notes and the associated detail drawings have been compiled to enable the elements of reinforced concrete structures to be detailed in a consistent and satisfactory manner for earthquake resistance. These details should be satisfactory in regions of medium and high seismic risk in so far as they reflect the present state-of-the-art. However considerable uncertainty exists regarding effective details for some members, particularly columns and beam-column connections. In low risk regions relaxations may be made to the following requirements, but the principles of lapping, containment and continuity must be retained if adequate ductility is to be obtained.

6.2.6.1 Laps

Laps in earthquake resisting frames must continue to function while the members or joints undergo large deformations. As the stress transfer is accomplished through the concrete surrounding the bars, it is essential that there be adequate space in a member to place and compact good quality concrete.

Laps should preferably not be made in regions of high stress, such as near beam-to-column connections, as the concrete may become cracked under large deformations and thus destroy the transfer of stress by bond. In regions of high stress, laps should be considered as an anchorage problem rather than a lap problem, i.e. the transfer of stress from one bar to another is not considered; instead the bars required to resist tension should be extended beyond the zone of expected large deformations in order to develop their strength by anchorage.

Tests have shown that contact laps perform just as well as spaced laps, because the stress transfer is primarily through the surrounding concrete. Contact laps usually reduce the congestion and give better opportunity to obtain well compacted concrete over and around the bars.

Laps should preferably be staggered but where this is impracticable and large numbers of bars are lapped at one location (e.g. in columns) adequate links or ties must be provided to minimize the possibility of splitting the concrete. In columns and beams even when laps are made in regions of low stress at least two links should be provided as shown in the details.

6.2.6.2 Anchorage

Satisfactory anchorage may be achieved by extending bars as straight lengths, or by using 90° and 180° bends, but anchorage efficiency will be governed largely by the state of stress of the concrete in the anchorage length. Tensile reinforcement should not be anchored in zones of high tension. If this cannot be achieved, additional reinforcement in the form of links should be added, especially where high shears exist, to help to confine the concrete in the anchorage length. It is especially desirable to avoid anchoring bars in the 'panel' zone of beam-column connections. Large amounts of the reinforcement should not be curtailed at any one section.

6.2.6.3 Bar bending

The use of BS 4466[17] will lead to standardization of bar shapes but due attention must be made to the bearing stresses in bends as follows. The bearing stress inside a bend in a bar which does not extend or is not assumed to be stressed beyond a point four times the bar size past the end of the bend need not be checked, as the longitudinal stresses developed in the bar at the bend will be small.

The bearing stress inside a bend in any other bar should be calculated from the equation

$$\text{bearing stress} = \frac{F_t}{r\phi}$$

where F_t is the tensile force due to ultimate loads in a bar or group of bars, r is the internal radius of the bend, and ϕ is the diameter of the bar or, in a bundle, the diameter of a bar of equivalent area.[18]

This stress should not exceed $1.5f_{cu}/(1 + 2\phi/a_b)$ where a_b for a particular bar or group of bars in contact should be taken as the centre to centre distance perpendicular to the plane of the bend between bars or groups of bars. For a bar or group of bars adjacent to the face of the member, a_b should be taken as the cover plus ϕ as defined above; f_{cu} is the characteristic cube crushing strength of the concrete.

6.2.6.4 Cover

Minimum cover to reinforcement should comply with local codes of practice. Where these do not exist, or in cases of doubt, reference may be made to British recommendations.[18]

6.2.6.5 Concrete quality

In some earthquake countries there may be a shortage of skilled labour and high quality building materials. Nevertheless the minimum recommended characteristic cube crushing strength for structural concrete is $20.0 \, \text{N/mm}^2$.

The use of lightweight aggregates for structural purposes in seismic zones should be very cautiously proceeded with, as many lightweight concretes prove very brittle in earthquakes. Appropriate advice should be sought in selecting the type of aggregate and mix proportions and strengths in order to obtain a suitably ductile concrete. It cannot be over-emphasized that quality control, workmanship and supervision are of the utmost importance in obtaining earthquake-resistant concrete.

6.2.6.6 Reinforcement quality

For adequate earthquake resistance, suitable quality of reinforcement must be ensured by both specification and testing. As the properties of reinforcement vary greatly between countries and manufacturers, much depends on knowing the source of the bars, and on applying the appropriate tests. Particularly in developing countries the role of the resident engineer may be decisive indeed onerous.

The following points should be observed (as amplified in Appendix B.2).

(a) An adequate minimum yield stress may be ensured by specifying steel to an appropriate standard, such as BS 4449,[19] or ASTM A615.[20]

(b) In California[21] it is required that for a given grade of steel, the actual yield stress should not exceed the minimum specified yield stress (characteristic strength) by more than $124 \, \text{N/mm}^2$ (18,000 psi). Retests should not exceed this value by more than an additional $21 \, \text{N/mm}^2$ (3000 psi). (These requirements are not strict enough for general application as discussed in Appendix B.2.)

(c) Grades of steel with characteristic strength in excess of 410 N/mm^2 (60,000 psi) are not permitted in some earthquake areas, e.g. California and New Zealand, but slightly greater strengths may be used if adequate ductility is proven by tests.

(d) Cold worked steels are not recommended in California or New Zealand, but steel to BS 4461[22] should be sufficiently ductile.

(e) Steel of higher characteristic strength than that specified should *not* be substituted on site.

(f) The elongation test is particularly important for ensuring adequate steel ductility. In BS 4461 and ASTM A615 appropriate requirements are set out for steels conforming to those standards. Steels to other standards require specific consideration.

(g) Bending tests are most important for ensuring sufficient ductility of reinforcement in the bent condition. In BS 4449, BS 4461 and ASTM A615 appropriate requirements are set out for steels conforming to those standards. Steels to other standards require specific consideration.

(h) The minimum bend radius for bars to ASTM A615 is sometimes greater than for British steels.

(i) Resistance to brittle fracture should be checked by a notch toughness test conducted at the minimum service temperature, where this is less than about 3–5°C.

(j) Strain–age embrittlement should be checked by rebend tests, similar to those for British steels.

(k) Welding of reinforcing bars may cause embrittlement and hence should only be allowed for steel of suitable chemical analysis and when using an approved welding process.

(l) Galvanizing may cause embrittlement and needs special consideration.

(m) Welded steel fabric (mesh) is unsuitable for earthquake resistance because of its potential brittleness. However, mesh to BS 4483[23] may be used for the control of shrinkage in non-structural elements such as ground slabs.

The above points are considered in more detail in Appendix B.2.

6.2.6.7 Notation

All notation in the design notes and details is in accordance with the British Concrete Code.[18]

6.2.6.8 Codes and standards

The reinforcing details recommended in this Manual are derived from a wide range of experience. Great reliance has been placed on Californian and New Zealand opinion, and their codes and leading research results have

210

been applied. The British Concrete Code[18] and British reinforcement standards have also been used where applicable.

In some earthquake countries, local codes may overrule some of the recommendations given in this document, but generally the requirements herein reflect the mainstream of current good aseismic detailing. As such they are imperfect and generalized and will need updating from time-to-time and at the discretion of earthquake-experienced engineers.

6.2.7 Foundation detailing

(See Details 6.1 and 6.2.)

6.2.7.1 Column bases and pile caps

The following rules apply;

(a) minimum percentage of steel = 0·15 percent each way;
(b) bars should be anchored at the free end as shown on the detail sheet;
(c) piles and caps should be carefully tied together to ensure integral action in earthquakes and sufficient reinforcement should be provided in non-tension piles to prevent separation of pile and cap due to ground movements.

6.2.7.2 Foundation tie-beams

In the absence of a thoroughgoing dynamic analysis of the substructure, tie-beams may be designed for arbitrary longitudinal forces of up to 10 percent of the maximum vertical column load into which the particular beam connects (Section 5.5.4.1). As the axial loads may be either tension or compression, the following rules are appropriate;

(a) minimum percentage of longitudinal steel = 1 percent;
(b) maximum percentage of longitudinal steel = 6 percent;
(c) minimum diameter of links = 8 mm;
(d) maximum and minimum spacing of links as for columns;
(e) minimum diameter of longitudinal steel = 12 mm;
(f) to enable foundation bases or footings to be cast before tie beams, beam starter bars from the footings should be detailed as shown on the detail sheet;
(g) the design check for the compressive case should be carried out as for design of columns with regard to such items as permissible compressive stresses, slenderness effects, and confining links.

6.2.7.3 Tie-beams taking bending

In some cases it may be required to transmit part of the bending moment at the column base into the tie-beams. Such tie-beams must therefore be designed for bending combined with axial compression or tension. The design should be carried out using the rules for beams or columns depending on the level of compressive stress. The minimum and maximum requirements (a) to (e) in Section 6.2.7.2 are applicable.

6.2.8 Retaining wall detailing

(See Detail 6.3.)

6.2.8.1

The minimum percentage of reinforcement should be 0·15 percent each face each way in both walls and footings. For adequate crack control more horizontal steel may be required especially in thin walls.

6.2.8.2

Top and bottom steel should be provided in the footings to provide for bending tensions which may not be apparent from a static analysis. Similarly steel should be provided in both faces of walls of thickness 150 mm and greater.

6.2.8.3

Footing bars should be anchored at the free end as shown on the detail sheet.

6.2.8.4

To ease fixing of wall bars the horizontal layer of bars on the face exposed to the air should wherever possible be in the outer layer. This also helps control of vertical cracking due to shrinkage.

6.2.8.5

The staggering of horizontal and vertical laps is desirable wherever possible as shown on the detail sheet.

6.2.8.6

Vertical construction joint positions along the length of wall should be selected to suit the detailing and should be shown on the drawings.

6.2.8.7

Basement walls or water-retaining structures may have detailing requirements which override the above recommendations.

6.2.9 Wall detailing

(See Details 6.4, 6.5 and 6.6.)

6.2.9.1

The minimum diameter of vertical and horizontal steel should be 10 mm.

6.2.9.2

In California it is recommended that for walls the minimum percentage of vertical and horizontal steel should be 0·125 percent each face.[4] This should be taken as the absolute minimum figure, and in many cases a more suitable minimum would be 0·2 percent each face.

6.2.9.3

The detailing around all openings is important. The details applicable to holes through suspended slabs are to be used for all walls.

6.2.9.4

Horizontal construction joints should be effectively cleaned and roughened.

6.2.9.5

For determining the reinforcement in highly stressed shear walls or coupling beams, see Section 5.7.

6.2.10 Column detailing

(See Details 6.7 and 6.8.)

6.2.10.1

The ratio of minimum to maximum column thickness should be not less than 0·4 nor should any dimension be less than 300 mm.

6.2.10.2

The minimum diameter for column longitudinal bars should be 12 mm.

6.2.10.3

The minimum diameter for column links should be 10 mm, and the minimum diameter for supplementary ties should be 8 mm.

6.2.10.4

The minimum content of longitudinal steel should be 1·0 percent of the gross cross-sectional area. The maximum content should be 6 percent.

6.2.10.5

Bundling of longitudinal bars is permitted providing the appropriate requirements for laps, curtailment and anchorage are complied with.

6.2.10.6

The longitudinal steel in columns should be of high tensile steel, to help ensure that beams fail before columns (Section 4.2.5).

6.2.10.7

The concrete core of a column shall be confined by special transverse reinforcement as described below when

$$N_e \geqslant 0.4 N_u$$

where N_e is the maximum axial load acting on a column during an earthquake, and N_u is the axial load capacity at simultaneous ultimate strain of concrete and yielding of tension steel (balanced conditions).

Confinement steel must be provided where the ductility demands are greatest in the confinement zones at the top and bottom of columns as shown on the column detail sheets. Either spiral reinforcement or links may be used. For spiral reinforcement the volumetric ratio of the transverse reinforcement provided shall not be less than the following:

$$\rho_{sp} = 0.38 \left(\frac{A_g}{A_c} - 1 \right) \frac{f_{cu}}{f_{yh}} \tag{6.12}$$

where

A_g = gross cross-sectional area of the concrete
A_c = area of core of column measured to the outside of the spiral,
f_{cu} = characteristic cube strength of concrete,
f_{yh} = characteristic strength of spiral or link reinforcement, and
ρ_{sp} = (volume of spiral reinforcement)/(volume of core measured to outside of spiral)

If links are used the required area of each leg of one link shall be

$$A_{h1} = \frac{l_h \rho_h s_{h1}}{2} \tag{6.13}$$

where

l_h = maximum unsupported length of link measured between perpendicular legs of the link or supplementary cross ties,
ρ_h = as for ρ_{sp}, except that for 'spiral' read 'hoop', and
s_{h1} = spacing of confining links. (See Detail 6.7).

6.2.10.8

In column/floor or column/beam connections special transverse steel is required as for the confinement zone described in Section 6.2.10.7, except that special transverse steel of one half of the amount required by Section 6.2.10.7 shall be used within the connection determined by the shallowest framing member, where such members frame into all four sides of a column and whose width is at least three quarters of the column width. But if a corner of such a column, unconfined by flexural members, exceeds 100 mm, the full special transverse steel shall be provided through the connection.

6.2.10.9 Column shear reinforcement

The transverse reinforcement in columns subject to bending and axial compression shall satisfy the following requirement:

$$A_h f_{yh}(d/s) = V_u - V_c \tag{6.14}$$

where

A_h = area of shear reinforcement within a distance s,
f_{yh} = yield stress of shear reinforcement,
d = the effective depth of the section,
s = spacing of shear reinforcement,
V_u = the ultimate shear force acting on the section,
$V_c = v_c bd$, and
v_c = nominal permissible shear stress (N/mm^2).

Various controls are placed on the value of v_c. In columns not subjected to axial tension v_c may be calculated from either of the two following equations:

$$v_c = 0 \cdot 144 \sqrt{f_{cu}} + \frac{138 A_s V_u}{b\{8M_u - N_e(4h - d)\}} \tag{6.15}$$

or

$$v_c = 0 \cdot 15 \left(1 + 0 \cdot 072 \frac{N_e}{bh} \right) \sqrt{f_{cu}} \tag{6.16}$$

where

f_{cu} = characteristic cube crushing strength (N/mm^2),
A_s = area of tensile reinforcement (mm^2),
b = width of column normal to V_u, (mm),
h = overall depth of column parallel to V_u, (mm),
M_u = ultimate moment acting simultaneously with V_u, (Nmm),
N_e = ultimate axial load acting simultaneously with V_u, (N).

But v_c shall be considered zero when

$$\frac{N_e}{bh} < 0 \cdot 10 f_{cu}$$

And notwithstanding equations (6.15) and (6.16), v_c must not exceed the following value:

$$v_c \not> 0\cdot26 \sqrt{\left\{ f_{cu}\left(1 + 0\cdot29 \frac{N_e}{bh} \right) \right\}} \tag{6.17}$$

For columns subjected to appreciable axial tension the permissible shear stress may be calculated as follows:

$$v_c = 0\cdot15\left(1 + 0\cdot29 \frac{N_e}{bh} \right) \sqrt{f_{cu}} \tag{6.18}$$

where N_e is negative for tension.

6.2.11 Beams

(See Details 6.9 and 6.10.)

Beams shall not have a width/depth ratio of less than 0·3 nor shall the width be less than 250 mm nor more than the supporting column width plus a distance on each side of the column of three quarters of the beam depth.

6.2.11.1 Beam longitudinal steel

(a)   The minimum diameter for any longitudinal bar = 12 mm.
(b)   To help ensure adequate ductility the longitudinal tensile steel content should not exceed 2·5 percent.
(c)   The minimum longitudinal steel content as a fraction of the gross cross-sectional area of the web ($h \times b$) should be $1\cdot4/f_y$ (f_y in N/mm^2) or $200/f_y$ (f_y in lb/in^2) where h is the overall depth of the beam and b is the width of the web.
(d)   Bent up bars are not recommended for aseismic construction.
(e)   Curtailment of longitudinal steel should allow for the most adverse loading conditions. Large numbers of bars should not be cut off at the same section.
(f)   Hooks and bends to be as BS 4466[17] except that in areas of high stress the ratio of bends may have to be increased to prevent local crushing; see the clause on bar bending in (Section 6.2.6.3).
(g)   The distance between bars is to be according to the code adopted (such as CP110[18]) but not less than 25 mm.
(h)   In beams forming part of a moment resisting framework, the positive moment capacity at columns must not be less than half the negative moment capacity provided. At least two bars should be provided both top and bottom throughout the length of the member.
(i)   Bundling of top and bottom bars is permitted providing the preceding requirements for laps, curtailment and anchorage are complied with.

(j)   Longitudinal bars in beams should be of mild steel, to help ensure that plastic hinges form in beams rather than columns.

6.2.11.2 Beam links

(a)   The minimum diameter for beam links should be 10 mm.

(b)   Within a distance equal to four times the effective depth from the end of the member, the minimum area of beam links A_v should not be less than that calculated from either of the following:

$$A_v \frac{h}{s_v} = 0 \cdot 15 A'_s \qquad (6.19)$$

or

$$A_v \frac{h}{s_v} = 0 \cdot 15 A_s \qquad (6.20)$$

where

A_v is the total area of all link legs within a distance s_v,
A'_s is the area of compressive reinforcement,
A_s is the area of tensile reinforcement; and
h is the overall depth of the beam.

(c)   Shear should not be resisted by bent up bars.

(d)   Links of the closed type are preferred wherever possible but open types with closing bars can be used provided sufficient anchorage is obtained.

(e)   Links should be provided to comply with the minimum requirements shown on the beam details, where design shears are not more onerous.

6.2.12 Slabs

(See Detail 6.11.)

Reinforcement designed for gravity loads in slabs forming part of a normal beam and slab system will generally be adequate to ensure that the slabs behave satisfactorily both as flexural members and as horizontal diaphragms transmitting earthquake forces. Certain elements such as flat slabs and waffle slabs, which may form part of the earthquake resisting framework must, of course, be designed and detailed accordingly. Their capacity to transfer earthquake moments between columns and slabs must be assured.

6.2.12.1

The minimum bar diameter should be 10 mm.

6.2.12.2

The minimum content of tension reinforcement in each direction should be 0·15 percent for high tensile steel, and 0·25 percent for mild steel. The minimum content of secondary reinforcement should be 0·15 percent.

6.2.12.3

For cantilevering slabs, bottom steel should be provided to counteract bending tensions which may occur during earthquakes.

6.2.12.4

Holes through slabs should be framed with extra steel as shown on the detail sheet because of the diaphragm action of slabs during earthquakes.

6.2.12.5

For ground floor or basement slabs which are designated as 'ground bearing', special seismic considerations may not exist and it is usual merely to place one layer of nominal steel in each direction to prevent cracking and shrinkage. This is usually placed in the top of the slab. Tie steel between column bases may also be placed in the ground slab in some instances, instead of in foundation tie beams (Section 6.2.7.2).

6.2.13 Staircases

(See Detail 6.12.)

6.2.13.1

Generally the rules for slabs apply. As shown on the detail sheet, top steel should be provided at each landing to provide for bending tensions which may not be apparent from a simple analysis.

6.2.13.2

If stairs are part of a horizontal diaphragm or moment resisting framework they should be reinforced accordingly. Due care must then be taken at the changes in slope to confine the longitudinal bars.

6.2.14 Upstands and parapets

(See Detail 6.13.)

6.2.14.1

Upstands and parapets should be carefully designed against seismic accelerations, which may considerably exceed those occurring elsewhere in the structure due to resonance or 'whip' effects.

6.2.14.2

The arrangement of reinforcement at corners and junctions should be as for walls.

218

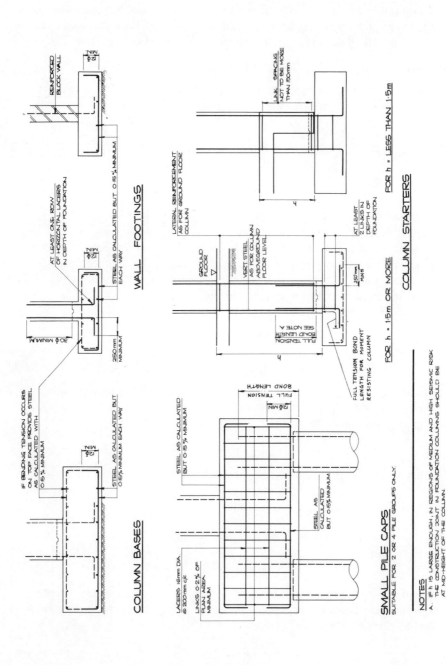

Detail 6.1 Foundations sheet 1

219

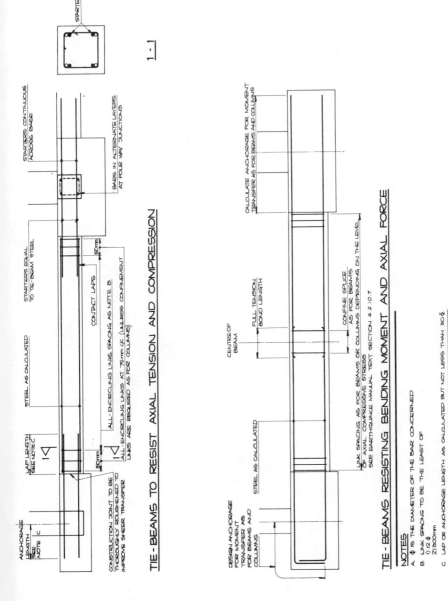

TIE - BEAMS TO RESIST AXIAL TENSION AND COMPRESSION

1 - 1

TIE - BEAMS RESISTING BENDING MOMENT AND AXIAL FORCE

NOTES

A. φ IS THE DIAMETER OF THE BAR CONCERNED

B. LINK SPACING TO BE THE LEAST OF
   1) 12 φ
   2) 300mm

C. LAP OR ANCHORAGE LENGTH AS CALCULATED BUT NOT LESS THAN 30 φ

Detail 6.2 Foundations sheet 2

220

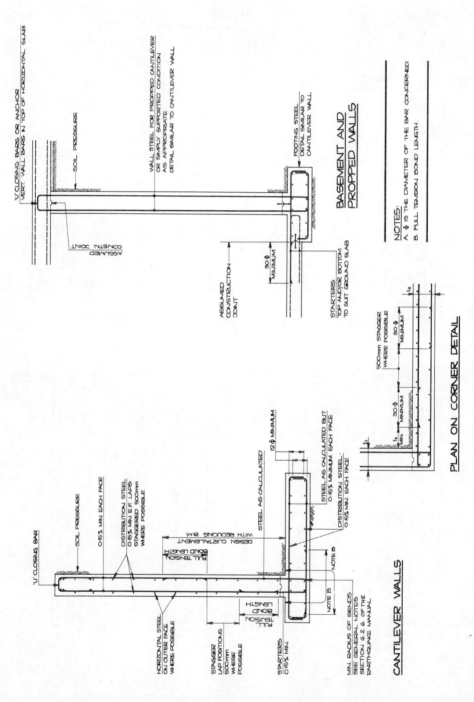

Detail 6.3 Retaining walls

221

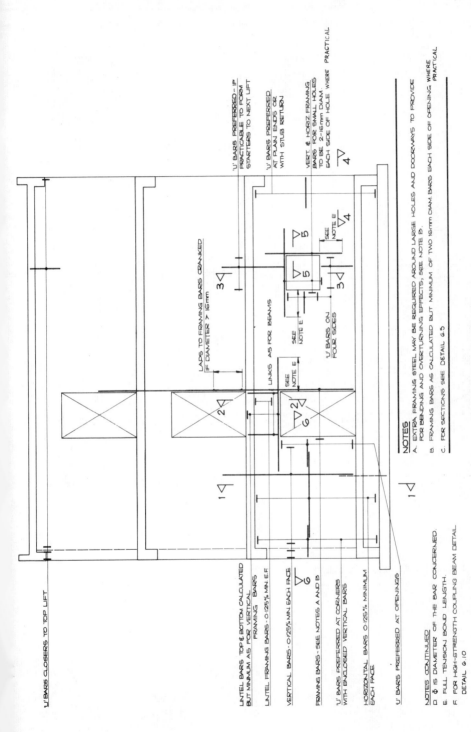

Detail 6.4 Walls sheet 1

'U' BARS CLOSERS TO TOP LIFT

LINTEL BARS TOP & BOTTOM CALCULATED BUT MINIMUM AS FOR VERTICAL FRAMING BARS

LINTEL FRAMING BARS - 0.125% MIN. E.F.

VERTICAL BARS - 0.125% MIN EACH FACE

FRAMING BARS - SEE NOTES A AND B

'U' BARS PREFERRED AT CORNERS WITH ENCLOSED VERTICAL BARS

HORIZONTAL BARS 0.125% MINIMUM EACH FACE

'U' BARS PREFERRED AT OPENINGS

'U' BARS PREFERRED - IF PRACTICABLE TO FORM STARTERS TO NEXT LIFT

'U' BARS PREFERRED AT PLAIN ENDS OR WITH STUB RETURN

VERT. & HORIZ FRAMING BARS FOR SMALL HOLES TO BE 2-16mm DIAM. EACH SIDE OF HOLE WHERE PRACTICAL

LADS TO FRAMING BARS CRANKED IF DIAMETER > 16mm

LINKS AS FOR BEAMS

SEE NOTE E

SEE NOTE E

SEE NOTE E

'U' BARS ON FOUR SIDES

NOTES

A. EXTRA FRAMING STEEL MAY BE REQUIRED AROUND LARGE HOLES AND DOORWAYS TO PROVIDE FOR BENDING AND OVERTURNING EFFECTS, SEE NOTE B.

B. FRAMING BARS AS CALCULATED BUT MINIMUM OF TWO 16mm DIAM BARS EACH SIDE OF OPENING WHERE PRACTICAL

C. FOR SECTIONS SEE DETAIL 6.5

NOTES CONTINUED

D. ϕ IS DIAMETER OF THE BAR CONCERNED.

E. FULL TENSION BOND LENGTH.

F. FOR HIGH-STRENGTH COUPLING BEAM DETAIL DETAIL 6.10

222

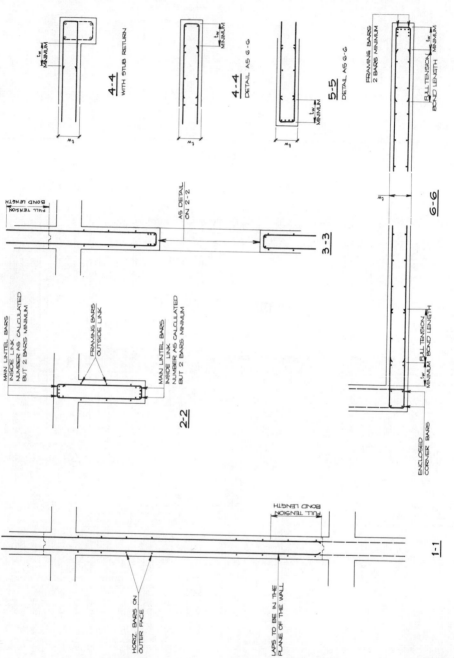

Detail 6.5 Walls sheet 2

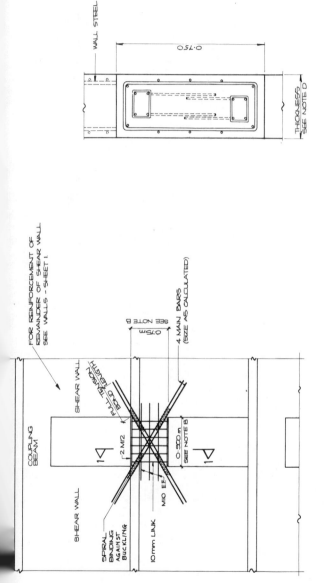

HIGH-STRENGTH COUPLING BEAM

NOTES

A. TO BE USED WHERE COUPLED SHEAR WALL ACTION IS ESSENTIAL.

B. THIS DETAIL IS APPROPRIATE FOR DEEP COUPLING BEAMS OF SIMILAR SPAN/DEPTH RATIO TO THAT SHOWN.

C. CARE IS NECESSARY IN DETAILING TO EASE THE STEEL FIXING AND CONCRETING PROBLEMS.

D. USE WITH WALLS THINNER THAN 0·300 m IS NOT RECOMMENDED.

Detail 6.6 Walls sheet 3

224

NOTES

A. l_c = LARGEST OF 1) h   (h > b)
   2) $l_o/6$
   3) 450mm

B. FOR COLUMNS REQUIRING SPECIAL CONFINEMENT STEEL

S_{h_1} = SPACING OF ALL CONFINEMENT LINKS & TIES IN CONFINEMENT ZONE USING A MAX SPACING OF 100mm. THE TOTAL CROSS-SECTIONAL AREA OF CONFINEMENT LINKS SHOULD BE DETERMINED ACCORDING TO SECTION 8.14.7

C. S_{h_2} = SPACING OF ALL LINKS & TIES IN INTERMEDIATE ZONE. SPACING TO BE LEAST OF : 1) 12 ϕ
   2) 300mm

ϕ = DIAMETER OF SMALLEST LONGITUDINAL BAR IN COLUMN.

D. SPACING AND TOTAL CROSS-SECTIONAL AREA OF LINKS & TIES THROUGH THE COLUMN/FLOOR CONNECTION (OR ANY OTHER CONNECTING MEMBER) SHOULD BE DETERMINED ACCORDING TO SECTION 8.14.8

E. ALL LINK ARRANGEMENTS MUST ALSO BE CAPABLE OF RESISTING THE APPLIED SHEARS THROUGHOUT THE WHOLE COLUMN LENGTH INCLUDING THE COLUMN / FLOOR CONNECTION ZONE.

F. IN MEDIUM AND HIGH RISK EARTHQUAKE ZONES LAPS SHOULD BE MADE AT MID COLUMN HEIGHT AND LAPS TO BE LARGER OF :
   1) CALCULATED BOND LENGTH
   2) 30 ϕ
   3) 600mm

G. FOR COLUMNS NOT REQUIRING SPECIAL CONFINEMENT OR SHEAR LINKS PROVIDE MINIMUM 10mm DIA LINKS AT 100mm PITCH THROUGH THE COL./FLOOR CONNECTION AND FOR DISTANCES l_c GIVEN IN NOTE A.

H. DIAM. OF SUPPLEMENTARY TIES TO BE SAME DIAM. AS LINKS.

J. ALL LAPS TO BE CONFINED BY MIN. OF 3 LINKS. TOPMOST LINK TO BE AT TOP OF LOWER SPLICE BAR.

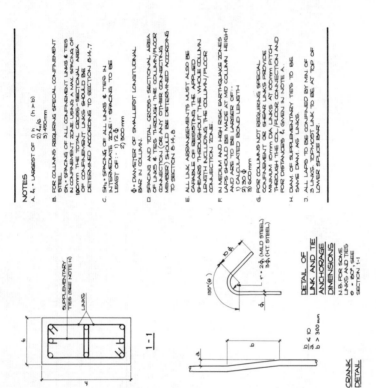

1 - 1

DETAIL OF
LINK AND TIE
ANCHORAGE
DIMENSIONS

N.B. FOR SOME LINKS AND TIES
θ = 180°, SEE SECTION 1-1

CRANK DETAIL

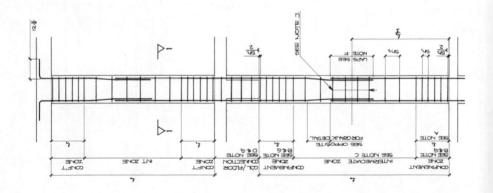

Detail 6.7 Columns sheet 1

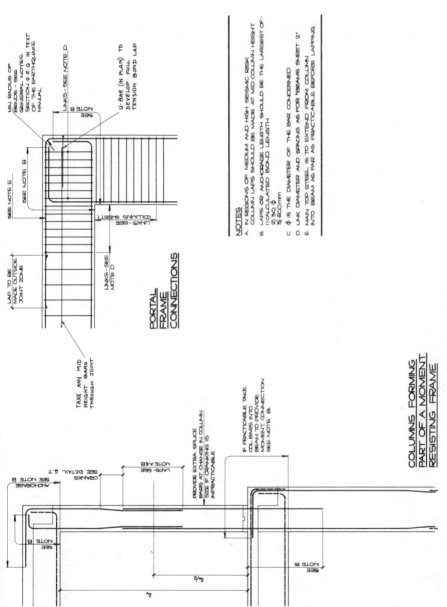

Detail 6.8 Columns sheet 2

226

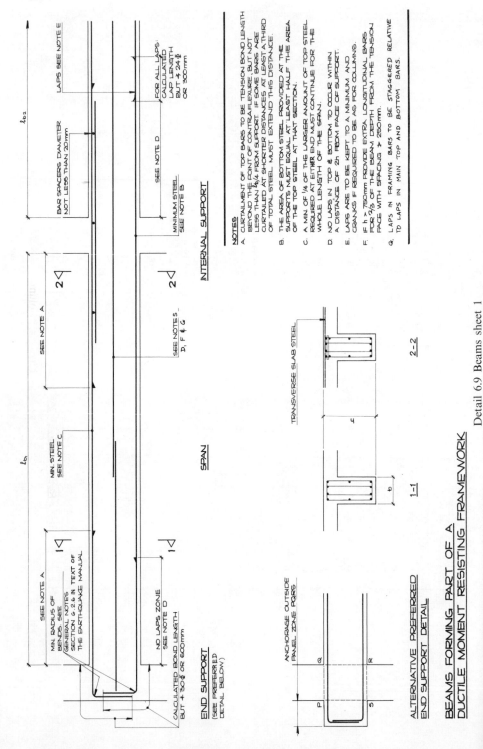

NOTES

A. CURTAILMENT OF TOP BARS TO BE TENSION BOND LENGTH BEYOND THE POINT OF CONTRAFLEXURE, BUT NOT LESS THAN $\ell_b/4$ FROM SUPPORT. IF SOME BARS ARE CURTAILED AT SHORTER DISTANCES AT LEAST A THIRD OF TOTAL STEEL MUST EXTEND THIS DISTANCE.

B. THE AREA OF BOTTOM STEEL PROVIDED AT THE SUPPORTS MUST EQUAL AT LEAST HALF THE AREA OF THE TOP STEEL AT THAT SECTION.

C. A MIN OF 1/4 OF THE LARGER AMOUNT OF TOP STEEL REQUIRED AT EITHER END MUST CONTINUE FOR THE WHOLE LENGTH OF THE SPAN.

D. NO LAPS IN TOP & BOTTOM TO OCCUR WITHIN A DISTANCE OF 2h FROM FACE OF SUPPORT.

E. LAPS ARE TO BE KEPT TO A MINIMUM AND CRANKS IF REQUIRED TO BE AS FOR COLUMNS.

F. IF h > 750mm PROVIDE EXTRA LONGITUDINAL BARS FOR 2/3 OF THE BEAM DEPTH FROM THE TENSION FACE WITH SPACING > 250mm.

G. LAPS IN FRAMING BARS TO BE STAGGERED RELATIVE TO LAPS IN MAIN TOP AND BOTTOM BARS.

LAPS SEE NOTE E

BAR SPACERS DIAMETER NOT LESS THAN 20mm

SEE NOTE A

FOR ALL LAPS CALCULATED LAP LENGTH BUT + 24 ϕ OR 300mm

SEE NOTE D

MINIMUM STEEL SEE NOTE B

MIN. STEEL SEE NOTE C

SEE NOTES D, F & G

INTERNAL SUPPORT

ℓ_{b2}

ℓ_b

SPAN

SEE NOTE A

MIN. RADIUS OF BENDS, SEE GENERAL NOTES SECTION G.2.6 IN TEXT OF THE EARTHQUAKE MANUAL

NO LAPS ZONE SEE NOTE D

CALCULATED BOND LENGTH BUT + 30 ϕ OR 600mm

END SUPPORT

(SEE PREFERRED DETAIL BELOW)

ANCHORAGE OUTSIDE PANEL ZONE PQRS

TRANSVERSE SLAB STEEL

2-2

1-1

ALTERNATIVE PREFERRED
END SUPPORT DETAIL

BEAMS FORMING PART OF A
DUCTILE MOMENT RESISTING FRAMEWORK

Detail 6.9 Beams sheet 1

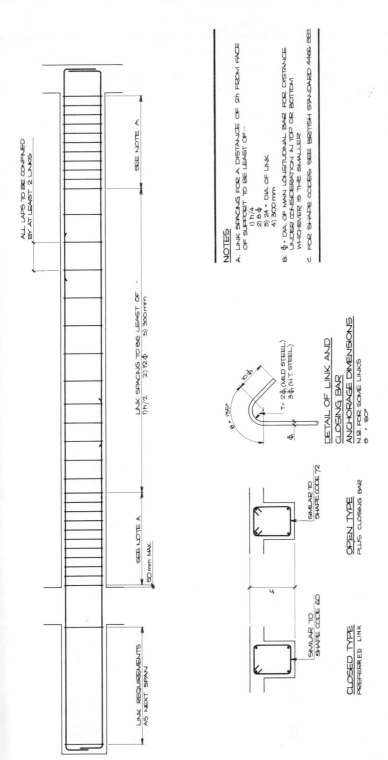

ALL LAPS TO BE CONFINED
BY AT LEAST 2 LINKS

SEE NOTE A

LINK SPACING TO BE LEAST OF :-
1) h/2   2) 12ϕ 3) 300mm

SEE NOTE A

50 mm MAX.

LINK REQUIREMENTS
AS NEXT SPAN

NOTES

A. LINK SPACING FOR A DISTANCE OF 2h FROM FACE
OF SUPPORT TO BE LEAST OF :-
1) h/4
2) 8ϕ
3) 24 × DIA OF LINK
4) 300 mm

B. ϕ = DIA. OF MAIN LONGITUDINAL BAR FOR DISTANCE
UNDER CONSIDERATION IN TOP OR BOTTOM
WHICHEVER IS THE SMALLER.

C. FOR SHAPE CODES SEE BRITISH STANDARD 4466 1969.

θ = 135°

T = 2ϕ (MILD STEEL)
3ϕ (H.T. STEEL)

ϕ

DETAIL OF LINK AND
CLOSING BAR
ANCHORAGE DIMENSIONS
N.B. FOR SOME LINKS
θ = 180°

SIMILAR TO
SHAPE CODE 72

OPEN TYPE
PLUS CLOSING BAR

SIMILAR TO
SHAPE CODE 60

CLOSED TYPE
PREFERRED LINK

ACCEPTED LINK TYPES

OTHER TYPES SUCH AS SHAPE CODES
73 TO 75 ARE PERMITTED PROVIDING
MINIMUM ANCHORAGE LENGTHS ARE USED.

Detail 6.10 Beams sheet 2

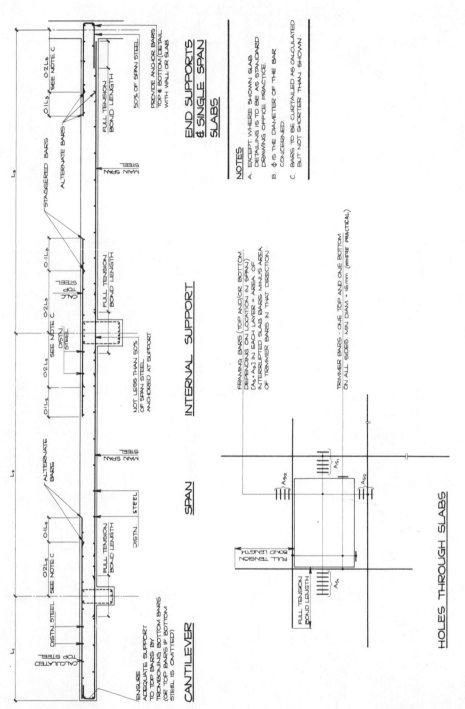

Detail 6.11 Slabs

229

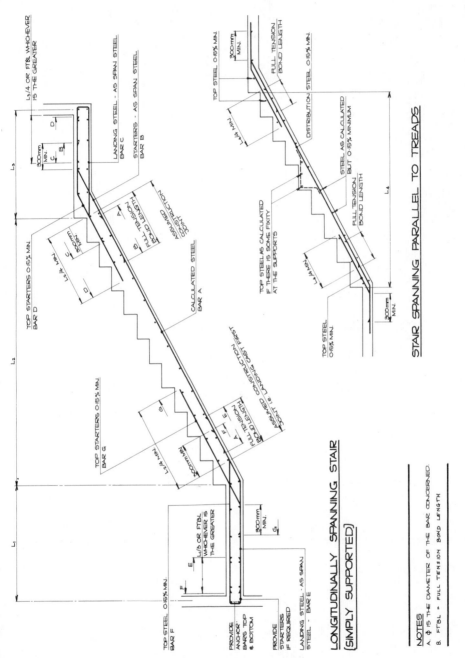

Detail 6.12 Staircases

230

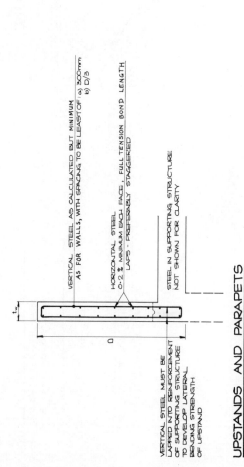

NOTES

A. Φ IS THE DIAMETER OF THE BAR CONCERNED

B. IF UPSTAND OR PARAPETS SPAN LONGITUDINALLY THEN DETAIL AS FOR BEAMS.

C. VERTICAL STEEL PREFERABLY IN FORM OF LINKS BUT IF HEIGHT OF LEG IS EXCESSIVE USE TWIN 'U' BARS.

VERTICAL STEEL AS CALCULATED BUT MINIMUM AS FOR WALLS, WITH SPACING TO BE LEAST OF: a) 300mm b) D/3

HORIZONTAL STEEL 0·2 % MINIMUM EACH FACE, FULL TENSION BOND LENGTH LAPS - PREFERABLY STAGGERED

STEEL IN SUPPORTING STRUCTURE NOT SHOWN FOR CLARITY

VERTICAL STEEL MUST BE LAPPED INTO REINFORCEMENT OF SUPPORTING STRUCTURE TO DEVELOP LATERAL BENDING STRENGTH OF UPSTAND

UPSTANDS AND PARAPETS

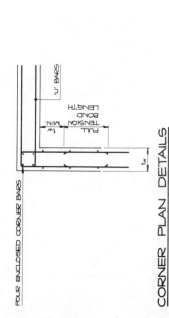

FOUR ENCLOSED CORNER BARS

'U' BARS

FULL TENSION BOND LENGTH

t_w MIN.

t_w

CORNER PLAN DETAILS

Detail 6.13 Upstands and parapets

6.3 STRUCTURAL PRECAST CONCRETE DETAIL

6.3.1 Introduction

Precast concrete structures have given mixed performance in earthquakes, difficulties mainly being experienced at connections between members. A negligible number of properly documented test results have been published on the behaviour of connections under cyclic loading. Much of the testing which has been done has been related to specific proprietary precast systems, and such results as have been published are usually either lacking in essential detail or are not readily applicable to other precasting assemblies. As a result the development of precasting has yet to reveal its full potential in aseismic construction.

Nevertheless precast concrete, either reinforced or prestressed, has been used to some extent in most forms of structure in earthquake areas, often in conjunction with a cautiously large amount of unifying *in situ* concrete. The nature of the seismic response of a precast structure must be inferred from the response of the reinforced or prestressed members involved (Sections 6.2 and 6.5). Allowance for the effect of the connection on the stress flow must also be made. This is particularly important when adapting proprietary precast products made for general purposes especially when intended originally for non-seismic areas. Without appropriate dynamic test results the effect of the connections may be difficult or impossible to assess, especially if they depart substantially from providing full continuity and homogeneity between adjacent members.

Dealing with building tolerances is a major problem in the design of connections. Constructional eccentricities may result in large secondary stresses in earthquakes, and should be either designed for or minimized by the manner of connection. It may be advantageous to design structural joints which permit generous constructional tolerances and restrict the expensive fine tolerance work to the cladding for visual or drainage purposes.

In order to overcome the connection problem, partial precasting is often done. For example precast beams may be used with *in situ* columns, or precast walls may be used with *in situ* floors, or *vice versa.*

As the basic detailing of reinforced and prestressed concrete has been discussed elsewhere in this document, only the essential problem of precast construction, that of *connection*, is considered in this section. For further reading see references 24–33.

6.3.2 Connections between bases and precast columns

The following typical details must be individually designed for the forces acting on the joint. The base considered may be at foundation level or on suspended members higher up the structure. Member reinforcement is not shown.

232

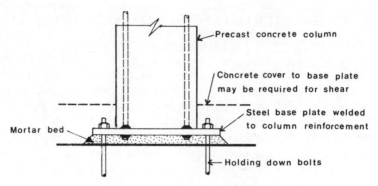

Precast concrete column

Concrete cover to base plate
may be required for shear

Steel base plate welded
to column reinforcement

Mortar bed

Holding down bolts

Detail 6.14 Site bolted. Moment transfer controlled by base plate

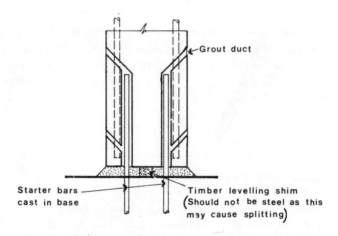

Grout duct

Starter bars
cast in base

Timber levelling shim
(Should not be steel as this
may cause splitting)

Detail 6.15 Site grouted. Effectiveness depends on grouting

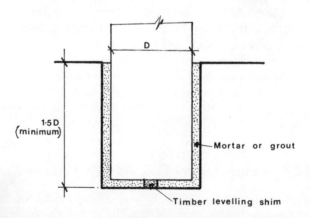

D

1·5 D
(minimum)

Mortar or grout

Timber levelling shim

Detail 6.16 Site grouted. Best all-round joint of this type. Method of transfer of vertical
load to base must be checked

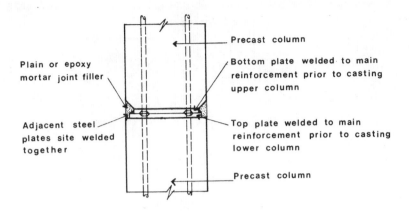

Precast column

Plain or epoxy
mortar joint filler

Bottom plate welded to main
reinforcement prior to casting
upper column

Adjacent steel
plates site welded
together

Top plate welded to main
reinforcement prior to casting
lower column

Precast column

Detail 6.17 Site welded

6.3.3 Connections between precast columns and beams

The following typical details must be individually designed for the forces
acting on the joint under construction. Variations on these connections may
be made to suit the circumstances. Member reinforcement is not shown.

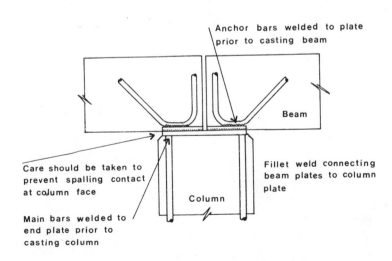

Anchor bars welded to plate
prior to casting beam

Beam

Care should be taken to
prevent spalling contact
at column face

Fillet weld connecting
beam plates to column
plate

Column

Main bars welded to
end plate prior to
casting column

Detail 6.18 Site welded. Low moment capacity

234

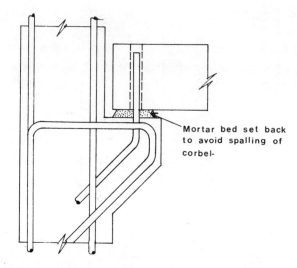

Mortar bed set back
to avoid spalling of
corbel-

Detail 6.19 Site grouted. Low moment capacity, poor in horizontal shear

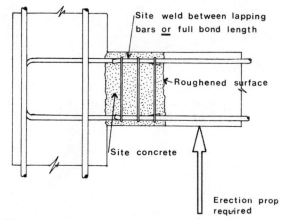

Site weld between lapping
bars or full bond length

Roughened surface

Site concrete

Erection prop
required

Detail 6.20 Site concreted and welded and links fixed

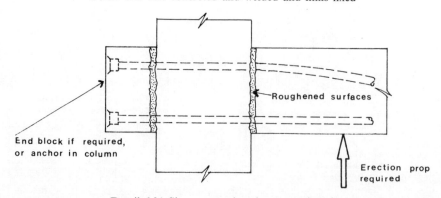

Roughened surfaces

End block if required,
or anchor in column

Erection prop
required

Detail 6.21 Site mortared and post-tensioned

6.3.4 Connections between precast floors and walls

The following typical details must be designed for the forces acting on the joint under consideration. Member reinforcement is not shown.

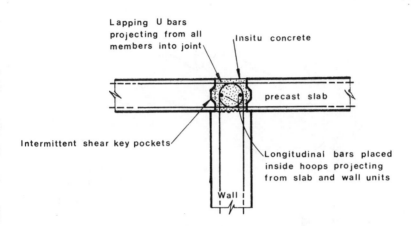

Detail 6.22 Site concrete and reinforcement

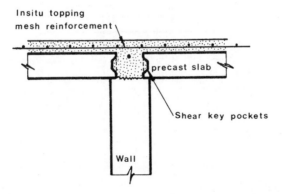

Detail 6.23 Site concrete and reinforcement

236

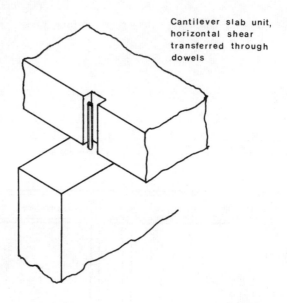

Cantilever slab unit, horizontal shear transferred through dowels

Detail 6.24 Site grouting

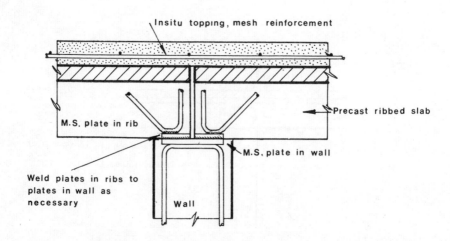

Insitu topping, mesh reinforcement

Precast ribbed slab

M.S. plate in rib

M.S. plate in wall

Weld plates in ribs to plates in wall as necessary

Wall

Detail 6.25 Site concrete, reinforcement and welding

6.3.5 Connections between adjacent precast floor and roof units

The following typical details must be designed for the forces acting on the joint under consideration. Floor slabs should be designed as a whole, to act as diaphragms distributing the shear between the vertical members of the structure. The unifying effect of perimeter beams should be taken into account. Member reinforcement is not shown.

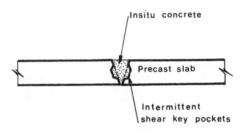

Detail 6.26 Site concreting. This joint depends on perimeter reinforcement to complete shear transfer system

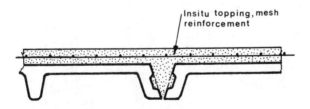

Detail 6.27 Site concreting and reinforcing

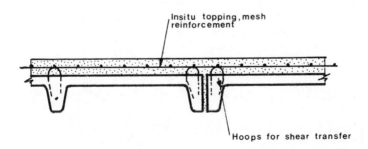

Detail 6.28 Site concreting and reinforcing

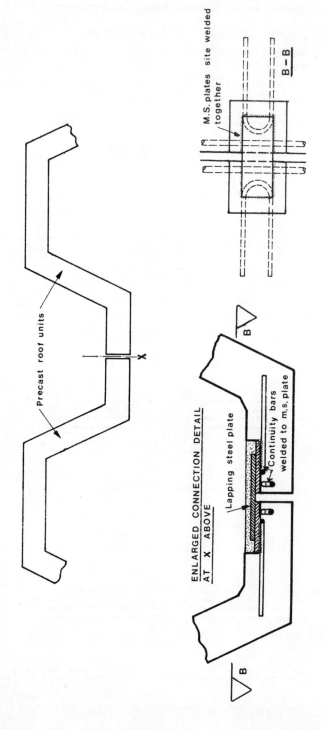

Detail 6.29 Site welding and mortaring. Lapping steel plate bent on site to suit differential camber of adjacent precast units

6.3.6 Connections between adjacent precast wall units

The following typical details should be designed for the forces acting on the joint under consideration. Great problems occur in producing a ductile and easily-erected precast shear wall, and no universal solution has as yet been evolved. The details below may be adapted for use with internal or external walls, i.e. cladding. Member reinforcement is not shown.

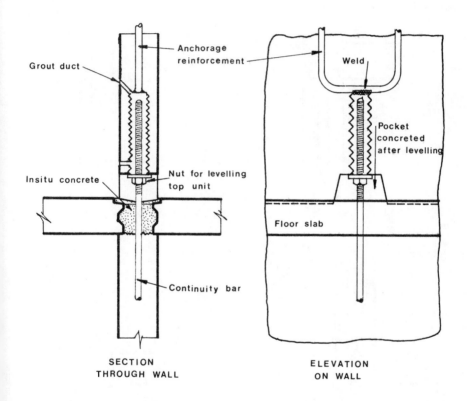

Detail 6.30 Site concreting and grouting

240

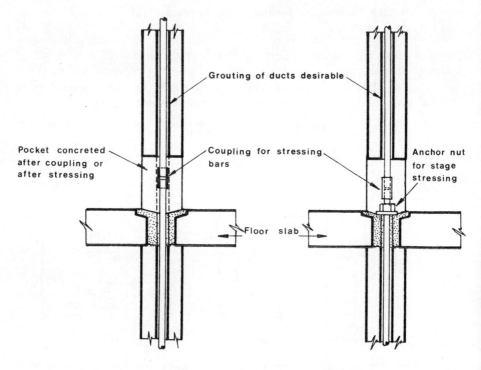

Grouting of ducts desirable

Pocket concreted
after coupling or
after stressing

Coupling for stressing
bars

Anchor nut
for stage
stressing

Floor slab

Detail 6.31 Site concreting and post-tensioning and grouting of ducts

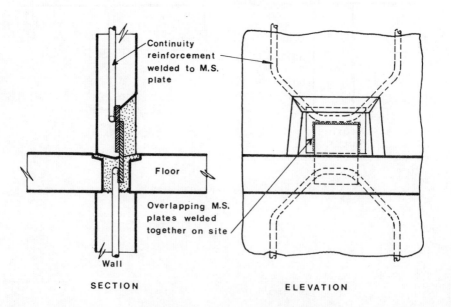

Continuity
reinforcement
welded to M.S.
plate

Floor

Overlapping M.S.
plates welded
together on site

Wall

SECTION

ELEVATION

Detail 6.32 Site welded and concreted

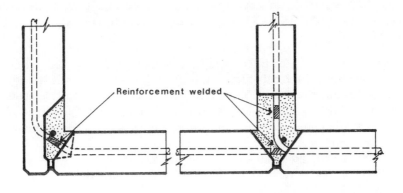

Detail 6.33 Site welded and concreted

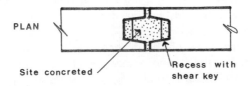

Detail 6.34 Site concreted

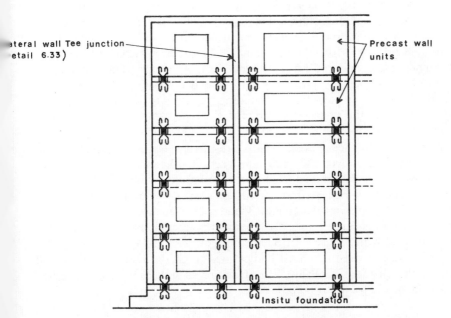

Detail 6.35 Layout of joints in wall elevation

6.4 PRECAST CONCRETE CLADDING DETAIL

Precast concrete cladding varies in its relationship to the building structure from being fully integrated to being fully separated from frame action. Ideally the cladding should be either fully integrated or fully separated, with no intermediate conditions. Fully integrated *structural* precast concrete cladding should be treated like any other precast structural element as discussed in Section 6.3. Cladding which is not considered as part of the structure is considered below.

In flexible beam and column buildings it is desirable to effectively separate the cladding from the frame action, both to protect the cladding from seismic deformations and also to ensure that the structure behaves as assumed in the analysis. For very flexible buildings in strong earthquakes the storey drift may be so large as to make full separation difficult to achieve, and some interaction of frame and cladding through bending of the connections may have to be accepted. Ductile behaviour of the cladding and of its connections to the structure is most important in such cases to ensure that the cladding does not fall from the building during an earthquake.[34,35]

In stiff (shear wall) buildings the storey drift will generally be small enough to significantly reduce the problem of detailing of connections which give full separation. On the other hand protection of the cladding from seismic motion is less necessary in stiff buildings, and connections permitting movement through bending may be satisfactory as long as the interaction between cladding and frame can be allowed for in the frame analysis.

There is a growing tendency for precast cladding to be fully separated from the frame in strong motion areas like California, Japan and New Zealand. This has been done on recent major buildings in Tokyo, such as the 47-storey Keio Plaza Hotel. Unfortunately little has been published regarding the connection details for separated cladding, although some reference is made to this problem by Uchida *et al.*[36] and Brooke-White.[37] In Uchida's structure, a 25-storey steel framed building, separation of the cladding was only partial, and the connections were designed so that the panels would not fall off if the storey drift was 50 mm.

Gaps between adjacent precast units are often specified to be 20 mm to allow for seismic movements and construction tolerances, but smaller or larger gaps may be determined from drift calculations. Waterproofing of gaps may be affected by baffled drain joints or mastic,[38] but the performance of mastic-filled joints in earthquakes is not known at the present time.

The principles of support for fully separated precast cladding are illustrated diagrammatically in Figure 6.11. Such connections should be made of corrosion-resistant materials, and must be designed to carry the gravity and wind loads of the cladding back into the structure as well as to allow the free movement of the frame to take place.

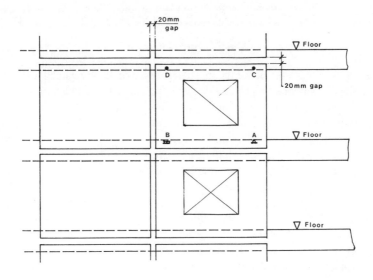

20mm
gap

▽ Floor

D   C

20mm gap

B   A

▽ Floor

▽ Floor

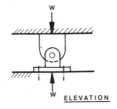

W

ELEVATION

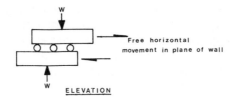

W

Free horizontal
movement in plane of wall

W   ELEVATION

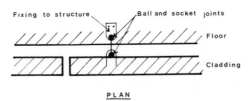

Fixing to structure   Ball and socket joints

Floor

Cladding

PLAN

Figure 6.11 Schematic illustration of supports for precast concrete cladding fully separated from frame action

6.5 PRESTRESSED CONCRETE DESIGN AND DETAIL

6.5.1 Introduction

Prestressed concrete elements in structures which have been subjected to earthquakes have mostly performed well. Failures have been mainly due to inadequate connection details or supporting structure.[39,40]

Although prestressed concrete is well established in bridge construction and various civil engineering applications, it is less widely used in building structures, and relatively few structures have been fully framed in prestressed concrete. This is true in both seismic and non-seismic areas. The comparative neglect of prestressed concrete for building structures has occurred partly for constructional and economic reasons, and in earthquake areas it has also occurred because of divergent opinions on the effectiveness of prestressed concrete in resisting earthquakes. Lack of suitable research data has prevented proper assessment of the seismic response characteristics of prestressed concrete, although recent research work should help to clarify the situation.

6.5.2 Official recommendations for seismic design of prestressed concrete

Some organisations interested in the use of prestressed concrete have published seismic design recommendations. For instance in 1966 the American P.C.I.[41] and the New Zealand P.C.I.[42] both suggested similar design principles, which made no attempt to describe the seismic response of prestressed concrete, and are too brief to be of practical use. In 1970 the F.I.P. Commission on Seismic Structures tabled recommendations[43] containing more design criteria but these are difficult to apply. In order to check the recommended curvature limits, a non-linear dynamic analysis would be necessary. Furthermore, Blakeley[39] suggests that the curvature limits are too severe as they would allow little reduction of response through inelastic deformations.

Mainly because of the lack of research data, codes of practice give litttle detailed guidance on the seismic design of prestressed concrete and official attitudes are therefore cautious towards its use in building structures. Nevertheless much can be said in favour of prestressed concrete as will be seen below.

6.5.3 Seismic response of prestressed concrete

The seismic response of structural materials has been discussed generally in Section 5.1, where some stress-strain diagrams were presented. The main characteristics of prestressed concrete under cyclic loading may be inferred from Figure 5.5(e).

Idealized forms of the hysteresis diagram are shown in Figure 6.12, which may be contrasted with that of steel (Figure 5.8). Blakeley and Park[39,44] give a more sophisticated idealization for prestressed concrete.

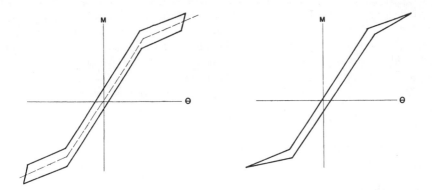

Figure 6.12 Idealized moment-rotation diagrams for prestressed concrete

It is evident from the narrowness of the hysteresis loops that the amount of hysteretic energy dissipation of prestressed concrete will be relatively small compared to steel or reinforced concrete. On the other hand the capacity of prestressed concrete to store elastic energy is higher than for a comparable reinforced concrete member.

Prestressed concrete suffers in comparison to reinforced concrete because of its lack of compression steel, so that its performance is poorer once concrete crushing begins. When compared to reinforced concrete, prestressed concrete undergoes relatively more uncracked deformation and relatively less deformation in the cracked state. This means that prestressed concrete structures should exhibit less structural damage in moderate earthquakes. In the event of structural repairs being necessary after an earthquake, there are obvious difficulties in restoring the prestress to sections of replaced concrete, and conversion of the failure zones to reinforced concrete may be necessary.[45]

It has been suggested that prestressed concrete buildings may be more flexible than comparable reinforced concrete structures, and that more non-structural damage may occur. However, differences in flexibility will be small in practical design terms, and structures in either material will generally be less flexible than steelwork. In any case proper detailing of the non-structure will be necessary regardless of the materials used in the structure.

For notes on the damping of prestressed concrete structures see Table 5.1.

6.5.4 Factors affecting ductility of prestressed concrete members

For the satisfactory seismic resistance of prestressed concrete members brittle failure must be avoided by the creation of sufficient useful ductility as discussed in Section 5.1.3. In the case of prestressed concrete the useful available section ductility may be defined as

$$\frac{\phi_{0.004}}{\phi_{cr}}$$

where $\phi_{0.004}$ is the curvature at a nominal maximum concrete strain of 0·004, and ϕ_{cr} is the curvature at first cracking. It can be seen in Figure 6.14 that $\phi_{0.004}$ is much smaller than the ultimate curvature, but the rotation after first crushing becomes increasingly less 'useful' because of physical degradation, especially under cyclic loading.

The ductility or rotation capacity of prestressed concrete is affected by:

(i)   the prestressing steel content;
(ii)   the transverse steel content;
(iii) the distribution of prestressing steel;
(iv)   the axial load.

Each of these variables is discussed below.

6.5.4.1 Prestressing steel content

From Figure 6.13 it may be seen that ductility decreases markedly with increasing prestressing steel content ρ. The value of $\rho = 0.0069$ corresponds to the A.C.I. 318–71 code[4] requirement for avoiding brittle failure under static loading:

$$\frac{A_{ps} f_{pu}}{0.85\, bd\, f_{cu}} \leqslant 0.3$$

In order to obtain adequate ductility for seismic design a reduced value of $\rho \leqslant 0.005$ is suggested as more appropriate.

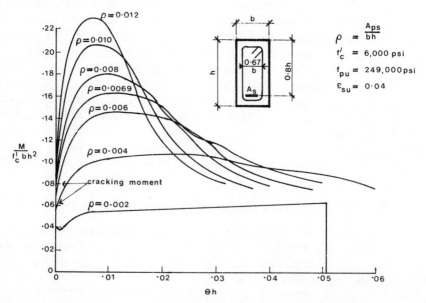

Figure 6.13 Moment-curvature relationship for rectangular prestressed concrete beams showing the effect of prestressing steel content on ductility (after Blakeley and Park[46])

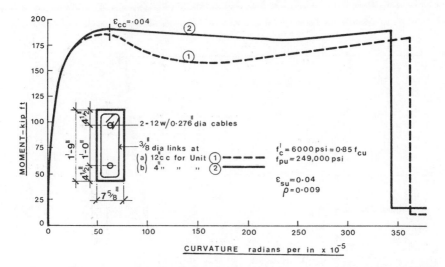

Figure 6.14 Effect of transverse steel content on ductility of prestressed beams (after Blakeley and Park[46])

6.5.4.2 Transverse steel content

Unfortunately comparatively little data exists on this subject, but from tests carried out by Blakeley and Park[46] it appears that the transverse steel content has relatively little effect on the useful ductility of beams with moderate prestress. In Figure 6.14 it can be seen that tripling the amount of transverse steel had only a modest effect on the strength retention at large curvatures, and barely affected the useful precrushing range of concrete strain $\epsilon_c \leqslant 0.004$. The small effect of confinement steel on ductility may also be seen in Figure 6.15.

6.5.4.3 Distribution of prestressing steel

At positions of moment reversal where the greatest ductility requirements exist, the required distribution of prestress will usually be nearly axial. Blakeley[39] demonstrated that a single axial tendon produced a less ductile member than that achieved by multiple tendons placed nearer the extreme fibres. At points in structures where stress reversals do not occur, eccentric prestress may be used. Where no unstressed reinforcement exists, an eccentrically prestressed beam is notably less ductile than a concentrically stressed beam with equal prestressing steel content (Figure 6.15).

The tendon distribution three in Figure 6.15 is not only as ductile as two, but has the advantage that the axial tendon will be practically unharmed by large rotations, and should hold the structure together after the tendons near the extreme fibres have failed.

248

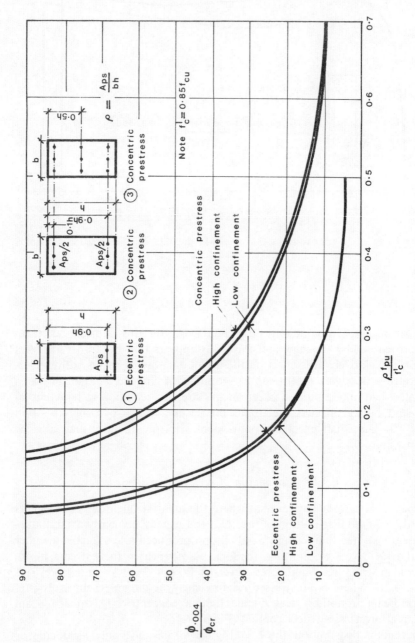

Figure 6.15 Variation of curvature ratio at crushing (section ductility) for prestressed concrete beams (after Blakeley[39])

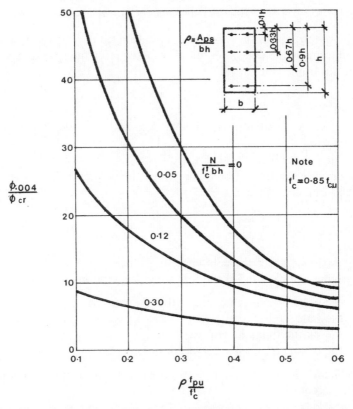

Figure 6.16 Variation of curvature ratio at crushing (section ductility) for columns with varying prestress and varying axial load (after Blakeley[39])

6.5.4.4 Effect of axial load on ductility of prestressed concrete columns

The section ductility $\phi_{0.004}/\phi_{cr}$ decreases rapidly with increasing column axial load N. This effect is seen in Figure 6.16 where ductility is plotted against the level of prestress for columns carrying varying axial loads. The curve for $N/f'_c bh = 0.12$ is of interest as this represents the load at which the ACI code[4] requires that the total ultimate moment capacity of the columns must be greater than the total ultimate moment capacity of the beams at a particular joint. It can be seen that the curvature ratio exceeds 10 for this axial load at practical levels of prestress.

6.5.5 Satisfying the ductility demands of prestressed concrete structures

The rotational capacity or section ductility as defined by the curvature ratio $\phi_{0.004}/\phi_{cr}$ may be found for rectangular beams or columns for Figures 6.15 and 6.16. The section ductility required may be determined from the frame analysis or possibly may be imposed arbitrarily.

6.5.5.1 Ductility demands of portal frames

For simple structures the required curvature ratio may be determined fairly simply in terms of the deflection ductility factor μ for the structure. Considering a portal frame (Figure 6.17),

$$\mu = \frac{x_{max}}{x_{cr}}$$

where x_{max} is the maximum lateral deflection, and x_{cr} is the lateral deflection at first cracking.

Following the method of Blakeley and Park,[44] the expression for the required column curvature ratio can be obtained for the column sidesway mechanism (Figure 6.17(a)) and is found to be

$$\frac{\phi_{max\ c}}{\phi_{cr\ c}} = 1\cdot5 + \frac{0\cdot37H^2}{h}\left(\frac{\mu - 1\cdot1}{H - 0\cdot6h}\right) \tag{6.21}$$

where h is the depth of the column.

In deriving equation (6.21) the following assumptions were made;

(a)   the point of contraflexture in the columns was at height $0\cdot6H$ above the base;

(b)   the plastic hinge length was $0\cdot6h$;

(c)   the maximum elastic curvature $= 1\cdot5\phi_{cr\ c}$.

For the beam sidesway mechanism (Figure 6.17b) the required column curvature ratio becomes

$$\frac{\phi_{max\ c}}{\phi_{cr\ c}} = 1\cdot5 + \frac{0\cdot37H^2}{h}\left(\frac{\mu - 1\cdot5}{H - 0\cdot3h}\right) \tag{6.22}$$

and the required beam curvature ratio is

$$\frac{\phi_{max\ b}}{\phi_{cr\ b}} = 1\cdot5 + \frac{0\cdot55H^2}{h}\left(\frac{\mu - 1\cdot5}{H - 0\cdot3h}\right) \tag{6.23}$$

In deriving equations (6.22) and (6.23) the additional assumptions were made that $L_b = 2L/3$, and that the column depth equalled the beam depth.

Consider an example of the beam sidesway mechanism in which the column height $H = 14h$. If the portal frame is required to have a deflection ductility $\mu = 4$, substituting in equations (6.22) and (6.23) it is found that the required column and beam curvature ratios are

$$\frac{\phi_{max\ c}}{\phi_{cr\ c}} = 21$$

and

$$\frac{\phi_{max\ h}}{\phi_{cr\ b}} = 15$$

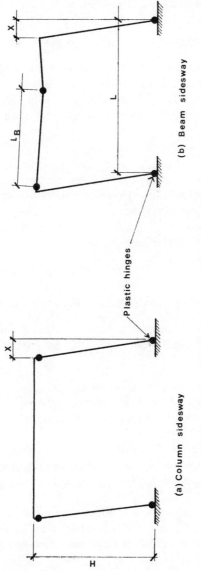

(a) Column sidesway

(b) Beam sidesway

Plastic hinges

Figure 6.17 Portal frame with two possible plastic hinge mechanisms

Referring to Figure 6.15 it can be seen that a curvature ratio of 15 is readily available for concentrically stressed beams with all practical amounts of prestress, i.e. for $\rho f_{pu}/f'_c \leqslant 0.5$. Referring to Figure 6.16, it is found that for columns a curvature ratio of 21 is available at an assumed column load of $N = 0.1 f'_c bh$, if the prestressing steel content $\rho \leqslant 0.2 f_{pu}/f'_c$.

6.5.5.2 Ductility demands of multi-storey prestressed concrete structures

For multi-storey prestressed concrete structures the ductility or curvature demands are clearly more difficult to determine than for the simple portal frame structures discussed above. Plastic hinge and rotation magnitudes may be determined by a fairly arbitrary selection of failure mechanisms, using an analysis similar to that underlying the portal frame method,[44] such as has been discussed for reinforced concrete in Section 6.2.4. However, this may lead to unconservative results. Alternatively in some cases an inelastic dynamic analysis may be warranted to find the curvature requirements, such as that described by Spencer[47] for a 20-storey building. Otherwise a practical solution would be to design the structure to the normal code loadings using load factors which encourage beam hinges to form before column hinges, and ensure that an arbitrary minimum section ductility is available at all likely hinge positions.

6.5.6 Detailing summary for prestressed concrete

For the adequate performance of prestressed concrete in earthquakes, its ductility and continuity should be maximized by careful consideration of the following items;

(i)   prestressing steel content (Section 6.5.4.1);
(ii)   prestressing steel distribution (Section 6.5.4.3);
(iii) continuity, ensured by adequate lapping of prestressing tendons or reinforcing bars;
(iv)   anchorages in post-tensioned construction, carefully positioned to avoid congestion and stress-raising in highly stressed zones. They should be situated as far from potential plastic hinge positions as possible;
(v)   joints between prestressed members involving ordinary reinforced concrete, properly designed as outlined in Sections 6.2 and 6.3;
(vi)   joints using mechanical details, as suggested in Section 6.3;
(vii) transverse steel appears to play a minor role in fully prestressed concrete (Section 6.5.4.2) but the lower the level of prestress, the more applicable is the use of confinement steel as in reinforced concrete beams or columns (Section 6.2);
(viii) grouting of post-tensioned tendons, mandatory in some seismic countries, based on fears of release of anchorage under reversed

loadings. However where anchorages are positive, the use of un-bonded tendons in earthquake areas is warranted. This condition is obviously met by thread and nut anchorages, or swaged systems, while some wedge systems of anchorage can be shown by testing to be secure under reversed loading.

6.6 MASONRY DESIGN AND DETAIL

6.6.1 Introduction

Masonry is a term covering a very wide range of materials such as adobe, brick, stone and concrete blocks; and each of these materials in turn varies widely in form and mechanical properties. Also masonry may be used with or without reinforcement, or it may be used in conjunction with other materials. As well as its use for primary structure, masonry is used for infill panels creating partitions or cladding walls.

The variety available in form, colour and texture makes masonry a popular construction material, as does its widespread geographic availability and in some cases its comparative cheapness. Properly used, it also has reasonable resistance to horizontal forces. However masonry has a number of serious drawbacks for earthquake resistance. It is naturally brittle; it has high mass and hence has high inertial response to earthquakes; its construction quality is difficult to control; and little research has been done into its seismic response characteristics.

Because of the poor performance of some forms of masonry in earthquakes, official attitudes towards masonry are generally cautious in most moderate or strong motion seismic areas. For example in Japan masonry is not permitted for buildings of more than three storeys in height.

By way of contrast, carefully designed apartment buildings of 15-storeys or more have been built in California. No doubt there is considerable inherent strength in the 'egg-crate' form of apartment buildings, but there has been little or no dynamic testing or seismic experience of tall masonry structures to date. The greater risk to human life in apartment blocks should also demand higher safety margins in such construction.

The other types of construction in which masonry is most popular are low-rise housing and industrial buildings (Appendices A.3 and A.4).

6.6.2 Seismic response of masonry

This discussion is supplementary to the general introduction to seismic response given in Section 5.1. As very little dynamic testing of masonry has been carried out to date, our knowledge of the seismic response of masonry is based largely on field observations after actual earthquakes and on inferences from static load testing. This scarcely provides a satisfactory design basis, especially as so many different types of masonry exist.

The tendency to fail in a brittle fashion is the central problem with masonry. While unreinforced masonry can be categorically labelled as brittle, uncertainty exists as to the degree of ductility which can be achieved in reinforced masonry. Based on static load-reversal tests, Meli[48] contended that 'for walls with interior reinforcement, where failure is governed by bending, behaviour is nearly elasto-plastic with remarkable ductility and small deterioration under alternating load except for very high deformation... If failure is governed by diagonal cracking, ductility is smaller and, when high vertical loads are applied, behaviour is frankly brittle. Furthermore...important deterioration (occurs) after diagonal cracking'.

Meli concluded that bending failure was the most favourable design condition for walls. In true dynamic tests, Williams and Scrivener[49] confirmed this conclusion. They also found that marked deterioration and stiffness degradation occurred with dynamic cycling in walls where flexural action predominated, while there was little difference between dynamic and static test results for walls failing in shear (Figure 6.18).

Reinforced hollow concrete blocks are popular in some areas. It is not clear whether this material can be made significantly more ductile than reinforced brick, but it certainly is far more brittle than normal reinforced concrete. Under cyclic loading the cover spalls off the grouted cores, and the use of confinement reinforcement is not practicable in masonry members of normal thickness. Clearly much more dynamic testing is required in order to establish the principles of the seismic response of masonry in its various forms. It appears wise to treat masonry essentially as a brittle material, and to use stiff rather than slender members until more information is available.

6.6.3 Structural form for masonry buildings

The general principles of earthquake-resistant structural form have been given in Section 4.2, but additional guidance peculiar to masonry is given here. Five interrelated criteria for consideration in masonry construction are that;

(i)   the aspect ratio H/B for the structure should be a minimum;

(ii)   the aspect ratio H'/B' for vertical members should be a minimum;

(iii)   the ratio of aperture area to wall area, $\Sigma A_a/HB$, should be a minimum;

(iv)   the distribution of apertures should be as uniform as possible;

(v)   stress-raising apertures should be located away from highly stressed zones.

Considering these criteria in relation to a single storey structure with zero or minimum reinforcement, Figure 6.19(a) represents a good structure and Figure 6.19(b) represents a bad one. Neither case has a bad overall aspect ratio, as is typical for single storey buildings. For unreinforced buildings with a maximum aperture area, a value of $H/B \ngtr 2/3$ might be taken.

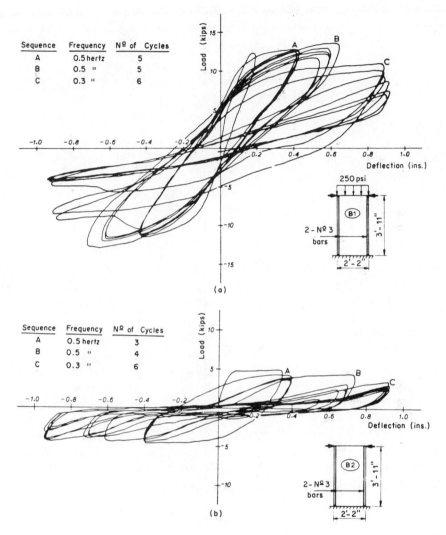

Figure 6.18 Response of reinforced brick walls to dynamic loading (after Williams and Scrivener[49])

The aspect ratio of vertical members, particularly those at the ends of walls (H_1'/B_1' and H_2'/B_4' in Figure 6.19(a)), should be little greater than unity in buildings of minimum reinforcement. This is clearly not so in Figure 6.19(b).

It is commonly recommended that the total area of holes should not exceed one third of the wall area, i.e. $\Sigma A_\mathrm{a}/HB \not> 1/3$. If criteria (ii) and (iii) above have been satisfied it is likely that the distribution of apertures will be reasonably uniform. Small holes, such as those used for the passage of services pipes and ducts, should be kept away from corners of load bearing members;

256

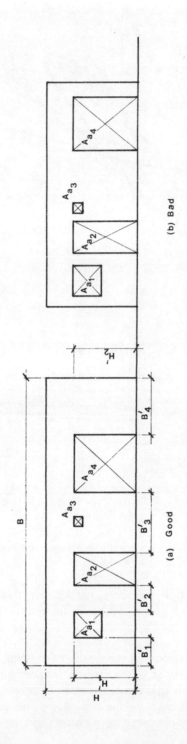

Figure 6.19 Structural form of low-rise masonry buildings

A_{a3} in Figure 6.19(b) is badly placed compared with that in Figure 6.19(a).

The main objective of the above criteria is to distribute the strength as uniformly as possible; in brittle structures the early failure of one member causes the remaining members to share the total load, and often leads to incremental collapse.

6.6.3.1 Structural form for reinforced masonry

The criteria set out above are also applicable to reinforced masonry, although some relaxation of the suggested limits may be made. The degree of relaxation will depend on the degree of protection against early brittle failure afforded by the reinforcement. In high quality construction, building aspect ratios $H/B = 2$, or $H/B = 3$ are reasonable, and in apartment block construction with small aperture ratios even higher values of H/B have been used.

The aspect ratios for vertical members again may be increased probably to values of two or three, but until the problem of whether dynamic response is better in shear or bending has been resolved (Section 6.6.2), safe maximum values of H'/B' are in doubt.

6.6.4 Design and construction details for reinforced masonry

On account of its questionable ductility, reinforced masonry (as well as unreinforced) should be designed and analysed as a linear elastic material (Section 5.2.4.1 and Table 5.3). Any ductility produced by reinforcement may be used to reduce the safety factors slightly against ultimate load. Design and construction procedures should conform to well-established codes of practice[50,51] as followed in the design handbook by Amrhein.[52] Seismic loadings set out in local regulations[21,53] should be used with larger load factors than those used for steel or concrete unless a specific factor for materials is used in finding the horizontal loads.

Some of the more important details of aseismic reinforced masonry construction are commented on below.

6.6.4.1 Minimum reinforcement

Although code recommendations on this subject vary somewhat, the following New Zealand practice[51] is fairly representative. At least one vertical bar, not less than 12 mm ($\frac{1}{2}''$) in diameter, should be placed at all corners, wall ends and wall junctions. Such bars should be anchored into the upper and lower wall beams or foundations, and adequately lapped at splices.

Walls should be reinforced both horizontally and vertically; a minimum of 10 mm diameter horizontal bars at 0·8 m (32") spacing, and 10 mm diameter vertical bars at 1·0 m (40") spacing, or the equivalent, should be provided.

At least one bar, not less than 12 mm ($\frac{1}{2}''$) in diameter, should be placed on all sides of apertures exceeding about 600 mm (2') in any direction. Such framing bars should extend not less than 600 mm (2') beyond each corner of the aperture or be equivalently anchored.

6.6.4.2 Horizontal continuity

Horizontal continuity around the perimeter of the building should be ensured at least at the levels of the base, the floors and the roof. The walls should be tied into an effective ring beam at these levels. Connections to floors or roofs other than of *in situ* concrete have proved especially vulnerable in earthquakes (Appendices A.3, A.4).

6.6.4.3 Grouting

The reinforcement of masonry depends for its effectiveness on transfer of stress through the grout from steel to masonry. Every effort should be made to ensure that compacted grout completely fills the cavities. A low-shrinkage grout is essential in order to minimize separation of the grout from the masonry. Grout for cavities of up to 60 mm ($2\frac{1}{2}''$) in width should contain aggregates up to 5 mm in size, while for larger cavities coarser aggregate may be suitable. Grout should have a characteristic cube strength of 20 N/mm^2 (3000 psi) at 28 days.

6.6.4.4 Hollow concrete blocks

Although the shape of hollow concrete blocks varies in detail in different countries, those shown in Figure 6.20 illustrate the principal types used in 150 mm (6") and 200 mm (8") thick walls.

6.6.4.5 Supervision of construction

In order to ensure adequate standards of construction, more supervision is required for reinforced masonry than on equivalent projects in other materials. The following points in particular need watching;

(a) cavities should be clean and free from mortar droppings;
(b) reinforcement should be placed centrally or properly spaced from the masonry;
(c) reinforcement should be properly lapped;
(d) the grouting procedure should be properly carried out;
(e) the grout mix should conform to the specification.

In multi-storey hollow concrete block construction, inspection holes at the bottom of walls on the line of vertical reinforcement are advisable to facilitate the checking of items (a) to (d) above (Figure 6.21).

259

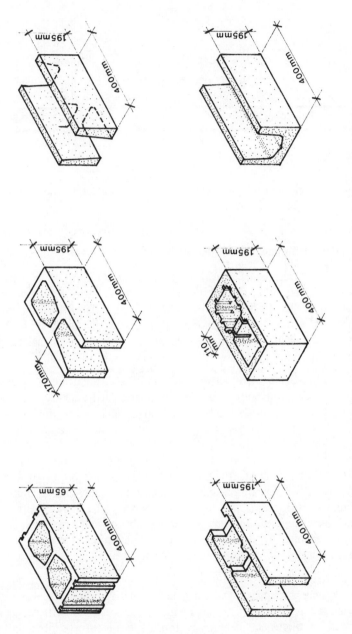

Figure 6.20 Principal structural concrete blocks used in New Zealand for reinforced masonry construction, 150 mm and 200 mm wall thicknesses

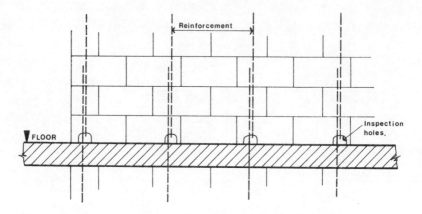

Figure 6.21 Inspection holes in hollow reinforced concrete block construction for checking reinforcement and grouting (see text)

6.6.5 Construction details for structural infill walls

Masonry is often used as structural infill, either as cladding or as interior partitions. It should be either effectively separated from frame action or fully integrated with it as discussed in Section 8.2.2. The analysis of the interaction between frames and integrated infill panels has been discussed in Section 5.8. Very little data on the detailing of masonry in this situation exists, but the following points may be made.

(i)   No gap should be left between the infill and the frame, so as to prevent accidental pounding damage during earthquakes.

(ii)   The top of the panel should be structurally connected to the structure above to ensure lateral stability of the infill in earthquakes.

(iii) Ideally the form of the structure and the strength of the infill panel would be such that shear failure of the masonry infill would not occur. When this is not the case, the reinforcement required in each infill panel for resisting shear is difficult to determine. It may be calculated in the same way as normal masonry shear walls, although it has been suggested by some workers such as Meli[48] that peripheral reinforcement (provided in this case by the frame) is more effective in creating ductile masonry than internal reinforcement.

Placing of full-height vertical reinforcement is obviously difficult in hollow block or brick construction of the form shown in Figure 6.20, when the infill is erected after the upper frame member has been constructed as is usually the case in this form of construction. This difficulty is obviated by the use of external reinforcement in the form of expanded metal sheets bonded to the sides of the block wall by means of a layer of mortar. Tso et al.[54] have reported favourable behaviour of this type of construction in cyclic loading tests, using washers and bolts through the wall to improve the bonding of the expanded metal to the wall.

6.7 STEELWORK DESIGN AND DETAIL

6.7.1 Introduction

Following the guidance on overall dynamic behaviour and analysis of structures given in Chapter 5, the detailed earthquake-resistant design of steelwork is discussed in this section together with further information on the seismic response of this material.

Although competently designed and fabricated structural steelwork has generally performed well in major earthquakes, careful detailing and control of material properties is necessary to ensure the development of its full ductility under earthquake loading. The main considerations in achieving adequate performance of steelwork in earthquakes are;

(i)   the specification of a sufficiently ductile grade of steel;
(ii)   the ductile design and fabrication of frame members and connections;
(iii) the avoidance of excessive lateral sway.

6.7.2 Materials

For the construction of earthquake-resistant ductile steel frames the basic steel material must, of course, be of good quality. While steels suitable for seismic resistance are found amongst those produced for general structural purposes, not all normal structural grades are sufficiently ductile. The main properties required are as follows;

(a)   adequate ductility;   } aseismic
(b)   consistency of mechanical properties;   } requirements
(c)   adequate *notch* ductility;   }
(d)   freedom from laminations;   } general
(e)   resistance to lamellar tearing;   } requirements
(f)   good weldability.   }

6.7.2.1 Ductility

Ductility may be described generally as the post-elastic behaviour of a material (Section 5.1.3). For steel it may be expressed simply from the results of elongation tests on small samples, or more significantly in terms of moment-curvature or hysteresis relationships as discussed later in this chapter.

Steels manufactured in various countries may have sufficient ductility, and earthquake codes of practice often recommend suitable steels. In California, for example, the SEAOC Code[21] requires that steels conform to the latest edition of the following ASTM Specification;

A36   (Structural carbon steel);
A440 (High strength, manganese-copper);
A441 (High strength, manganese-vanadium);
A572 (Grade 42, 45, 50 and 55), (High strength, columbium or vanadium);
A588 (High strength multiple alloy).

6.7.2.2 Consistency of mechanical properties

In economically designing so that beams fail before columns it is desirable that the maximum and minimum strengths of members are as nearly equal in magnitude as possible. This means that the standard deviation of strengths should be as small as possible. While it is satisfactory for non-seismic design, it is unfortunate for earthquake-resistant design that steel manufacturers have been more concerned with simply achieving their minimum guaranteed yield strengths, than in producing consistent ultimate strengths.

6.7.2.3 Notch ductility

Notch ductility is a measure of the resistance of a steel to brittle fracture and is a separate property from that of general ductility discussed in Section 6.7.2.1. Adequate notch ductility is required in *all* structural steelwork, not only in seismic areas of the world. It is generally expressed as the energy required to fracture a test piece of particular geometry. Three widely used tests are the Charpy V-notch test, the Izod test, and Charpy keyhole test. The results although quantitative are generally empirical and are not comparable between tests and between materials. While many national steel standards specify levels of resistance to brittle fracture, American standards do not.

6.7.2.4 Laminations

Laminations are large areas of unbonded steel found in the body of a steel plate or section. This implies a layering of the steel with little structural connection between the layers. The laminated areas originate in the casting and cropping procedures for the steel ingots, and may be as much as several square metres in extent. Steel may be screened ultrasonically for lamination before fabrication, and some guidance may be found on this procedure in a British Standards Institution publication.[55]

6.7.2.5 Lamellar tearing

It should first be pointed out that lamellar tearing should not be confused with laminations, the two being different phenomena.

Lamellar tearing is a tear or stepped crack which occurs under a weld where sufficiently large shrinkage stresses have been imposed in the through thickness direction of susceptible material. It commonly occurs in **T**-butt welds and in corner welds and is caused by inclusions which act as 'perforations' in the steel. Lamellar tearing has been discussed in some detail by Farrar and Dolby[56] and Jubb.[57] Unfortunately no non-destructive method of screening for susceptibility to lamellar tearing is as yet available. The usual method of checking is by measuring the ductility in a through plate tensile test.

Electric arc steelmaking incorporating vacuum degassing can produce steels with reduced (but not eliminated) susceptibility to lamellar tearing— although at some extra cost.[57] The risk of lamellar tearing can also be reduced by the use of suitable welding techniques and details.[56,57]

6.7.2.6 Weldability

Weldability may be considered simply as the capacity of the parent metal to be joined by sound welds. The weld metal should be able to closely match the properties of the parent plates, and few material defects should arise. To some extent weldability will be assured by the use of steels produced according to major national standards, such as those referred to in Section 6.7.2.1, or British Standard 4360:1972, 'Weldable structural steels'. The weldability of a steel is often assessed by means of a formula based on the chemical analysis of the steel. Such methods determine the preheating temperature necessary to avoid hydrogen cracking.[58] In general the higher the tensile strength of the steel, the lower is its weldability.

6.7.3 Ductile design of steel structures

Ductile design involves making proper use of the inelastic properties of building materials. A general discussion of inelastic behaviour has been given in Section 5.1.3, while the means of analysis available for the inelastic state are discussed in Section 5.2.4.1. In order to obtain adequate ductility from a steel structure, both the members and the connections must be considered. This will involve observing the following requirements;

(i)   excessive alternating stresses in the flanges of beam and column members must be avoided;

(ii) inelastic buckling of columns and diagonals must be prevented;

(iii) connections must be designed to allow extensive yielding of the frame members.

In principle the use of plastic design of sections is appropriate to seismic resistance, but care must be taken to ensure that the above three criteria are complied with and that excessive lateral sway is avoided. Generalized plastic collapse mechanism theories will not be appropriate if they lead to column hinges forming before beam hinges.

In the following sections the ductile design of the main components of steel structures is briefly discussed.

6.7.4 Steel beams

In this section the behaviour and design of beams acting primarily in bending will be considered. In most beams axial forces are small enough to be neglected, but where large axial forces may occur column design procedures should be employed.

6.7.4.1 Moment-curvature relationships for beams

For the adequate aseismic design of steel beams, and the associated connections and columns, the moment-curvature or moment-rotation relationship should be known. A long stable plastic plateau is required which is not terminated too abruptly by lateral or local buckling effects, such as indicated by terminating at points *A*, *B* and *C* in Figure 6.22. The curves terminating at *D* and *E* are typical of the desired behaviour achieved by well designed beams under moment gradient and uniform moment respectively.

Moment gradient is the usual loading condition to be considered with plastic hinges forming at the ends of beams in laterally loaded frames. The

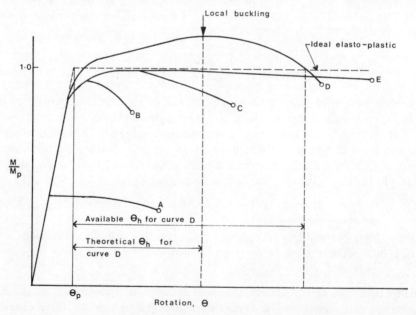

Figure 6.22 Behaviour of steel beams in bending

localization of high stresses produced by the moment gradient causes strain-hardening to occur during plastic rotation, resulting in an increase in moment capacity above the ideal plastic moment M (curve D in Figure 6.22). Strain-hardening may increase the plastic moment by as much as 40 percent. Local buckling and lateral buckling arising from plastic deformation of the compression flanges generally produces a reduction of moment capacity in the later stages of rotation[59] as illustrated in curve D of Figure 6.22.

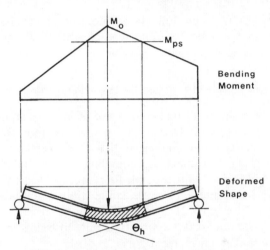

Figure 6.23 Beam under moment gradient with plastic hinge deformations and the hinge rotation θ_h of equation (6.24) as defined by Lay and Galambos[60]

In order to predict the rotation capacity of a plastic hinge, the following expressions presented by Lay and Galambos[60] for the inelastic hinge rotation θ_h (Figure 6.23) of a beam under *moment gradient* may be used:

$$\theta_h = 2 \cdot 84 \, \epsilon_y (\beta - 1) \frac{b}{h} \frac{t_f}{t_w} \left(\frac{A_w}{A_f} \right)^{1/4} \left(1 + \frac{V_1}{V_2} \right) \tag{6.24}$$

where

b = flange width,
h = overall depth of section,
t_f = flange thickness,
t_w = web thickness,
A_f = flange area,
A_w = web area,
$V_1 \text{ and } V_2$ = absolute values of shears acting either side of the hinge, arranged so that $V_1 \leqslant V_2$,
β = ratio of strain at onset of strain-hardening to strain at first yield,
ϵ_y = strain at first yield.

θ_h represents a substantial proportion of the total rotation capacity of the beam (Figure 6.22). For the American section 10WF25 (A36 steel), equation (6.24) predicts that $\theta_h = 0\cdot07$ radians.[60] It should be noted that equation (6.24) incorporates simplifications which lead to underestimations of θ_h of 20 percent or more.

6.7.4.2 Shear force and plastic moment

The effect of shear on the full plastic moment of members will usually be negligible in steel frames,[59] and no requirements for shear in steel beams specifically relating to earthquake resistance arise.

6.7.4.3 Behaviour of steel beams under cyclic loading

In steel frames designed to make good use of inelastic resistance in earthquakes, several reversals of strain of 1·5 percent or more may have to be withstood. Several tests have indicated that steel members can have very stable hysteretic behaviour under cyclic loading such as that shown in Figure 5.5(b). Bertero and Popov[61] for example found that a steel cantilever strained to $\pm1\cdot0$ percent failed after 607 cycles, while strain amplitudes of $\pm2\cdot5$ percent produced failure after 21 cycles. These latter results should be used carefully as they are apparently appropriate only to beams with considerable stiffening in the maximum moment zone.

More recent tests[62,63,64] have examined beams under more normal conditions of restraint, and in some cases earlier hysteretic degradation was observed. In the example shown in Figure 6.24 the decay was mainly due to web buckling. Flange buckling and lateral torsional buckling were also observed to influence the loss of strength and stiffness of the beams.

It is clear, however, that with suitable geometry and restraint to control buckling, sufficient hysteretic stability for seismic resistance can be obtained. To this effect Popov[64] has recommended that more conservative local and lateral buckling criteria should be used for earthquake resistance than has been normal practice, i.e. the web and flange thickness and lateral bracing requirements of normal static design manuals such as Ref. 59 are not sufficiently conservative for strong earthquake conditions. Until further experimental evidence becomes available no specific new anti-buckling recommendations may be made with great certainty, but modest extra caution would seem advisable (see further comments in Section 6.7.7).

6.7.5 Steel columns

In this section the seismic deformation behaviour of steel columns is compared with that of beams with appropriate allowance for the effect of axial loads. Some consequent design implications are mentioned.

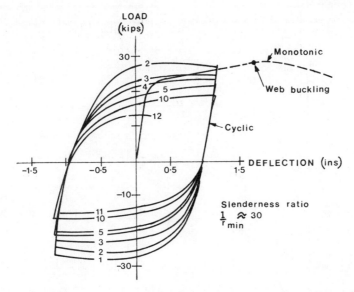

Figure 6.24 Hysteresis loops for a steel beam tested under moment gradient (after Vann et al.[62])

6.7.5.1 Moment-curvature relationship for columns

Columns are commonly required to resist appreciable bending moments as well as axial forces. The moment-curvature relationships for the so-called 'beam-column' are similar to those for beams under uniform moment, except that the value of the plastic moment M_p is reduced by the presence of axial load as shown in Figure 6.25.

The full plastic moment for a column M_{pc} may be taken as equal to that of the pure bending case M_p for columns with small axial loads; the ASCE[59] recommends that for

$$0 \leqslant N \leqslant 0\cdot15\,N_y$$

use

$$M_{pc} = M_p \tag{6.25}$$

but for

$$0\cdot15\,N_y \leqslant N \leqslant N_y$$

use

$$M_{pc} = 1\cdot18\left(1 - \frac{N}{N_y}\right)M_p \tag{6.26}$$

where N is the maximum expected axial compressive load, and $N_y = f_y A$ is the section capacity at yield stress in axial compression.

As indicated in Figure 6.25, the available plastic moment of columns may not be developed because of local buckling or lateral torsional buckling as for beams.

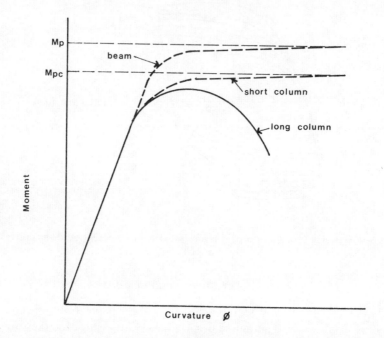

Figure 6.25 Typical moment-curvature relationships for short and long columns compared with pure beam behaviour

6.7.5.2 Behaviour of steel columns under cyclic loading

Although columns should generally be protected against inelastic cyclic deformations by prior hinging of the beams, some column hysteretic behaviour is likely in strong earthquakes in most structures.

The behaviour of steel columns under cyclic bending is similar to that of beams without axial load, except that the axial force added to the bending moment concentrates the yielding in the regions of larger compressive stress. This leads to a more rapid decay of load capacity owing to more extensive buckling, as may apparently be inferred by comparing Figure 6.26 with Figure 6.24. Second-order bending ($P \times \Delta$ effect) may also be important in the inelastic range.

Unfortunately too little experimental work has been done on the hysteretic behaviour of steel columns to allow definitive statements to be made yet on the adequacy of present-day design requirements. For further reading, see elsewhere.[62,64]

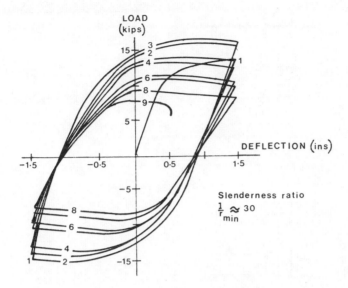

Figure 6.26 Hysteresis loops for a steel member under cyclic bending, and with a constant axial force of $N = 0.3 \, N_y$ (after Vann *et al.*[62])

6.7.6 Steel connections

In order to take full advantage of the strength and ductility of the members of a steel frame, the connections should be able to develop the full plastic capacity of the members. Because the behaviour of connections is not as well understood as that of members, some conservativeness in the design of connections relative to members is required.

Connections should also be designed to make fabrication and erection of the framework as simple and quick as possible. They should not be too sensitive to factory or field tolerances, and should minimize the use of highly skilled crafts. Connections should also permit adequate inspections to be made at the time of construction.

Butt welding, fillet welding, bolting and rivetting may be employed for aseismic connections, either individually or in combination. As fully bolted or rivetted connections tend to be very large and expensive, fully welded connections or a combination of welding and bolting are most frequently used.

6.7.6.1 Behaviour of steel connections under cyclic loading

Compared with beam and column elements, relatively few cyclic load tests have been carried out on steel connections, and conclusive design criteria are not yet available for seismic conditions. However, Popov and Pinkney[65] tested five types of joint, two involving minor-axis bending of the column

270

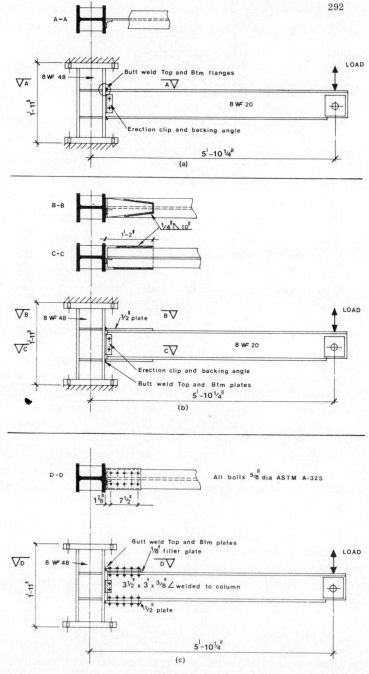

Figure 6.27 Beam-column connections with major axis column bending tested by Popov and Pinkney.[65] (a) Butt welded beam-column joint, (b) fillet welded beam-column joint, (c) bolted beam-column joint

and three involving major-axis bending of the column. The latter three joints (Figure 6.27) were of the following types;

(a)   a butt-welded joint;
(b)   a fillet-welded joint using flange plates;
(c)   a joint using high-strength bolts and flange plates.

In the tests it was found that the butt-welded joints were superior to the other two types in terms of total energy absorption. In the bolted joints the hysteresis loops were reduced in area considerably by slippage, although the use of smaller than normal oversize holes reduced this effect. All the joints sustained loads in excess of their design limit values until the onset of cracking.

In tests on connections using fully welded and flange welded-web bolted joints by Popov and Stephen,[66] very large increases in bending strength (up to 69 percent) due to strain-hardening were observed. Despite some favourable aspects of these tests, the SEAOC Steel Subcommittee[67] considered that the flange welded-web bolted connections should not be recommended without further research data.

6.7.6.2 Detailing of steel connections for seismic loading

As discussed earlier, the connections of a steel structure should be strong enough to allow the adjacent members to develop their full strength. In assessing the full yield strength of beams, allowance should be made for the probability of the actual strength of the beams exceeding the guaranteed minimum strength. Typical increases are indicated below.

Guaranteed f_y minimum	Average f_y
248 N/mm^2 (36 Ksi)	$1.15 f_y$ min.
345 N/mm^2 (50 Ksi)	$1.10 f_y$ min.

Allowance should also be made for the effects of strain hardening; under moment gradient the maximum moment of beams may be as much as 40 percent greater than the plastic moment M_p (Section 6.7.5.1).

As mentioned in the previous section *butt welded joints* should give the best earthquake resistance. Suitable design methods for such joints are given in the ASCE Manual.[59]

Fillet welded joints are capable of developing the full plastic moment of the members joined when the loading is monotonic. The tests by Popov *et al.* referred to in the previous section are the only tests of fillet welds under cyclic loading so far published, and indicate reasonable behaviour. Fillet weld stresses suitable for plastic design have been given in the ASCE Manual[59] or similar sources, but no recommendations specifically for seismic stress conditions have as yet been made.

Bolted joints are capable of developing the full plastic moment of the connected members, although with local loss of stiffness and energy absorption (Section 6.7.6.1). The loss of section area at the bolt holes should be considered. In California the SEAOC Code[21] recommends that when using steel for which the specified ultimate strength is less than 1·5 times the specified yield strength, plastic hinges in beams should not occur at positions where the beam flange area has been reduced such as by bolt holes. As with welded joints, plastic design criteria for bolted connections such as given by the ASCE Manual[59] are appropriate.

6.7.6.3 *Deformation behaviour of steel panel zones*

The panel zone of a connection between two members is the intersection zone common to the two members. This zone is assumed to deform in shear as indicated in Figure 6.28(a). Kato and Nakao[68] have suggested a tri-linear relationship between the shear stress and shear strain as a good approximation to the results of their monotonic tests on Japanese H-Section connections (Figure 6.28b).

Although little is known of the deformation characteristics of panel zones, specially under cyclic loading, it has been demonstrated[69] that the deformation of beam-column connections may contribute up to about one third of the interstorey deflection in multistorey buildings, and of this deformation about half may arise from the shear deformation of the panel zone itself.[70]

The large influence of panel zone behaviour on overall frame strength and stiffness has also been indicated by Kato and Nakao.[68,71] If bilinear hinges were assumed in the panel zones, the ultimate shear resistance of the frame was found to be half that achieved if the panels were kept elastic and hinges were formed only in the frame members. In practice it is difficult to prevent some yielding in the panel zone, and limited yielding in the panel may beneficially reduce the amount of plasticity and plastic instability which occurs in the adjacent beam hinges.

6.7.6.4 *Detailing panel zones for seismic resistance*

In the absence of design criteria based on cyclic testing, the panel zone should be designed to allow the adjacent members to reach their full strength while avoiding excessive shear deformation of the panel zone itself. The following simplified design method has been suggested by Kato[71] for frames in which;

(a) joint panels have the same sectional shape as that of adjoining columns;

273

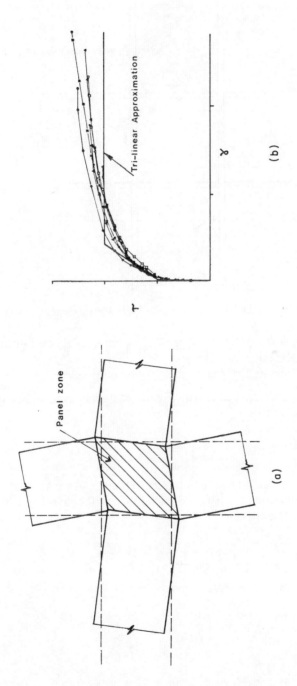

Figure 6.28 Idealized shear deformation of beam-column panel zones

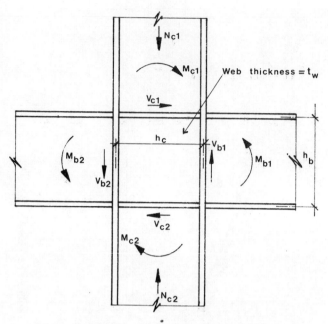

Figure 6.29 Forces acting on a typical panel zone

(b)   joint panels are stiffened by pairs of diaphragms located at the levels of the adjoining beam flanges;

(c)   axial compression of the column is not unusually large.

Referring to the notation defined in Figure 6.29, the average shear stress $\bar{\tau}$ in the panel is given by

$$\bar{\tau} = \frac{M_{b1} + M_{b2}}{h_b h_c t_w} - \frac{V_{c1} + V_{c2}}{2 h_c t_w} \tag{6.27}$$

Applying the condition that shear stresses remain elastic, i.e. $\bar{\tau} \leqslant \tau_y$, and Von Mises criterion that $\tau_y = f_y/\sqrt{3}$, where f_y is the tensile yield stress, equation (6.27) may be written

$$\bar{\tau} = \frac{M_{b1} + M_{b2}}{h_b h_c t_w} - \frac{V_{c1} + V_{c2}}{2 h_c t_w} \leqslant \frac{f_y}{\sqrt{3}} \tag{6.28}$$

Kato[71] simplified the above expression to

$$\bar{\tau} = \frac{M_{b1} + M_{b2}}{h_b h_c t_w} \leqslant \frac{4}{3} \cdot \frac{f_y}{\sqrt{3}}$$

or

$$\bar{\tau} = \frac{M_{b1} + M_{b2}}{V_{sp}} \leqslant 0 \cdot 77 f_y \tag{6.29}$$

The conservative approximation of neglecting the column shears V_{c1} and V_{c2} of equation (6.28) has been approximately compensated for in equation (6.29) by increasing the permissible stress. It should be noted that in equation (6.29) the quantity $h_b h_c t_w$ has been written as V_{sp} which may be described as the effective shear volume of the panel.

Attention should also be paid to the possible need for stiffeners to the panel zone.[59]

The panel zone design criterion given by equation (6.29) may be applied to joints of different configuration, as described in Figures 6.30 to 6.32.

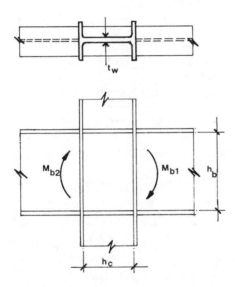

Figure 6.30 Typical H shape joint as applied to equation (6.29). $V_{sp} = h_b h_c t_w$, where h_b and h_c are measured as the clear distances between the inner faces of the flanges

6.7.7 Seismic response of complete steel structures

Little experimental work has as yet been done on the hysteretic behaviour of complete steel frames in order to determine the applicability of measured individual member behaviour to complete design problems. The largest scale testing so far reported is that of Carpenter and Lu,[72] the results of which indicate a 40 percent increase in maximum load capacity over the maximum monotonically applied lateral load predicted by second-order elastic-plastic analysis. Unfortunately the lateral restraint conditions for the above work were not published (by Carpenter and Lu),[72] but low slenderness ratios seem to be implied in the light of comparison with more recent tests[62,63,64] as discussed in Section 6.7.4.3.

Comments on other problems associated with overall frame design such as the avoidance of excessive lateral sway, have been made earlier in this chapter.

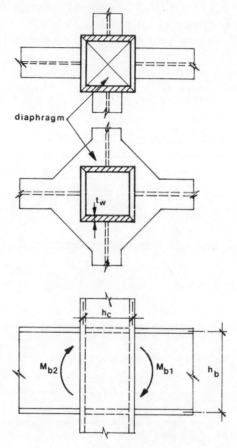

Figure 6.31 Typical joint at a box shape column as applied to equation (6.29). $V_{sp} = 2h_b h_c t_w$, where h_b and h_c are measured between the centre lines of the flanges as shown

6.7.7.1 Frames with diagonal members

This chapter generally has been devoted to design of frames containing vertical and horizontal members. Frames containing diagonal members have a number of aseismic problems peculiar to themselves which have only begun to be studied comparatively recently. Some tests have indicated the possibility of failure after very few load cycles, of diagonals in which the axial loads vary from high compression to some tension in earthquake loading. A short discussion of triangulated frames may be found in Appendix A.2.

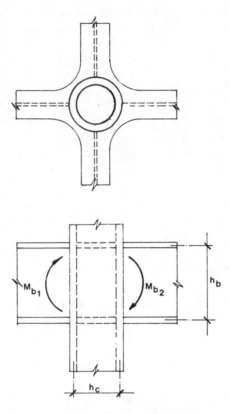

Figure 6.32 Typical joint at a tubular column as applied to equation (6.29). $V_{sp} = \frac{1}{2}V_{op}$, where V_{op} is the total volume of the connection surrounded by the beam and column ends

6.8 TIMBER DESIGN AND DETAIL

6.8.1 Introduction

The performance of timber construction in earthquakes can be very good mainly by virtue of its high strength to weight ratio. Timber also has the favourable property that its ultimate strength under dynamic loading is of the order of 25 percent higher than under static design conditions (Figure 6.33).

Unlike steel and reinforced concrete, timber does not have the advantage of inelastic behaviour (Figure 6.34) and therefore must be designed as a brittle material (Section 5.2.4.1). However timber shows little degradation of strength or stiffness under cyclic loading and timber construction has high damping. Test results of Medearis[74] for timber diaphragms indicated an equivalent viscous damping of 8–10 percent independent of amplitude (Figure 6.34).

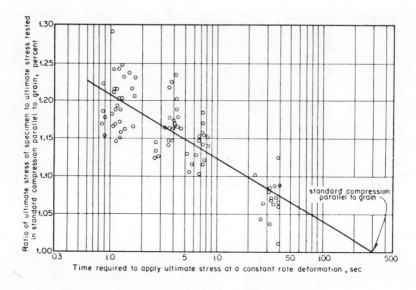

Figure 6.33 Effect of loading rate on ultimate strength of timber (after Brokaw and Foster,[73] by permission of U.S. Forest Products Laboratory Forest Service, U.S. Department of Agriculture)

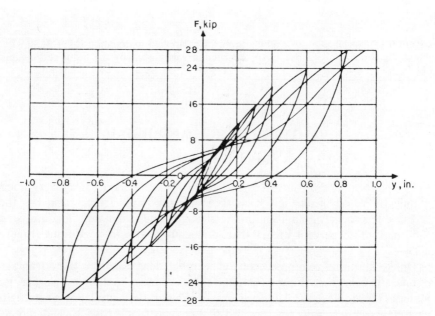

Figure 6.34 Hysteretic behaviour of timber diaphragms under cyclic loading (after Medearis[74])

The figure of 15 percent of critical damping in Table 5.1 for timber shear wall construction makes allowances for the additional damping which occurs in all parts of a timber building.

In earthquakes[75,76,77,78] the main causes of inadequate performance of timber construction have been as follows;

(i)   large response on soft ground;
(ii)   lack of integrity of substructure;
(iii)   asymmetry of the structural form;
(iv)   insufficient strength of chimneys;
(v)   inadequate structural connections;
(vi)   use of heavy roofs without appropriate strength of supporting frame;
(vii)   deterioration of timber strength through decay or pest attack;
(viii) inadequate resistance to post-earthquake fires.

Within certain limitations, means are available for dealing with all these aspects of earthquake resistance of timber construction as discussed below.

6.8.2 Site responses and timber buildings

It has been variously reported[78,79] that timber buildings suffer more earthquake damage when sited on soft ground than when on hard ground. The reasons for this occurrence are uncertain. For instance the possible role of resonance is obscure. As most one and two storey timber buildings have fundamental periods of vibration in the range 0·1 to 0·6 seconds, resonance with the ground would seem likely on thin rather than thick layers of soft ground. After heavy shaking a timber building will loosen at the joints and its natural periods are likely to lengthen, but the manner of its vibration is uncertain and it is unlikely to have well-defined modes in which resonance can occur.

It is possible that timber houses on soft ground are weakened by seasonal ground movements, making them more vulnerable to earthquakes. Also the differential earthquake ground movements in softer soils are larger than in firm soils and this is likely to affect timber buildings with light foundations more than stiffer forms of construction. If timber buildings are to be built on soft ground in an earthquake area, extra measures should be taken to ensure structural integrity, particularly at foundation level.

6.8.3 Foundations of timber buildings

For the design of foundations of timber buildings, the guidance for commercial-industrial buildings and housing given in Sections A.3.4 and A.4.6 is appropriate. It should also be noted that the requirements of codes of practice[80,81] for holding down the timber structure on to the foundation should be treated as a minimum precaution. The size of holding-down bolts for details such as shown in Figure 6.41 will of course have to be determined by calculation in many cases.

The type of foundation provided by pole construction (Figure 6.35) overcomes some of the weaknesses of orthodox substructures to timber building, as the poles themselves provide vertical continuity and the pole frameworks develop the necessary resistance to horizontal forces.[82,83,84] For some species of timber, preservative treatment such as tanalizing will of course be essential for durability below ground.

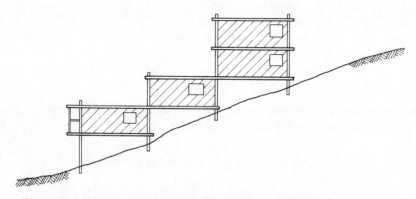

Figure 6.35 Pole frame apartments as built at Lugunda Beach, California[84]

6.8.4 Structural form of timber buildings

The form of the superstructure of timber buildings should be chosen according to the requirements of low-rise construction as described in Sections A.3.2 and A.4.2.

6.8.5 Chimneys and roofs of timber buildings

Chimneys and roofs are the source of a great deal of earthquake damage in timber buildings. Chimneys are often themselves structurally inadequate, or there are interaction problems between the stiff chimneys and the more flexible timber construction (Section A.4.8). Timber frames have often proved to have been inadequately designed for the forces deriving from heavy roof construction (Section A.4.7).

6.8.6 Timber shear panel construction

Most timber buildings derive their strength and stiffness from shear panels or diaphragms which may constitute walls, floors, ceilings or roof slopes. Individual shear elements are built up from planks, plywood, metal, plaster or other sheeting which is fixed to the basic timber framework by nails, screws or glue. The effectiveness of different types of diaphragm for resisting in-plane shears depends on;

(a) its overall size and shape;
(b) the size, shape and position of any apertures;
(c) the nature of the timber framework;
(d) the nature and disposition of the diagonal or sheeting members;
(e) the connections between elements (c) and (d).

A useful study of some of the above factors was carried out by the U.S. Forest Products Laboratory in 1946, the results of which are shown in Figure 6.36. The superiority of plywood for the panelling and gluing for the means of connection are obvious. In the field however problems arise in obtaining reliable glues of suitable strength. The use of plywood in seismic areas has been discussed in some detail elsewhere,[85] while the good performance of metal clad timber houses under dynamic loading has been demonstrated in New Zealand.[86] Although nails are moderately effective for connecting plywood to frames, this form of fixing has not been entirely satisfactory at the perimeter of major shear elements, such as the connection between roof diaphragms and walls of industrial buildings (Section A.3.5).

The qualitative information given in Figure 6.36 is turned into design data for horizontal and vertical plywood diaphragms by the Uniform Building Code[80] as set out in Figure 6.37. Here the effect of nail size, nail spacing, plywood thickness and quality have been taken into account. Care should be taken in translating Figure 6.37 to other areas where different grades of timber are in use. For example in New Zealand the timber code[81] permits loads on plywood panels which are about half those allowed in the U.S.A. This appears to arise from the lower lateral shear force allowed on edge nails driven into the comparatively soft New Zealand radiata pine framing.[85]

From Figure 6.36 it is also clear that diagonals are much more effective when continuous between opposite framing members of a panel, rather than when broken by apertures. In domestic buildings it is of course common for only one or two diagonals to be used within any individual wall unit, and such diagonals should clearly be inclined between 30° and 60° to the horizontal for greatest effectiveness. In timber which is likely to split, nail holes near the ends should be predrilled slightly smaller than the nail diameter. In framing up shear panels, care should be taken that the perimeter members, and diagonals if used, are made from sound timbers. The framing members for door and window apertures should similarly be of good quality timber.

External timber framed walls are often clad with plaster, and the earthquake performance of such walls has been greatly improved using expanded metal lath. In Japan and California expanded metal is now commonly used in conjunction with cement, lime and sand plaster (mixes of between 1:1:3 and 1:1:4$\frac{1}{2}$ are found to be durable), the lime reducing the brittleness of

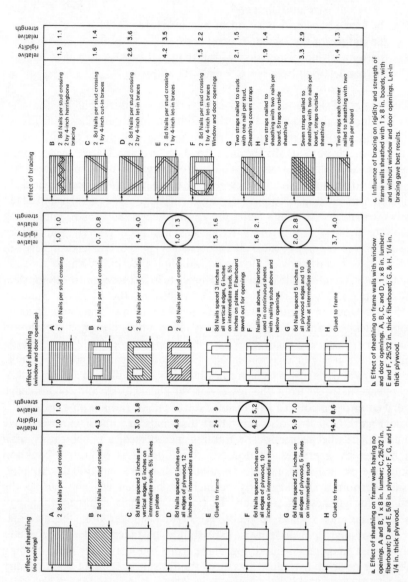

Figure 6.36 Tests on timber framed walls with various forms of sheeting and fixing carried out by the U.S. Forest Products Laboratory

the plaster. The application of two or three coats of plaster, giving a total thickness of about 20 mm, is normal practice.

6.8.6.1 Deflection of plywood diaphragms

Excessive deflection of plywood diaphragms is to some extent controlled by limiting the aspect ratio of diaphragms. Closely representing international practice, in New Zealand[81] the length of horizontal diaphragms may not exceed 4 times the width, while the height of vertical diaphragms may not exceed 3·5 times the width.

As horizontal diaphragms may deflect sufficiently to endanger supporting or attached wall components (Figure 6.38), a means of calculating their deflections under in-plane loading is desirable. This deflection involves bending and shear deformation and nail slippage as expressed by

$$\Delta = \frac{52vL^3}{EAB} + \frac{vL}{4Gt} + 0·308Le_\text{n} \tag{6.30}$$

where

Δ = deflection (mm),
v = applied shear loading (N/m),
L = length of diaphragm (m),
B = width of diaphragm (m),
A = cross-sectional area of chord (mm²),
E = modulus of elasticity of chords (N/mm²),
G = shear modulus of plywood (N/mm²),
t = thickness of plywood (mm),
e_n = nail deformation (mm).

The nail deformation e_n may be derived from Figure 6.39 when using timbers equivalent to those grown in North America, as indicated. In New Zealand it has been recommended[85] that a value of $e_\text{n} = 0·50$ mm be used regardless of plywood thickness.

6.8.7 Connections in timber construction

Connections between timber members may be formed in the timber itself, or may involve glue, nails, screws, bolts, metal straps, metal plates or toothed metal connectors. Under earthquake loading, joints formed in the timber are inferior to most other forms of joint. In light timber construction such as smaller dwellings, Evans[88] has recommended the use of metal corner plates (Hurricane braces) or toothed steel connectors.

Allowable shear in pounds per foot for horizontal plywood diaphragms[1]

Plywood Grade	Common Nail Size	Minimum Nominal Penetration in Framing (in Inches)	Minimum Nominal Plywood Thickness (in Inches)	Minimum Nominal Width of Framing Member (in Inches)	Blocked Diaphragms — Nail Spacing at diaphragm boundaries (all cases) and continuous panel edges parallel to load (cases 3 & 4): 6 / Nail spacing at other plywood panel edges: 6	4 / 6	2½ / 4	2 / 3	Unblocked — Load perpendicular to unblocked edges and continuous panel joints (case 1)	Unblocked — All other configurations (cases 2, 3 & 4)
Structural I	6d	1¼	5/16	2	185	250	375	420	165	125
				3	210	280	420	475	185	140
	8d	1½	3/8	2	270	360	530	600	240	180
				3	300	400	600	675	265	200
	10d	1⅝	½	2	320	425	640²	730²	285	215
				3	360	480	720	820	320	240
Structural II, C-C Exterior, Standard Sheathing and Other Grades Covered in U.B.C. Standard No. 25-9	6d	1¼	5/16	2	170	225	335	380	150	110
				3	190	250	380	430	170	125
			3/8	2	185	250	375	420	165	125
				3	210	280	420	475	185	140
	8d	1½	3/8	2	240	320	480	545	215	160
				3	270	360	540	610	240	180
			½	2	270	360	530	600	240	180
				3	300	400	600	675	265	200
	10d	1⅝	½	2	290	385	575²	655²	255	190
				3	325	430	650	735	290	215
			5/8	2	320	425	640²	730²	285	215
				3	360	480	720	820	320	240

[1] These values are for short time loads due to wind or earthquake and must be reduced 25 per cent for normal loading. Space nails 10 inches on centre for floors and 12 inches on centre for roofs along intermediate framing members.

[2] Reduce tabulated allowable shears 10 percent when boundary members provide less than 3-inch nominal nailing surface.

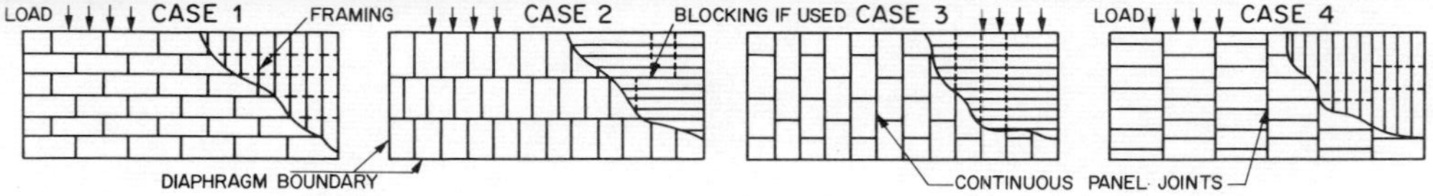
LOAD
CASE 1
FRAMING
CASE 2
BLOCKING IF USED
CASE 3
LOAD
CASE 4
DIAPHRAGM BOUNDARY
CONTINUOUS PANEL JOINTS

NOTE: Framing may be located in either direction for blocked diaphragms.

Allowable shear for wind or seismic forces in pounds per foot for plywood shear walls[1]

Plywood grade	Nail size (Common or Galvanized Box)	Minimum Nail Penetration in Framing (Inches)	Minimum Nominal Plywood Thickness (Inches)	Plywood applied direct to framing: Nail Spacing at Plywood Panel Edges 6	4	$2\frac{1}{2}$	2	Nail size (Common or Galvanized Box)	Plywood applied over $\frac{1}{2}$-inch Gypsum Sheathing: Nail Spacing at Plywood Panel Edges 6	4	$2\frac{1}{2}$	2
Structural I	6d	$1\frac{1}{4}$	$\frac{5}{16}$	200	300	450	510	8d	200	300	450	510
	8d	$1\frac{1}{2}$	$\frac{3}{8}$	230[3]	360[3]	530[3]	610[3]	10d	280	430	640[2]	730[2]
	10d	$1\frac{5}{8}$	$\frac{1}{2}$	340	510	770[2]	870[2]	—	—	—	—	—
Structural II, C-C Exterior, Standard Sheathing, Panel Siding Plywood and Other Grades Covered in U.B.C. Standard No. 25-9	6d	$1\frac{1}{4}$	$\frac{5}{16}$	180	270	400	450	8d	180	270	400	450
	8d	$1\frac{1}{2}$	$\frac{3}{8}$	220[3]	320[3]	470[3]	530[3]	10d	260	380	570[2]	640[2]
	10d	$1\frac{5}{8}$	$\frac{1}{2}$	310	460	690[2]	770[2]	—	—	—	—	—
	Nail size (Galvanized Casing)							Nail size (Galvanized Casing)				
Plywood Panel Siding in Grades Covered in U.B.C. Standard No. 25-9	6d	$1\frac{1}{4}$	$\frac{5}{16}$	140	210	320	360	8d	140	210	320	360
	8d	$1\frac{1}{2}$	$\frac{3}{8}$	130[3]	200[3]	300[3]	340[3]	10d	160	240	360	410

[1] All panel edges backed with 2-inch nominal or wider framing. Plywood installed either horizontally or vertically. Space nails at 6 inches on centre along intermediate framing members for $\frac{3}{8}$-inch plywood installed with face grain parallel to studs spaced 24 inches on centre and 12 inches on centre for other conditions and plywood thicknesses. These values are for time loads due to wind or earthquake and must be reduced 25 percent for normal loading.

[2] Reduce tabulated allowable shears 10 percent when boundary members provide less than 3-inch nominal nailing surface.

[3] The values for $\frac{3}{8}$-inch thick plywood applied direct to framing may be increased 20 percent provided studs are spaced a maximum of 16 inches on centre or plywood is applied with face grain across studs.

Figure 6.37 Allowable shear in plywood diaphragms according to the Uniform Building Code.[80] Reproduced from the 1973 edition of the Uniform Building Code by permission of the International Conference of Building Officials

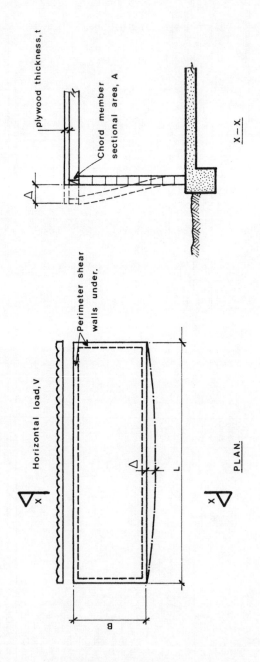

plywood thickness, t

Chord member
sectional area, A

X – X

Perimeter shear
walls under.

Horizontal load, V

PLAN.

Figure 6.38 A typical horizontal timber diaphragm showing the effect on supporting walls of deflections under horizontal loading

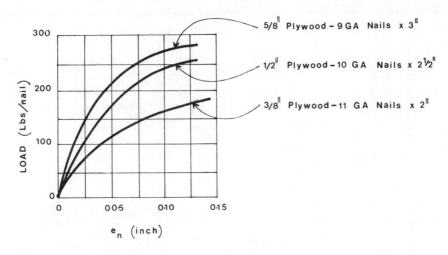

Figure 6.39 Nail deformation in plywood diaphragms framed on to Douglas fir as derived in the USA[87]

For nailed joints, the nail load, size and spacing require careful attention. A nail driven parallel to the timber grain should be designed for not more than two-thirds of the lateral load which would be allowed for the same size of nail driven normal to the grain. Nails driven parallel to the grain should not be expected to resist withdrawal forces. Edge or end distance of nails should not be less than half the required nail penetration.

In diaphragms, perimeter framing may need jointing capable of carrying the longitudinal forces arising from wind or seismic loading. A simple method of connection is shown in Figure 6.40.

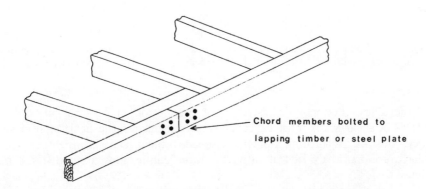

Figure 6.40 Method of jointing chord members of timber diaphragms

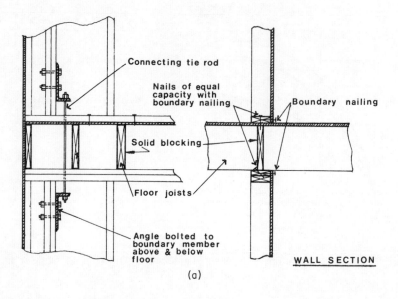

(a)

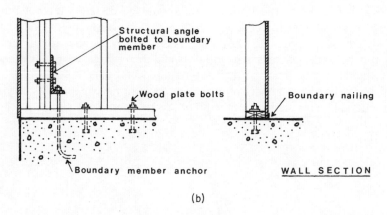

(b)

Figure 6.41 Connection details for plywood shear walls.[85] (a) Interstory connections in timber buildings, (b) connection of timber members to concrete foundations

Connections between shear walls and foundation or between successive storeys of shear walls must be capable of transmitting the horizontal shear forces and the overturning moments applied to them. Details which are considered good practice in California for these connections are illustrated in Figure 6.41.

For some comments on the connections of timber roof diaphragms to walls of other materials refer to Sections 6.8.6.1 and A.3.5.

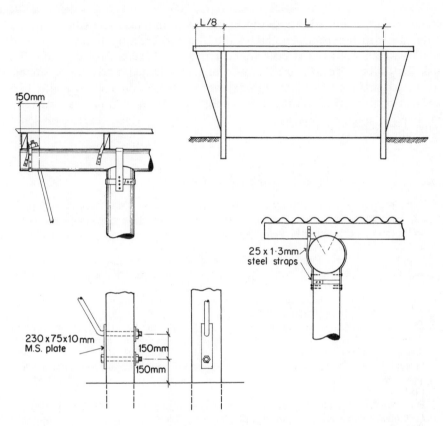

Figure 6.42 Tied rafter pole building showing typical connection details[82]

Pole frame buildings are usually jointed using bolts, steel straps and clouts (Figure 6.42) as described in detail elsewhere.[83,84] An effective means of obtaining resistance to lateral shear forces is to create moment-resisting triangles at the knees of portals (Figure 6.42) using steel rods as the diagonal member.

6.8.8 Fire resistance in timber construction

The danger from fires after earthquakes is very great and timber construction is of course particularly vulnerable in this respect. Most fire retarding chemicals are considered to reduce the strength of timber, and under the Uniform Building Code[80] design stresses must be reduced by 10 percent in timber treated for fire resistance with approved chemicals. Plywood so treated should be designed with a 16 percent reduction in permissible stresses. When using new fire retarding chemicals, design stresses should be determined from strength tests carried out before and after treatment. Fire stops are also of great importance in timber buildings and local regulations should have fire stop provisions similar to those in the Uniform Building Code.[80]

It is worth noting that pole frame structures have a relatively low fire risk for timber construction. Because of their comparatively large volume to surface area ratio and smooth exterior, poles are difficult to ignite, and because of their wide spacing, fire cannot spread easily from one structural member to another. The loss of strength due to surface charring will not generally be critical.

REFERENCES

1. Park, R., and Paulay, T. *Reinforced concrete structures*, Wiley, New York, 1975.
2. Park, R. 'Theorisation of structural behaviour with a view to defining resistance and ultimate deformability', *Bulln. New Zealand Society for Earthquake Engineering*, **6**, No. 2, 52–70 (June 1973).
3. Krawinkler, H., and Popov, E. P. 'Hysteretic behaviour of reinforced concrete rectangular and T-beams', *Proc. 5th World Conference on Earthquake Engineering, Rome*, **1**, 249–258 (1973).
4. A.C.I. Committee 318. *Building code requirements for reinforced concrete (A.C.I. 318-71)*, American Concrete Institute, 1971.
5. Park, R., and Paulay, T. 'Behaviour of reinforced concrete external beam-column joints under cyclic load', *Proc. 5th World Conference on Earthquake Engineering, Rome*, **1**, 772–781 (1973).
6. Megget, L. M. 'Cyclic behaviour of exterior of reinforced beam-column joints', *Bulln. New Zealand National Society for Earthquake Engineering*, **7**, No. 1, 27–47 (March 1974).
7. Hollings, J. P. 'Reinforced concrete seismic design', *Bulln. New Zealand Society for Earthquake Engineering*, **2**, No. 3, 217–250 (Sept. 1969).
8. Park, R. 'Ductility of reinforced concrete frames under seismic loading', *New Zealand Engineering*, **23**, No. 11, 427–435 (Nov. 1968).
9. Blume, J. A., Newmark, N. M., and Corning, L. H. *Design of multi-storey reinforced concrete buildings for earthquake motions*, Portland Cement Association, Skokie, Illinois, 1961.
10. Base, G. D., and Read, J. B. 'Effectiveness of helical binding in the compression zone of concrete beams', *ACI Jnl*, **62**, 763–781 (July 1965).
11. Bertero, V. B., and Felippa, C. Discussion to paper by Roy, H. E. H., and Sozen, M. A. 'Ductility of concrete', *Proc. of the International Symp. on Flex. Mechanics for Reinforced Concrete, ASCE-ACI, Miami*, 213–235 (Nov. 1964).
12. Nawy, E. G., Danesi, R. F., and Grosko, J. J. 'Rectangular spiral binders, effect on plastic hinge rotation capacity in reinforced concrete beams', *ACI Jnl*, **65**, 1001–1016 (Dec. 1968).
13. Baker, A. L. L., and Amarakone, A. M. N. 'Inelastic hyperstatic frame analysis', *Proc. Intl. Symp. on the Flex. Mechanics of r.c., ASCE-ACI, Miami*, 85–142 (Nov. 1964).
14. Soliman, M. T. M., and Yu, C. W. 'The flexural stress-strain relationship of concrete confined by rectangular transverse reinforcement', *Magazine of Concrete Research*, **61**, 223–238 (Dec. 1967).
15. Corley, W. G. 'Rotational capacity of reinforced concrete beams', *Jnl of the Structural Division, ASCE*, **92**, No. ST5, 121–146 (Oct. 1966).
16. Pfrang, E. O., Siess, C. P., and Sozen, M. A. 'Load-moment-curvature characteristics of r.c. cross sections', *A.C.I. Jnl*, **61**, 763–778 (July 1964).

17. British Standards Institution. *Bending dimensions and scheduling of bars for reinforcement of concrete*, British Standard 4466: 1969.
18. British Standards Institution. *The structural use of concrete*, British Standard Code of Practice, CP 110: 1972.
19. British Standards Institution. *Hot rolled steel bars for the reinforcement of concrete*, British Standard 4449: 1969.
20. American Society for Testing and Materials. *Deformed and plain billet-steel bars for concrete reinforcement*, ASTM A615, 1974.
21. Seismology Committee, S.E.A.O.C. *Recommended lateral force requirements and commentary*, Structural Engineers Association of California, 1973.
22. British Standards Institution. *Cold worked steel for the reinforcement of concrete*, British Standard 4461: 1969.
23. British Standards Institution. *Steel fabric for the reinforcement of concrete*, British Standard 4483: 1969.
24. Forrest, E. J. 'Seismic resistance of industrialised building', in 'Concrete for the 70's, Industrialisation', *Proc. National Conference Wairakei New Zealand, Portland Cement Association, New Zealand*, 70–75 (1972).
25. Wood, B. J. 'Structural jointing', published in 'Concrete for the 70's, Industrialisation', *Proc. National Conference Wairakei New Zealand, Portland Cement Association, New Zealand*, 76 79 (1972).
26. Mast, R. F. 'Seismic design of 24-storey building with precast elements', *Jnl Prestressed Concrete Institute*, **17,** 45–59 (July/August 1972).
27. Walocha, H. J. 'Parking structure solutions in a seismic zone', *Jnl Prestressed Concrete Institute*, **17,** No. 4, 60–64 (July/August 1972).
28. Seminar under the Japan–U.S. Co-operative Science Program—Construction and Behaviour of Precast Concrete Structures, *Reports by Japanese Participants, Parts I and II, August 23–27*, Seattle, U.S.A. (in English), 1971.
29. Petrovic, B., Muravljov, M., and Dimitrievic, R. 'The IMS assembly framework system and its resistance to seismic influences', *Earthquake Engineering, Proc. 3rd European Symposium on Earthquake Engineering, Sofia*, 513–520 (1970).
30. P.C.I. Committee on Connection Details, '*Connection details for precast–prestressed concrete buildings*', Prestressed Concrete Institute, U.S.A., 1963.
31. Building Research Station. *Digest No. 114*, Garston, Watford, Gt. Britain, Feb. 1970.
32. Building Research Station. *Digest No. 137*, Garston, Watford, Gt. Britain, Jan. 1972.
33. Tankersley, R. N. and Sewell, J. M. F. 'Design and construction of a precast multistorey parking station', *New Zealand Engineering*, **25,** No. 11, 291–299 (Nov. 1970).
34. N.Z.M.O.W. 'Design of public buildings', (Clause 11.5.4, Precast Concrete Claddings) *Code of Practice PW 81/10/1*, 1970.
35. City of Los Angeles. *Building Code, Division 23, Item 7*, Exterior Elements, Los Angeles, California.
36. Uchida, N., Aoyagi, T., Kawamura, M., and Nakagawa, K. 'Vibration test of steel frame having precast concrete panels', *Proc. 5th World Conference on Earthquake Engineering, Rome*, **1,** 1167–1176 (1973).
37. Brooke-White, C. J. 'An architectural application of precast concrete', in 'Concrete for the 70s—Industrialisation', *Proc. National Conference, Wairakei, New Zealand, Portland Cement Association, New Zealand* (Oct. 1972).
38. Miller, M. S. 'Joints in architecture', in 'Concrete for the 70's—Industrialisation', *Proc. National Conference, Wairakei, New Zealand, Portland Cement Association, New Zealand* (Oct. 1972).
39. Blakeley, R. W. G. 'Prestressed concrete seismic design', *Bulln. New Zealand Society for Earthquake Engineering*, **6,** No. 1, 2 21 (March 1973).
40. Pond, W. F. 'Performance of bridges during San Fernando earthquake', *Jnl Prestressed Concrete Institute*, **17,** No. 4, 65–75 (July/August 1972).

41. P.C.I. Seismic Committee. 'Principles of the design and construction of earthquake resistant prestressed concrete structures', Journal Prestressed Concrete Institute, Vol. 11, No. 3, June, 1966, pp 18–22.
42. N.Z.P.C.I. 'Seismic design recommendations for prestressed concrete', New Zealand Prestressed Concrete Institute, Sept. 1966.
43. F.I.P. 'Report of the F.I.P. Commission on Seismic Structures', Proc. Sixth Congress of La Federation Internationale de la Precontrainte, Prague, 1970.
44. Blakeley, R. W. G., and Park, R. 'Response of prestressed concrete structures to earthquake motions', New Zealand Engineering, Vol. 28, No. 2, Feb. 1973, pp 42–54.
45. Park, R., and Thompson, K. J. 'Behaviour of prestressed, partially prestressed and reinforced concrete interior beam-column assemblies under cyclic loading', University of Canterbury, Oct. 1973.
46. Blakeley, R. W. G., and Park, R. 'Ductility of prestressed concrete members', Bulln. New Zealand Society for Earthquake Engineering, 4, No. 1 145–170 (March 1971).
47. Spencer, R. A. 'The non-linear response of a multi-storey prestressed concrete structure to earthquake excitation', Proc. 4th World Conference on Earthquake Engineering, Chile II, A4, 139–154 (1969).
48. Meli, R. 'Behaviour of masonry walls under lateral loads', Proc. 5th World Conference on Earthquake Engineering, Rome, 1, 853–862 (1973).
49. Williams, D., and Scrivener, J. C. 'Response of reinforced masonry shear walls to static and dynamic cyclic loading', Proc. 5th World Conference on Earthquake Engineering, Rome, 2, 1491–1494 (1973).
50. International Conference of Building Officials, Uniform Building Code, Clause 2418, Reinforced Masonry Design, I.C.B.O., Pasadena, California, 1971.
51. Standards Association of New Zealand. NZSS 1900: Model Building Bylaws, 1964, Chaps 6.2 and 9.2.
52. Amrhein, J. E. Reinforced masonry engineering handbook, Masonry Institute of America, Los Angeles, 1972.
53. Standards Association of New Zealand. NZSS 1900: Model Building Bylaws, 1964, Chap. 8.
54. Tso, W. K., Pollner, E., and Heidebrecht, A. C. 'Cyclic loading on externally reinforced masonry walls', Proc. 5th World Conference on Earthquake Engineering, Rome 1, 1177–1186 (1973).
55. British Standards Institution. 'Quality grading of steel plate from 12 mm to 150 mm thick by means of ultrasonic testing', B.S.I. Draft for Development, DD21, 1972.
56. Farrar, J. C. M., and Dolby, R. E. Lamellar tearing in welded steel fabrication, The Welding Institute, Cambridge, 1972.
57. Jubb, J. E. M. 'Lamellar tearing', Welding Research Council, U.S.A. Bulletin 168, 1968.
58. Coe, F. R. Welding steels without hydrogen cracking, The Welding Institute, Cambridge, 1973.
59. Welding Research Council and the American Society of Civil Engineers, 'Plastic design in steel—a guide and commentary', ASCE Manual No. 41, 2nd ed., ASCE, New York, 1971.
60. Lay, M. G., and Galambos, T. V. 'Inelastic beams under moment gradient', Jnl. Structural Division, ASCE, 93, No. ST1, 381–399 (Feb. 1967).
61. Bertero, V. V., and Popov, E. P. 'Effect of large alternating strains on steel beams', Jnl. Structural Division, ASCE, 91, No. ST1, 1–12 (Feb. 1965).
62. Vann, W. P., Thompson, L. E., Whalley, L. E., and Ozier, L. D. 'Cyclic behaviour of rolled steel members', Proc. 5th World Conference on Earthquake Engineering, Rome 1, 1187–1193 (1973).
63. Takanashi, K. 'Inelastic lateral buckling of steel beams subjected to repeated and reversed loadings, Proc. 5th World Conference on Earthquake Engineering, Rome 1, 795–798 (1973).

64. Popov, E. P. 'Experiments with steel members and their connections under repeated loads', *Preliminary Report of the Symposium on Resistance and ultimate deformability of structures acted on by well defined repeated loads*, IABSE, Lisbon, 1973, pp. 125–135.

65. Popov, E. P., and Pinkney, R. B. 'Cyclic yield reversal in steel building connections', *Jnl. Structural Division, ASCE*, **95**, No. ST3, 327–353 (March, 1969).

66. Popov, E. P. and Stephen, R. M. 'Cyclic loading of full size steel connections', *American Iron and Steel Institute, Steel Research for Construction Bulletin No. 21* (Feb. 1972).

67. Building Code Steel Subcommittee, SEAOC. *Report in Bulln. Structural Engineers Association of California*, Nov. 1971.

68. Kato, B., and Nakao, M. 'The influence of the elastic plastic deformation of beam-to-column connections on the stiffness, ductility and strength of open frames', *Proc. 5th World Conference on Earthquake Engineering, Rome* **1**, 825–828 (1973).

69. Teal, E. J. 'Structural steel seismic frames—drift ductility requirements', *Proc. 37th Annual Convention Structural Engineers Association of California* (1968).

70. Surtees, J. O., and Mann, A. P. 'End plate connections in plastically designed structures', *Conference on Joints in Structures, Institution of Structural Engineers and the University of Sheffield* (July 1970).

71. Kato, B. 'A design criteria of beam-to-column joint panels', *Bulln. New Zealand National Society for Earthquake Engineering*, **7**, No. 1, 14–26 (March, 1974).

72. Carpenter, L. D., and Lu, L. W. 'Repeated and reversed load tests on full-scale steel frames', *Proc. 4th World Conference on Earthquake Engineering, Chile*, **I**, B—2, 125–136 (1969).

73. Brokaw, M. P., and Foster, G. W. *Effect of rapid loading and duration of stress on the strength properties of wood tested in compression and flexure*, U.S. Department of Agriculture, Forest Products Laboratory, Madison, Wisc., 1946.

74. Medearis, K. 'Static and dynamic properties of shear structures', *Proc. International Symp. Effects of Repeated Loading on Materials and Structures, RILEM—Inst. Ing., Mexico*, **VI** (1966).

75. Falconer, B. H. 'Preliminary comments on damage to buildings in the Inangahua earthquake', *Bulln. New Zealand Society for Earthquake Engineering*, **1**, No. 2, 61–71 (Dec. 1968).

76. 'The San Fernando California earthquake of February 9, 1971', *NBS Report 10556*, U.S. Department of Commerce, National Bureau of Standards, March, 1971.

77. Earthquake Engineering Research Laboratory, California Institute of Technology. 'Engineering Features of the San Fernando Earthquake', *Bulln. New Zealand Society for Earthquake Engineering*, **6**, No. 1, 22–45 (March, 1973).

78. Soil Research Team, Earthquake Research Institute. 'Earthquake damage and subsoil conditions as observed in certain districts of Japan', *Proc. 2nd World Conference on Earthquake Engineering, Japan*, **1**, 311–325 (1960).

79. Tsai, Zuei-Ho. 'Earthquake and architecture in Japan', *Proc. 3rd World Conference on Earthquake Engineering, New Zealand*, **III**, V–114 (1965).

80. International Conference of Building Officials. *Uniform Building Code*, I.C.B.O., Pasadena, California, 1973, Chap. 25.

81. New Zealand Standards Institute, Wellington. 'Design and Construction—Timber', *Model Building Bylaw, NZSS 1900*, 1964, Chap. 9.1.

82. New Zealand Timber Research and Development Association. 'Pole frame buildings', *T.R.A.D.A., Timber and Wood Products Manual*, Section 16-1, 12 pages (August, 1972).

83. New Zealand Timber Research and Development Association. 'Engineering design data for radiata pine poles', *TR.A.D.A., Timber and Wood Products Manual*, Section 26–2, 5 pages (Nov. 1972).

84. New Zealand Timber Research and Development Association. 'Pole type construction', *T.R.A.D.A.*, *Timber and Wood Products Manual*, Section 2f—1, 6 pages (Dec. 1973).

85. New Zealand Timber Research and Development Association. 'Plywood design for seismic areas', *T.R.A.D.A.*, *Timber and Wood Products Manual*, Section 5f—1, 12 pages (Dec. 1973).

86. Tracey, W. J. 'A simulated earthquake test for a timber house', *Bulln. New Zealand Society for Earthquake Engineering*, **2**, No. 3, 289–294 (Sept. 1969).

87. Timber Engineering Company. *Timber design and construction handbook*, F. W. Dodge Corporation, New York, 1956.

88. Evans, F. W. 'Earthquake engineering for the smaller dwelling', *Proc. 5th World Conference on Earthquake Engineering, Rome*, **2**, 3010–3013 (1973).

Chapter 7

Earthquake resistance of services

7.1 SEISMIC RESPONSE AND DESIGN CRITERIA

7.1.1 Introduction

This chapter sets out to advise services engineers on the earthquake-resistant design of services components and installations. Much of the background information on the earthquake problem is contained in other chapters of this book or in the literature of structural engineering and seismology. Up till about 1970 only a comparatively small effort had been made in this field by services engineers on their own account. Recently, however, there has been a growing awareness that services equipment needs its own specialist aseismic design and detailing. The following points are worthy of attention

(i) Aseismic design of services is a problem of dynamics, which cannot be treated adequately with equivalent static methods alone.

(ii) Earthquake accelerations applied in the design of services equipment generally should be much larger than the corresponding values used in the design of the buildings housing the equipment.

(iii) The response spectrum method (See Section 7.1.4) provides ready-worked solutions of the equations of motion, and is a powerful aid to understanding the true dynamic nature of the earthquake problem.

(iv) In many cases a high level of earthquake resistance can be provided at relatively small extra cost.[1]

7.1.2 Earthquake motion—accelerograms

Strong motion earthquakes are most commonly recorded by accelerographs which produce a plot of the variation of acceleration in a given direction with time. Perhaps the most widely used accelerogram is that obtained at El Centro, California during the Imperial Valley earthquake of 18th May, 1940. Figure 5.2 shows the north–south accelerogram of this earthquake, with a peak acceleration of $0.33\,g$. By integration of the acceleration record, the ground velocity was deduced, showing a maximum value of 34 cm/s. Similarly by integrating the velocity, the displacement of the ground was inferred, showing a maximum of 21 cm. The record of acceleration in the east-west direction was similar, with a maximum value of $0.22\,g$. The vertical component showed considerably more rapid variations, reaching a maximum of $0.2\,g$.

7.1.3 Design norms—design earthquakes

In the past some attempts have been made to establish the worst earthquake forces which could occur, so that equipment could be designed for them and made completely earthquake-proof. Unfortunately the likely maximum ground motions of earthquakes are usually above practical design levels. It is therefore common for a 'standard' or 'design' earthquake to be taken in a given region for certain types of construction. The standard earthquake may be thought of as the worst earthquake officially likely to occur in that region in a given time interval (say 50–100 years), and may be specified in terms of peak ground accelerations, a response spectrum, or an earthquake (Richter) magnitude. Although there is a risk of a worse real earthquake occurring, the standard for officially acceptable minimum risk is set by the design earthquake. Individual structures or equipment items may be designed to some authorized or discretionary fraction (greater or less than unity) of the standard earthquake.

7.1.4 The response spectrum design method

A direct analytical approach to the problem of earthquake strength is to make a mathematical model of the structure and subject it to accelerations as recorded in actual earthquakes. Many structures approximate to the single degree-of-freedom model shown in Figure 7.1, where a mass is supported by a spring and is connected to a damping device. If linear material behaviour is assumed, the ratio of spring stiffness to horizontal shear is constant. For mathematical convenience the damping force is usually taken as proportional to velocity, which is generally a satisfactory approximation.

Figure 5.3 is a typical 'response spectrum' diagram. It shows the maximum acceleration response to a given earthquake motion of a linear single degree-of-freedom structure, with any fundamental period in the range $T = 0$ to $T = 4.0$ s. Note the considerable reduction in response resulting from an increase in damping.

As individual earthquakes give different irregular responses dependent on local ground conditions, a design criterion is sought by averaging the response curves for a number of earthquakes (Figure 7.2).

The curves in Figure 7.2 clearly show that for 'Californian-type' earthquakes;

(i)   extremely rigid structures (i.e. period less than 0.05 s) experience peak accelerations not much larger than the maximum applied acceleration;

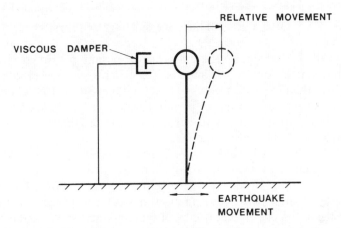

Figure 7.1 Single-degree-of-freedom model for studying earthquake response of equipment

(ii)   structures with small flexibility (periods of 0·2–0·6 s) act as mechanical amplifiers and experience accelerations up to 4 or 5 times the peak applied (ground) acceleration;

(iii) structures with large flexibilities experience accelerations less than the peak applied acceleration;

(iv) structures with strong damping, whatever their natural period, experience greatly reduced response to ground motion.

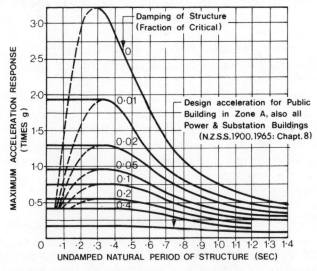

Figure 7.2 Response spectra proposed as a design standard for New Zealand (after Skinner.[2] Reproduced from *New Zealand Engineering* by permission of the N.Z. Institution of Engineers)

7.1.5 Design accelerations for buildings and services

It is important to recognize the difference in seismic performance criteria for buildings on the one hand, and for some items of services equipment on the other. Building codes assume that buildings are able to absorb a large amount of earthquake energy by non-elastic deflection; it is also economically preferable to accept some damage in each earthquake rather than to attempt to prevent damage completely.

On the other hand, it is vital that key services installations such as power stations and communications facilities should survive the strongest earthquakes in full working order. Therefore seismic motions should be accepted by elastic rather than plastic deformation in most components. Because of this difference in the design philosophy of buildings and services, the accelerations (often expressed as coefficients of gravity) for the elastic design of buildings are generally much lower than the equivalent design accelerations for services.

7.1.6 Services equipment mounted in buildings

Equipment mounted in a building should be designed to withstand the earthquake motions to which it will be subjected by virtue of its dynamic relationship to the building. The design of the equipment and its mountings should take into account the dynamic characteristics of the building, both as a whole and in part.

A building tends to act as a vibration filter and transmits to the upper floors mainly those frequencies close to its own .natural frequencies. Thus on the upper floors there will be a reduction in width of the frequency band of the vibrations affecting the equipment. As a rough guide, the fundamental period of a flexible building may be taken as $0.1 N$ seconds where N is the number of storeys, but individual parts of the structure such as floors (on which the equipment is mounted) may have lower fundamental periods. Also the magnitude of the horizontal accelerations will generally increase with height up the building; and hence amplification of the ground accelerations usually takes place.

The accurate prediction of the vibrational forces occurring in equipment mounted in a building is a complex dynamical problem which at present is only attempted on major installations such as nuclear power plants. In ordinary construction a simpler approach has to suffice, such as the response spectrum technique. With such methods, however, it is difficult to make realistic allowance for the filtering and amplification characteristics of the building.

7.2 ELECTRICAL EQUIPMENT

7.2.1 Acceptability of damage

In the seismic design of buildings it is usually considered satisfactory if a building can be evacuated without loss of life after a major earthquake; the damage sustained by the building varies from minor to irreparable. A higher standard is essential for electrical and other equipment providing key public services because there is an enormous dependence on electricity supply for the functioning of important or vital services such as hospitals, food treatment and cold storage services, trains and trolley buses, petrol pumps, water supply and fire-fighting pumps, sewage pumps, lifts and ventilating equipment in large buildings.

The continuation of electricity supply is a major factor in the success of emergency plans after earthquakes. A notable failure of electricity supply equipment occurred in 1971 in San Fernando, California, where there was $30 million of damage at the new $110 million Pacific Intertie Electric Convertor Station.[3]

7.2.2 Fundamental design considerations

The following illustrate three fields in which it is essential to derive seismic loads from first principles in order to obtain sufficiently accurate design criteria.

7.2.2.1 Brittle materials

Whereas structural engineers generally try to avoid the use of brittle materials, electrical engineers have no choice but to use one of the most brittle of all materials, namely porcelain, in most of their structures. Because there is no ductility, any failure of such a material is total. Therefore seismic design accelerations for these structures should be ten to twenty times those used for ordinary buildings. Diagonally braced structures carrying heavy loads, even though made of steel, may also have to be designed for large accelerations to avoid failure by buckling of struts.

7.2.2.2 Ductility

It is often necessary to make sure that the ductility is uniformly distributed throughout the structure. If ductility is confined to a few members, the structure as a whole can still fail by brittle fracture of the remaining members.

For massive rigid bodies such as transformers, all the energy imparted by an earthquake has to be absorbed in holding-down bolts or clamps, which are very small compared with the mass of the whole structure. Thus if reliance is to be placed on the ductility of these fastenings to justify using reduced

300

seismic accelerations for elastic design, considerable knowledge of the post-yield behaviour and energy absorbing capacity of the fastenings is essential.

7.2.2.3 Damping

In the design of buildings, as only a fairly small amount of damping is readily available, survival in a large earthquake depends largely on post-yield energy dissipation and ductility. On the other hand, with much electrical equipment the provision of high damping becomes a practical possibility because of the smaller masses involved. Such damping may be in the form of rubber pads, stacks of Belleville washers, or true viscous damping units.

7.2.3 Cost of providing earthquake resistance

Many smaller items of electrical equipment can withstand horizontal accelerations of $1{\cdot}0\,g$ as currently designed and installed. Even a 10 tonne transformer with the height of its centre of gravity equal to the width of its base could be secured against a horizontal acceleration of $1{\cdot}0\,g$ with four 20 mm diameter holding down bolts without exceeding the yield stress—an inexpensive protection for such a valuable piece of equipment.

Provided the nature of earthquake loading is understood and taken into account from the beginning of a design, earthquake resistance can often be obtained at little cost. The introduction of additional earthquake strength into an existing design is bound to be more costly.

7.2.4 Design procedures using dynamic analyses

7.2.4.1 Single degree-of-freedom structures

For structures which can be thought of as having effectively only one mode of vibration in a given direction, the following simple *response spectrum* design procedure will usually prove to be both easy to carry out and seismically realistic.

(i) Ascertain the natural period of vibration in the direction being studied (by calculation or by measurement of similar structures). This should be done for its condition as installed, including the effects of supports and foundations.
(ii) Determine an appropriate value of equivalent viscous damping by measurement of similar structures or by inference from experience, or by calculation if special dampers are provided.
(iii) Read the acceleration response to the standard earthquake from the appropriate spectrum on Figure 7.2. For natural periods less than $0{\cdot}3$ s, the maximum value for the damping concerned should generally be used, unless there is convincing proof that the structure concerned will remain very rigid throughout strong shaking, i.e. that it will

always retain a natural period less than 0·1 s. This latter condition is very difficult to prove, and should not normally be used.

(iv) Combine the stresses from this earthquake loading with other stresses such as those from dead loads, and working pressures, including short-circuit loads, but not with stresses due to wind loads.

(v) For ductile structures, design to meet total loadings with stresses not exceeding normal working stresses, or such higher stresses as may be shown to meet the specification for survival in a major earthquake.

(vi) For brittle structures, design to meet total loading with stresses that allow a factor of safety of at least 2·0 on the guaranteed breaking load of brittle components, or at least 2·5 if the breaking load is not based on statistically adequate information.

7.2.4.2 Multi-degree-of-freedom structures

For structures that have more than one mode of vibration two main methods of dynamic analysis exist as described below.

(a) A *response spectrum* technique similar to that described in the preceding section can be used, but it is more complex in that the responses due to a number of modes must be combined in a *modal analysis*. The background mathematics is described in Sections 5.2 and 5.4 of this manual and examples of this type of calculation may be seen in Ref. 4.

(b) A more powerful dynamic analysis involves the application of *time-dependent forcing functions* directly to structures, rather than using response spectra. The equations of motion for the structure are solved using *full modal analysis* or *direct integration*, as described in Sections 5.2 and 5.4 of this book. This approach is being increasingly used in the dynamic analysis of a wide range of engineering structures, and is now being used by some Japanese manufacturers of high-voltage circuit breakers.

7.2.5 Design procedures using equivalent-static analyses

Where dynamic analysis is not feasible, it is desirable to establish suitable equivalent-static forces expressed as coefficients of gravity. Such coefficients (sometimes called seismic design factors) should preferably be determined only for structures that fall into well-defined groups within which dynamic characteristics do not vary greatly. Each group must have its coefficients derived from fundamental principles in such a way as to cover reasonable variations from the chosen dynamic characteristics. Hitchcock[1] suggested three such groups: (i) basemounted free-standing equipment, (ii) equipment mounted on suspended floors, and (iii) equipment that would fail in a brittle

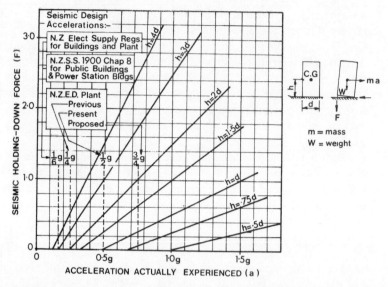

Figure 7.3 Relationship between seismic holding-down bolt loads, applied accelerations, and geometry of basemounted free-standing equipment (after Hitchcock.[1] Reproduced from *New Zealand Engineering* by permission of the N.Z. Institution of Engineers)

manner. The following discussion of these groups is largely based on Hitchcock's paper.

7.2.5.1 Basemounted free-standing equipment

Transformers are the chief members of this group of equipment. Figure 7.3 shows the forces acting on such equipment; the graph shows how the calculated holding-down force, expressed as a fraction of the weight, varies with the maximum acceleration experienced and with the ratio of height of centre of gravity to effective width of base.

This graph shows the weakness of the widely used $0.25g$ earthquake specification, under which any equipment with a height of centre of gravity less than twice the effective base width would have a calculated holding-down requirement of zero. When such equipment is subjected to strong earthquake accelerations, which generally will be larger than $0.25g$, the required holding-down forces will be much greater than zero.

As an example of equipment in this group, consider the power transformer mounted on a concrete pad shown in Figure 7.4. Some field measurements with small amplitude vibrations in the transverse direction gave the damping as 0.9 percent of critical, and the fundamental period as 0.24 s. Plotting these values of damping and period on Figure 7.5 shows that the acceleration response of this transformer to the proposed New Zealand standard earthquake would be nearly $2.0g$ if the period and the damping remain unchanged. It

303

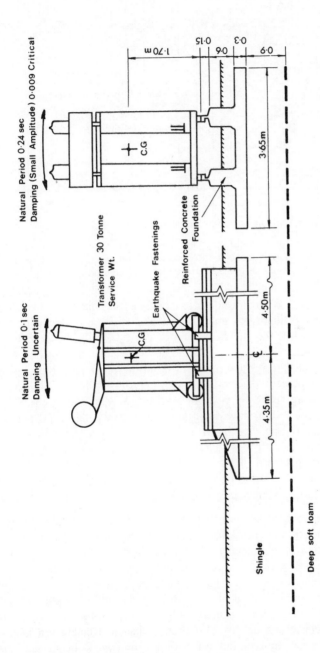

Figure 7.4 110/33 kV transformer on concrete pad (after Hitchcock.[1] Reproduced from *New Zealand Engineering* by permission of the N.Z. Institution of Engineers)

304

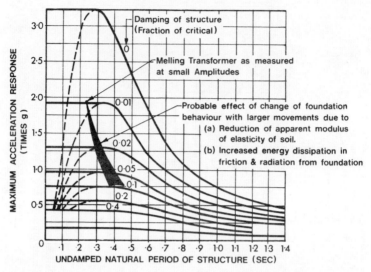

Figure 7.5 Response of 30 tonne transformer on concrete pad on soft subsoil to design earthquake of Figure 7.2 (after Hitchcock.[1] Reproduced from *New Zealand Engineering* by permission of the N.Z. Institution of Engineers)

is known, however, that when foundations rock in this manner, the subsoil properties may be modified; its modulus of elasticity (and hence natural frequency) decreases while the energy dissipated per cycle (and hence equivalent damping) increases.

It has been estimated[5] that the equivalent damping of rocking foundations can reach about 10 percent of critical as compared with about 20 percent for foundations moving vertically without rocking. As the overall equivalent damping factor for this example will probably lie in the range 2–10 percent, it can be seen from Figure 7.5 that this particular transformer would be subjected to a peak acceleration of $1 \cdot 3g$ to $0 \cdot 75g$ in an earthquake corresponding to this response spectrum.

Therefore the equivalent-static design method for basemounted freestanding equipment (including transformers and fastenings) should be as follows.

(a)   If the natural period of vibration of the equipment as finally installed on its foundations is not known or is known to be larger than $0 \cdot 1$ s, then a design acceleration of $0 \cdot 75g$ should be used, in conjunction with normal working stresses and with properly designed ductile material behaviour in the weakest part of the fixings.

(b)   If the natural period of vibration of the equipment as finally installed on its foundations can be shown to be less than $0 \cdot 1$ s (and to remain so for accelerations up to $0 \cdot 4g$) then a design acceleration of $0 \cdot 4g$ should be used, in conjunction with normal working stresses and properly designed ductile material behaviour. As mentioned previously it is very difficult to be sure that the lower portion of the response spectrum for very small values of the period (T) can be safely used, and this provision should seldom be applied in practice.

Two final points concerning the equivalent-static force design of basemounted equipment may be made. Firstly, equipment for unspecified locations should be provided with fixings capable of resisting an acceleration of $0.75g$ applied to the equipment. Secondly, the difference between working stress and yield point provides a margin of safety if the damping is less than that assumed in the design, and ductility beyond yield point would provide further protection against collapse should a larger earthquake than the design one occur.

7.2.5.2 Electrical equipment mounted on suspended floors of buildings or other structures

Equipment mounted in buildings is generally subjected to modified earthquake effects; this can mean amplification especially in the upper floors of buildings. This amplification of the ground and building motions is worse when the building and the equipment resonate, i.e. when they have equal periods of vibration. Fortunately damping between the building and the equipment can be used to drastically reduce amplification, as it is not always possible to avoid the resonance effect. Table 7.1 illustrates the effect of resonance and damping as obtained in a simple analysis by Shibata et al.[6]

Any rule of thumb for seismic design should require all equipment items in a building above ground floor to be designed and fastened for $1.0g$ acceleration without exceeding working stresses (as the New Zealand Code recommends for exterior and interior ornaments, veneers and appendages; see Table 7.2 of this book).

7.2.5.3 Electrical equipment that would fail in a brittle manner in earthquakes

Hitchcock[1] reports on some tests on the dynamic characteristics of porcelain-supported equipment, some ground-mounted, some supported on concrete posts. The natural periods of vibration were found to be in the range

Table 7.1. Response of plant mounted in buildings. Plant resonant with building with fundamental periods in range 0·2–0·4 s

Fraction of critical damping for building	0·07	0·07	0·07	0·07
Fraction of critical damping for plant item	0·007	0·02	0·1	0·2
Peak response of plant item to 1940 El Centro earthquake.	8–10g	5–7g	3g	2g

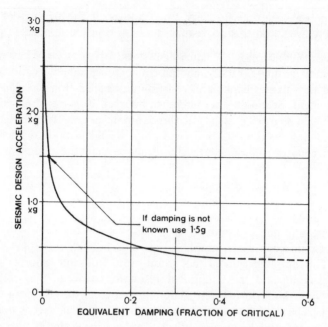

Figure 7.6 Seismic design accelerations for brittle structures with fundamental period · less than 0·4 second (after Hitchcock.[1] Reproduced from *New Zealand Engineering* by permission of the N.Z. Institution of Engineers)

0·2–0·4 s, corresponding to the peak of the response spectra in Figure 7.2. As the damping ranged from 0·018–0·006 of critical, the expected response varied from about 1·5*g* to greater than 2·0*g*. Most of the items of equipment involved had strengths appreciably less than those required to withstand such accelerations. In order to deal with this situation Hitchcock[1] suggested three alternative procedures as follows.

(a) Provide the required strength with factors of safety of the order of 2–3 to cover uncertainties in the assessment of the strength of brittle materials.

(b) Provide ductile components that yield early enough to prevent the brittle components reaching breaking load.[7]

(c) Provide additional damping. This solution is quite practicable when dealing with small masses of electrical equipment.[8]

Equipment used in an electrical installation must in general be suitably rigid to avoid variations in clearance between live parts, and to limit the amount of flexibility to be provided in electrical connections. In fact any acceptable structure of equipment is unlikely to have a period longer than about 0·4 s. From Figure 7.2 it can be seen that for periods less than about 0·4 s acceleration is taken as constant for any given damping. Hence the relationship between acceleration response and damping can be plotted as in Figure 7.6.

Assuming that the New Zealand design earthquake is to be used, the following design rules for this type of brittle equipment may be adopted.

(i)   If the amount of damping in the equipment is not accurately known the equivalent-static acceleration for equipment that fails in brittle components under horizontal loading should be 1·5g. This should be used with a factor of safety of 2·0 on the guaranteed breaking strength of the brittle portions, and with ordinary working stresses in the ductile parts of the structure.

(ii) Alternatively, if satisfactory evidence is available of the amount of damping inherent in the equipment, the seismic coefficient may be that read from Figure 7.6 for that amount of damping.

These rules would be suitable for the design of standard items of equipment installed in any part of a seismic country, because any type of foundation, from extremely rigid to highly flexible, could be used without invalidating the underlying assumptions.

7.2.6 Design examples

For examples of design calculations using the preceding equivalent-static design coefficients on a variety of electrical equipment, readers are referred to the appendices of Hitchcock's paper.[1]

7.3 MECHANICAL EQUIPMENT AND PLUMBING

7.3.1 Standards for seismic resistance of mechanical services and plumbing

Until recently there have been few spectacular failures of mechanical equipment during earthquakes, and it seemed that existing design practice was mainly adequate. However, the extensive damage which occurred to all types of construction in the San Fernando earthquake[3,9] has provoked reviews of existing design standards and of how they are being applied.

A reasonable basic design requirement is that the services should not fail before the building fails in an earthquake. If a building received minor structural damage, but could readily be made habitable by makeshift repairs, such as covering broken windows and relieving jammed doors, then the services installation could reasonably be expected to be in no worse condition, i.e. it should have suffered no more than a few broken joints in unimportant services.

If however, a building became uninhabitable for a prolonged period because of substantial damage to the structure or cladding, then it could be argued that the condition of the services installation would be less important. This would always be true provided that any failures in the services had not in themselves caused secondary damage or casualties—for example, by the release of dangerous substances such as steam, oil, or gas. Safety equipment, such as that used for fire fighting or emergency ventilation, must remain usable after strong earthquakes, and hazardous equipment, such as that containing flammable fluids, should remain intact after the building has failed.

Within the above philosophy should be incorporated the rule that high costs due to earthquake damage should be avoided, especially if only a small extra initial capital outlay is necessary.

7.3.2 Earthquake design fundamentals for mechanical services and plumbing

Much of the remainder of this chapter is based on a review paper by Blackwell.[10] Blackwell worked in terms of the New Zealand earthquake regulations which are also adopted here as they provide a reasonable basis for discussing seismic design generally.

Two alternative methods of structural calculations for earthquake forces can be used. The first one is the traditional procedure using seismic coefficients (such as that specified in NZS 4203:1976[11]) and the second one is 'a more precise form of dynamic analysis which may be required for special structures and may be accepted for any structure'. If the second method is used, then the Code's seismic coefficients for pipework and plant (Table 7.2) need to be reviewed and possibly amended by the structural engineer before they are used. In fact the mechanical engineer should consult the structural engineer at the beginning of the design in order to establish the vibrational characteristics of the building. Of particular interest will be the natural periods of vibration, the maximum accelerations expected on all floors housing equipment, and relative movements between adjacent structures or parts thereof.

Table 7.2 and Figure 7.7 of this document reproduce the seismic acceleration coefficients recommended for buildings and parts of buildings by the New Zealand Code[11], which forms the basis of the following suggestions.

The seismic coefficients for the fastenings of mechanical equipment can be found from Table 7.2. The severity of the requirements compared with those for buildings is clear; for example pipework for sprinkler systems in normal-use multi-storey buildings in Zone A of New Zealand should be designed for horizontal accelerations of $1·0g$.

The above equivalent-static force values should be adequate unless equipment with low damping has a natural period of vibration close to one of the important periods of the building. Although it is not feasible to avoid all building frequencies, there should be little trouble if the equipment's fundamental period is less than $0·1$ s as most buildings are much more flexible than this.

Vertical accelerations of the same order as the horizontal ones can occur. Indeed near the epicentre of an earthquake the vertical acceleration may

Table 7.2. Seismic coefficients for parts or portions of buildings (extract from New Zealand Code[11])

Item	Part or portion	C_p*
(7)	Towers not exceeding 10 percent of the mass of the building. Tanks and full contents, not included in item (8); chimneys and smoke stocks and penthouses connected to or part of the building except as provided for by clause 3.4.5:	
	(a) Single storey buildings where the height to depth ratio of the horizontal force resisting system is:	
	(i) Less than or equal to 3	0·2
	(ii) Greater than 3	0·3
	(b) Multistorey buildings where the height to depth ratio of the horizontal force resisting system is:	
	(i) Less than or equal to 3	0·3
	(ii) Greater than 3	0·5
(8)	Containers and full contents and their supporting structures, pipelines and valves:	
	(a) For toxic liquids and gases, spirits, acids alkalis, molten metal, or poisonous substances, including containers for materials that could form dangerous gases if released:	
	(i) single storey buildings	0·6
	(ii) multistorey buildings	1·3
	(b) For sprinkler systems:	
	(i) single storey buildings	0·5
	(ii) multistorey buildings	1·0
	(c) Other:	
	(i) single storey buildings	0·3
	(ii) multistorey buildings	0·7
(9)	Furnaces, steam boilers, and other combustion devices, steam or other pressure vessels, hot liquid containers, transformers and switchgear, shelving for batteries and dangerous goods:	
	(i) single storey buildings	0·6
	(ii) multistorey buildings	1·3
(10)	Machinery, shelving not included in item (9), trestling, bins, hoppers etc, other fixtures:	
	(i) single storey buildings	0·3
	(ii) multistorey buildings	0·7

*Note: Base Shear $F = C_p W_p$ where W_p is the weight of the item concerned.

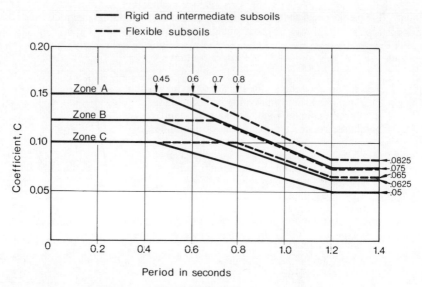

Figure 7.7 Basic seismic coefficients from the New Zealand Code[11]

be of the order of twice the horizontal. In a severe earthquake the vertical seismic accelerations could exceed gravity. As the seismic accelerations reverse in direction, equipment may need to survive for vertical accelerations between zero and $2 \cdot 0g$. This will obviously greatly affect the stability of equipment. For example friction between a tank and the floor supporting it could not be relied on to locate the tank in such circumstances.

7.3.3 Types of earthquake protection required for mechanical services and plumbing

There are two main problems affecting earthquake protection in a mechanical services or plumbing installation. The first concerns movements, and the second energy absorption. Both problems are worsened if resonance or quasi-resonance exists. Movements can be dealt with in either of the following ways;

(a)   by preventing serious relative displacement during an earthquake by anchoring the components of the installation to the building structure;

(b)   by accommodating the relative movements of components without fracture of pipelines, ducts, cables and other connections. (These relative movements may result from movements of either the building fabric or the mechanical services components themselves.)

The energy absorption problem means dealing with the seismic stresses occurring in the equipment, its mountings, and its fastenings to the structure.

The equipment may have to be strengthened, and damping devices may have to be fitted. Mountings should not be made too strong because, apart from the expense, this may cause the equipment to fail somewhere else. Also the resulting lower period of vibration sometimes leads to higher stresses; for example, it can be seen from Figure 7.2 that if equipment with a natural period of 1·4 s and 5 percent damping is stiffened so that its natural period decreases to 0·5 s, the inertial force will have increased three times.

It would be impracticable to make connections at positions of maximum sway on equipment whose natural period is above 1·0 s. Fitting limit stops might overcome this problem but such stops would have to be designed to limit shock loading (Section 7.3.5). Energy is absorbed by the deformation of fasteners, springs, and rubber mountings, but, if the materials have little natural damping, and deformation remains within the limits of elasticity, dissipation of energy may be insufficient for earthquake protection. In this respect springs and even rubber mountings may prove unsatisfactory. Hydraulic or friction dampers could be added to increase the energy absorption but this would be expensive and require detailed design.

Simpler methods of absorbing energy are usually possible, including the plastic deformation of supports and holding-down bolts, and the frictional work done when units slide about on the floor. Once plastic deformation has taken place, bolts will be slack on the return movement and this is when floor friction is useful. Floor friction is free from back-lash and shock effects apart from the deceleration at the end of the slide, and is generally free of cost; the unit of course must be designed not to tip over.

When fastenings are designed for plastic deformation, they should be proportioned and sized so that the stresses are evenly distributed throughout the whole volume of the material, because the amount of energy dissipated is directly proportional to the stress developed and to the volume of material developing stress. Fastenings should be free of weak links or stress concentrations which would result in early fracture of the fastening without much dissipation of energy.

7.3.4 Rigidly-mounted equipment

7.3.4.1 Boilers, calorifiers, control panels, batteries, air-conditioners, kitchen equipment and hospital equipment

(i)   The first requirement is to prevent the equipment sliding across the floor. The coefficient of friction rarely exceeds 0·3, and the effectiveness of friction can be greatly reduced by the upwards component of the earthquake, so friction alone is unlikely to be sufficient. Mounting on bituminous-felt or lead would increase the friction and may be sufficient for less important equipment. The use of a suitable glue with neoprene pads would also give increased security against sliding.

(ii)   The next requirement is to ensure stability against overturning. In the first instance a simple geometric calculation will show whether the equipment is inherently stable or not. This is a function of the base width and the height of centre of mass of the equipment (Section 7.2.5.1 and Figure 7.3). Referring to the New Zealand Code requirements (Table 7.2) it may be noted that there would not be any upthrust due to overturning, with a horizontal acceleration of $0.6g$, if the centre of gravity of the equipment is not higher than 0.84 times the width of the base.

(iii)   Where overturning stability cannot be obtained from geometric considerations, the equipment will have to be fastened to the building structure in some way. If this is done by holding-down bolts fixed into the floor, the bolts should be the weakest part of the system so that they protect the equipment by yielding first. This is particularly desirable when the equipment itself is not very strong. Fine-thread bolts with a length not less than ten times the diameter should be used, and the thread should be designed such that the ultimate strength of the bolt based on the thread root area exceeds the yield strength of the gross bolt area (See also Section 7.3.3). Restraint against overturning can in some cases be obtained by fastening the top of the equipment to walls or columns; but the walls in particular must be seen to be strong enough for this purpose.

(iv)   Pipework and electrical wiring connections are vulnerable and therefore must be strong. It would also be wise to allow some flexibility in the pipes and wires away from the equipment in case of relative seismic movement between the items on either side of the connections.

(v)   Doors to control-panels should be hinged to prevent them being dislodged in earthquakes; loose covers can fall against live contacts, shorting out the equipment.

(vi)   Mercury switches should be avoided, as should essential instruments that have heavy movable components likely to break away from their supports.

(vii)   Boilers with extensive brickwork are undesirable as it is very difficult to reinforce the fire brick.

7.3.4.2. Chimneys

Chimneys should be subjected to a thorough seismic structural design (Appendix A.2). Lightweight double wall sheet-metal flues should be used where possible and prefabricated stacks should be avoided or used with great care.

7.3.4.3 Tanks

As well as considerations of sliding and overturning as discussed above, the following points are peculiar to tanks.

(i) Corrugations of copper tanks are liable to collapse with subsequent failure of the bottom joint. This can be remedied by making a stronger joint and possibly reducing the number of corrugations. Alternatively,welded stainless steel tanks can be used to increase the tank strength while retaining corrosion resistance.

(ii) Where there will be a possibility of a tank sliding, severance of the connections can be avoided by flexibility in the pipes (ten diameters on each side of a bend should be adequate) and by provision of strong connections between the pipes and the tank. The bottom connection can be strengthened by passing it right through the tank and welding it at each end. The top connection could be similarly treated unless a large arm ball valve were required, when extra strengthening at the connection would be satisfactory.

(iii) Suspended tanks should be strapped to their larger systems, and provided with lateral bracing.

(iv) Because of the build-up of surface waves in liquid during earthquakes, some protection against liquid spillage may be desirable. This may be either in the form of a lid, or a spill tray with a drain under the tank. The effects of pressures on the tank due to the liquid oscillation may have to be taken into account in the design of larger tanks. For a treatment of the dynamics of liquids in tanks see elsewhere,[4,12] and for elevated liquid containers, see Section A.2.6.1.

7.3.5 Flexibly-mounted equipment

This refers to virtually any mountings that are not completely rigid, such as those obtained with holding down bolts and metal or concrete mountings. Flexible mountings may be divided into two groups relating to the predominant forcing periods of most earthquakes (Figure 7.2). These two groups of mountings are discussed below.

Group 1. Mountings with a natural period less than the predominant earthquake period (i.e. below about 0·07 s)—felt, cork, and most rubber mountings would usually come into this category. Provided the mountings will not permit sliding to occur (e.g. by gluing) and the connections to the equipment are reasonably flexible, no further mounting precautions should be necessary.

Group 2. Mountings with a natural period corresponding to the predominant earthquake periods (i.e. above about 0·07 s). Spring mountings fall into this category, such as those used for low-speed fans, engines, compressors, and

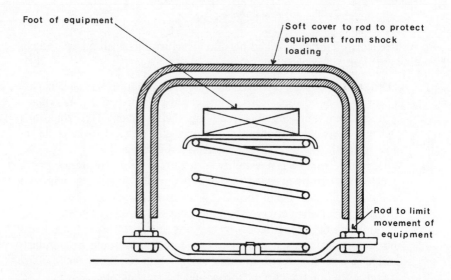

Figure 7.8 Detail of a flexible mounting with a resilient restraint against excessive movements (after Blackwell.[10] Reproduced from *New Zealand Engineering* by permission of the N.Z. Institution of Engineers)

possibly electric motors. As resonance is likely, some method must be provided which limits the movement and transfers the forces directly to the floor instead of through the mounts. Steel rods or angles would be suitable and should be designed to yield at the design acceleration. An example of such a fastener is shown in Figure 7.8. Note the covering round the rod to reduce shock loading. Alternatively the rods could be replaced by multiple-strand steel wire. The flexibility normally provided in pipe, duct, and electric wire connections would be adequate for an earthquake.

7.3.6 Light fittings

Pendant fittings can have a wide range of natural frequencies. Wire-supported fittings may not fail but could swing and smash if brittle. Metal-supported fittings, with friction damping or ductility in the supports, should survive if the components are designed for a horizontal acceleration of $0.32g$ (twice the maximum basic coefficient for New Zealand as given in Figure 7.7). Heavy fittings and brittle materials for supports should be avoided, as should any combination of low damping and low fundamental periods (in the range 0.2–1.0 second). In such cases accelerations in excess of $1.5g$ can occur during an earthquake.

All components of an emergency lighting system should be designed for about six times the maximum coefficient given in Figure 7.7.

7.3.7 Ductwork

Ductwork is usually quite strong in itself, and despite relatively flexible hangers, it is usually susceptible to earthquake damage only where it crosses seismic movement gaps in buildings. At these points flexible joints should be provided which are long enough to take up the seismic movements. Canvas joints may be suitable, asbestos if there is a fire risk, or lead-impregnated plastic if noise is a problem. Wherever possible, seismic movement gaps in buildings should not be crossed. It may be possible to locate fire walls at seismic movement gaps and to design pipe and duct systems to be separate on each side of the gap, thus avoiding crossing the gap as well as keeping the number of systems down to a minimum.

The other most vulnerable position in ductwork is at its connection to machines (e.g. fans). At these positions flexible duct connections should be installed in a semifolded condition with enough material to allow for the expected differential deflection between the machines and the ductwork.

Duct openings and pipe sleeves through walls or floors should be large enough to allow for the anticipated movement of the pipes and ducts.

7.3.8 Pipework

7.3.8.1 Flexibility requirements

Flexibility is required in pipework to allow for building and equipment movement. Seismic flexibility requirements are different from those for accommodating thermal expansion, as seismic movements take place in three dimensions. Sliding joints or bellows cannot be used as they do not have the required flexibility and introduce a weakness which could cause early failure without making use of the ductility of the pipework as a whole. Accordingly, those expansion joints which are installed to accommodate thermal expansion must be fully protected from earthquake movements.

The movements should be taken up by bends, off-sets or loops which have no local stress concentrations and which are so arranged that if yielding occurs there will not be any local failure. Note that short radius bends can cause stress concentrations. Anchors adjacent to loops must also be strong, and connections to equipment must be able to resist the pipe forces caused by earthquake movements. Connections using screwed nipples and some types of compression fittings should be avoided, unless they can be arranged so as to be unaffected by seismic movements.

U-bends and Z-bends as shown in Figure 7.9 can be used to obtain flexibility. The dimensions L should be determined by calculation, so as to give safe stresses in the pipes and at the supports for the applied seismic movements.

Where the laying of pipes across seismic movement gaps in buildings cannot be avoided, details as shown in Figures 7.9, 7.10 or 7.11 can be used.

316

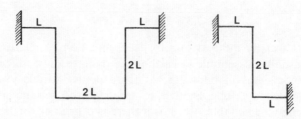

Figure 7.9 Suggested pipe arrangements for crossing movement gaps. (Reproduced from *New Zealand Engineering* by permission of the N.Z. Institution of Engineers)

Such crossings should be made at the lowest floor possible, in order to minimize the amount of movement which has to be accommodated.

Pipework should be tied to only one structural system. Where structural systems change, and relative deflections are anticipated, flexible joints should be provided in the pipework to allow for the same amount of movement. Suspended pipework systems should have consistent degrees of freedom throughout. For example, branch lines should not be anchored to structural elements if the main line is allowed to sway. If pipework is allowed to sway, flexible joints should be installed at equipment connections.

For further information on installation details applicable to any piping systems reference may be made to the Standards of the U.S. National Board of Fire Underwriters for earthquake protection to fire sprinkler systems.

7.3.8.2 Methods of supporting pipework

Simple hangers will allow the pipe to swing like a pendulum. With usual support spacings, pipes will have a fundamental period of about 0·1 s if sideways movement is prevented at every support, and the period will in-

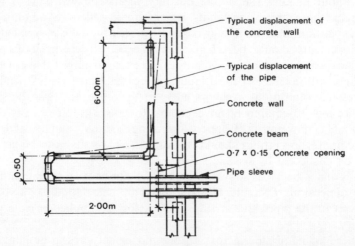

Figure 7.10 Plan view of pipework crossing a seismic movement gap (After Berry[13])

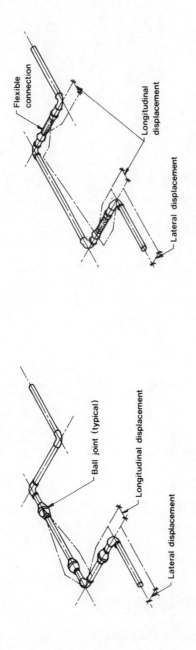

Figure 7.11 Pipework details for crossing seismic movement gaps where space limitations prevent use of pipe loop shown in Figure 7.10 (after Berry[13])

318

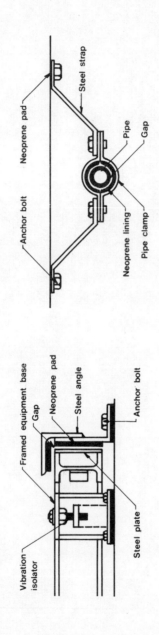

Figure 7.12 Combined earthquake mountings and vibration isolation for machine bases and pipework (after Berry[13])

crease to 0·2 s with twice this spacing, and to about 1·0 s with three times the spacing. The latter two periods are very close to common building periods and the resulting resonance would cause large movements, considerable noise and possible failure. This can be avoided by the provision of horizontal restraints or by the use of two hangers in a V-formation.

7.3.9 Vibration isolation and earthquake protection

Vibrating and noisy equipment should be located remote from critical occupancies so that vibration isolation is not required and the equipment can be anchored directly to the structure. Avoid mounting heavy mechanical equipment on the top or upper floors of tall buildings, unless all vibration isolation mounts and supports are carefully designed to be earthquake resistant.

Where equipment and the attached piping must be separated from the structure by vibration isolators, supports of the type shown in Figure 7.12 should be used. To reduce inertia forces avoid the use of heavy bases under equipment that is mounted on vibration isolators. All vibration isolators for equipment should be anchored to the floor and to the equipment.

REFERENCES

1. Hitchcock, H. C. 'Electrical equipment and earthquakes', *New Zealand Engineering*, **24**, No. 1, 3–14 (Jan. 1969).
2. Skinner, R. I. 'Earthquake-generated forces and movements in tall buildings', *Department of Scientific and Industrial Research Bulletin 166*, 1964.
3. Housner, G. W., and Jennings, P. C. 'The San Fernando California earthquake', *Earthquake Engineering and Structural Dynamics*, **1**, No. 1, 5–32 (1972).
4. Thomas, T. H. 'Nuclear reactors and earthquakes', *U.S. Atomic Energy Commission, T1D-7024*, National Technical Information Service, Virginia (Aug. 1963).
5. Whitman, R. V. 'Analysis of foundation vibrations', in *Proc. Symp. on Vibration in Civil Engineering, organized by the Int. Assn. of Earthquake Engineering, 1965*, Butterworths, London, 1966, pp. 157–179.
6. Shibata, H., Sata, H., and Shigeta, T. 'Aseismic design of machine structure', *Proc. 3rd World Conference on Earthquake Engineering, New Zealand*, **2**, II—552 (1964).
7. Gilmour, R. M., and Hitchcock, H. C. 'Use of yield ratio response spectra to design yielding members for improving earthquake resistance of brittle structure', *Bulln. New Zealand Society for Earthquake Engineering*, **4**, No. 2, 285–293 (April 1971).
8. Winthrop, D. A., and Hitchcock, H. C. 'Earthquake design of structures with brittle members and heavy artificial damping by the method of direct integration', *Bulln. New Zealand Society for Earthquake Engineering*, **4**, No. 2, 294–300 (April 1971).
9. U.S. Department of the Interior and U.S. Department of Commerce. 'The San Fernando California Earthquake of February 9th, 1971', (A preliminary report), *Geological Survey Professional Paper 733*. U.S. Govt. Washington, 1971.
10. Blackwell, F. N. 'Earthquake protection for mechanical services', *New Zealand Engineering*, **25**, No. 10, 271–275 (October 1970).
11. Standards Association of New Zealand. 'General structural design and design loadings for buildings', *Standard Code of Practice NZS 4203:1976*.

320

12. Housner, G. W. 'Dynamic behaviour of water tanks', *Bulln. Seismological Society of America*, **53**, No. 2, 381–387 (Feb. 1963).
13. Berry, O. R. 'Architectural seismic detailing', *State of the Art Report No. 3*. Technical Committee No. 12, Architectural-Structural Interaction IABSE-ASCE International Conference on Planning and Design of Tall Buildings, Lehigh University, August, 1972, (Conference Preprints, Reports Vol. 1a–12).

Chapter 8

Architectural detailing for earthquake resistance

8.1 INTRODUCTION

A large part of the damage done to buildings by earthquakes is non-structural. For instance in the San Fernando earthquake of February 1971, a total of £200 million worth of damage was done of which over half was non-structural. The importance of sound anti-seismic detailing in earthquake areas should need no further emphasizing. The choice of a suitable structural form is crucial (Section 4.2).

Buildings in their entirety should be tailored to safely ride through an earthquake and the appropriate relationship between structure and non-structure must be logically sought. For the effect of non-structure on the overall dynamic behaviour of a building see Section 4.4, where the question of full separation or integration of infill panels into the structure is discussed.

Architectural items such as partitions, doors, windows, cladding and finishes need proper seismic detailing; many non-seismic construction techniques do not survive strong earthquake motion as they do not provide for the right kinds or size of movements. Detailing for earthquake movements should however be considered in conjunction with details for the usual movements due to live loads, creep, shrinkage and temperature effects. As with so many other problems it is worth saying that good planning can provide the right framework for practical aseismic details.[1]

An ironic example of the inadequacy of a non-structural item comes from the San Fernando earthquake; a modern firestation withstood the earthquake satisfactorily with regard to its structure, but the main doors were so badly jammed that all the fire engines were trapped inside.

Unfortunately there is little literature available giving specific guidance on aseismic architectural detailing. Indeed few countries seem to have Codes of Practice on this subject, though there are helpful clauses in a New Zealand Ministry of Works engineering code[2] and in the City of Los Angeles building bylaws. Virtually no basic research had been done in this area and it appears that architects in earthquake areas to date have largely relied on details considered to be 'good practice', without discussing their experience.[3] We are forced to start almost from square one, observe what goes wrong with architectural details in earthquakes, and try to prevent repetitions.[4] The San

321

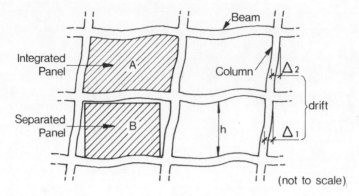

Figure 8.1 Diagrammatic elevation of structural frame and non-structural infill panels

Fernando earthquake caused numerous failures, many on photographic record,[5] from which we can learn.

8.2 NON-STRUCTURAL INFILL PANELS AND PARTITIONS

8.2.1 Introduction

The recommendations of this section should be applied in conjunction with normal design considerations regarding creep, shrinkage and temperature effects which overlap, but are generally less exacting than the seismic design requirements for infill panels.

In earthquakes all buildings sway horizontally producing differential movements of each floor relative to its neighbours. This is termed storey drift (Figure 8.1), and is accompanied by vertical deformations which involve changes in the clear height h between floors and beams.

Any infill panel should be designed to deal with both these movements. This can be done by either (i) integrating the infill with the structure or (ii) separating the infill from the structure. A discussion of both systems of constructing infill panels follows.

8.2.2 Integrating the infill panels with the structure

In this case the panels will be in effective structural contact with the frame such that the frame and panels will have equal drift deformations (Panel A in Figure 8.1). Such panels must be strong enough (or flexible enough) to absorb this deformation, and the forces and deformations should be computed properly. Where appreciably rigid materials are used the panels should be considered as *structural* elements in their own right as discussed in Sections 5.8 and 6.6.5. Reinforcement of integrated rigid walls is usually necessary if seismic deformations are to be satisfactorily withstood.

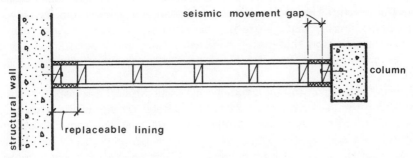

Figure 8.2 Lightweight partition detailed so that earthquake hammering by the structure will damage limited end strips only

Integration of infill and structure is most likely to be successful when very flexible partitions are combined with a very stiff structure (with many shear walls). Attention is drawn to the fact that partitions not located in the plane of a shear wall may be subjected to deformations substantially different from those of the shear wall. This is particularly true of upper storey partitions.

Light partitions may be dealt with by detailing them to fail in controlled local areas thus minimizing earthquake repairs to replaceable strips (Figure 8.2).

Finding suitably flexible construction for integral infill may not be easy, especially in beam and column frames of normal flexibility. These may experience a storey drift of as much as 1/100 of the storey height in a moderate earthquake.

8.2.3 Separating infill panels from the structure

See Figure 8.1, Panel B. For important structural reasons this method of dealing with non-structural infill is likely to be preferable to integral construction when using flexible frames in strong earthquake regions. The size of the gap between the infill panels and the structure is considerably greater than that required in non-seismic construction. In the absence of reliable computed structural movement, it is recommended that horizontal and vertical movements of between 20 mm and 40 mm should be allowed for. The appropriate amount will depend on the stiffness of the structure, and the structural engineer's advice should be taken on this.

This type of construction has two inherent detailing problems which are not experienced to the same extent in non-seismic areas. Firstly awkward details may be required to ensure lateral stability of the elements against out-of-plane forces. Secondly sound-proofing and fire-proofing of the separation gap is difficult. Moderate sound-proofing of the movement gap can be achieved with cover plates or flexible sealants, but where stringent fire-proofing and sound-proofing requirements exist, the separation of infill panels from the structure is inappropriate. Designers should be careful in the choice of so-called 'flexible' materials in movement gaps; the material must be not only sufficiently soft, but also permanently soft. Both polysulphide and foamed polyethylene are *not* flexible enough (or weak enough) in this situation.

It is in fact difficult to find a suitable material; Mono-Lasto-Meric is both permanently and sufficiently soft, but is not suitable for gap widths exceeding 20 mm. Foamed polyurethane is probably the best material from a flexibility point of view and will provide modest sound-insulation, but may have little fire resistance. A fire-resistant possibility is Declon 156, a polyester/polyurethene foam which intumesces in fire conditions.

Figures 8.3 to 8.6 show some details used for separated infill panels. Note that great care has to be taken during both detailing and building to prevent the gaps being accidentally filled with mortar or plaster. Figure 8.6 shows a detail which helps prevent plaster bridging the gap. Further details suitable for small seismic movements may be found elsewhere.[6]

8.2.4 Separating infill panels from intersecting services

Where ducts of any type penetrate a full-height partition, the ducts should not be tied to the partition for support. Support should occur on either side of the partition from the building structure above. If the opening is required to be sealed because of fire resistance or acoustics, the sealant should be of a resilient non-combustible type to permit motion of the duct without affecting the partition or duct. It is important for both seismic and acoustic considerations that the duct be independently supported by hangers and horizontal restraints from the building structure.

Further discussion of ducts is to be found in the chapter on Mechanical Services (Section 7.3), and for some remarks on the required properties of gap sealants around ducts, see discussion on infill panels in Section 8.2.3.

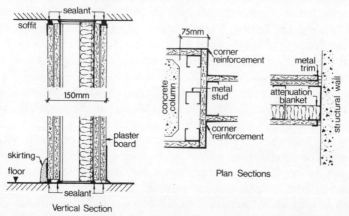

Figure 8.3 Light partition details for small seismic movements i.e. suitable for stiff framed buildings or small earthquakes

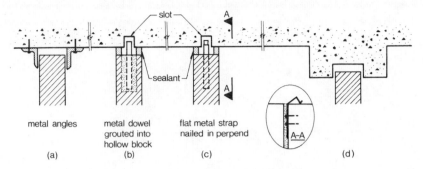

Figure 8.4 Separated stiff partitions; top details for lateral stability of brick or block walls (See section 8.2.2)

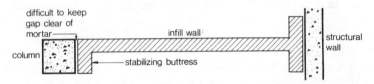

Figure 8.5 Separated stiff partition; plan view of stabilizing buttress systems

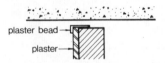

Figure 8.6 Plastering detail to ensure preservation of gap between partition and structure

8.3 CLADDING, WALL FINISHES, WINDOWS AND DOORS

8.3.1 Introduction

The problems involved in providing earthquake-proof details for these items are the same in principle as those for partitions as discussed in the preceding section. Their in-plane stiffness renders them liable to damage during the horizontal drift of the building, and the techniques of integral or separated construction must again be logically applied.

8.3.2 Cladding and curtain walls

Precast concrete cladding is discussed in Section 6.4. Suffice it here to point out that in flexible buildings, non-structural precast concrete cladding should be mounted on specially designed fixings which ensure that it is fully separated from horizontal drift movements of the structure. Brick or other

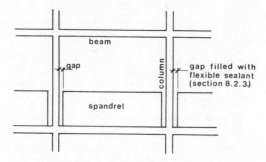

Figure 8.7 Detail of external frame showing separation of spandrel or parapet from columns to avoid unwanted interaction

rigid cladding should be either fully integral and treated like infill walls (Sections 5.8, and 6.6.5), or should be properly separated with details similar to those for rigid partitions (Figures 8.4, 8.5) or for parapets such as shown on Figure 8.7.

External curtain walling may well be best dealt with as fully-framed pre-fabricated storey-height units mounted on specially-designed fixings capable of dealing with seismic movements in a similar way to precast concrete cladding as mentioned above. Increasing numbers of new high-rise buildings in the U.S.A. are using this type of fabrication, because of both design and cost considerations.

8.3.3 Weather seals

Weather seals that may be damaged in severe earthquakes should be accessible and suitable for replacement.

8.3.4 Wall finishes

Brittle or rigid finishes should be avoided or specially detailed on any walls subjected to shear deformations, i.e. drift as applied to panel A, Figure 8.1. This applies to materials such as stone facings or most plasters. In Japan it is recommended that stone facings should not be used on walls where the storey drift is likely to be more than $1/300$.

Brittle veneers such as tiles, glass or stone, should not be applied directly to the inside of stairwells, escalators or open wells. If they must be used, they should be mounted on separate stud walls or furrings. Preferably the stairwells should be free of material which may spall or fall off and thus clog the exit way or cause injury to persons using the area.

Heavy ornamentation such as marble veneers should be avoided in exit lobbies. If a veneer of this type must be used, it should be securely fastened to structural elements using appropriate structural fastenings to prevent the veneers from spalling off in the event of seismic disturbance.

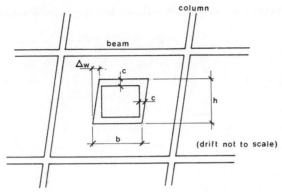

Figure 8.8 Detail of external frame with window glazing set in soft putty

Plaster on separated infill panels must be carefully detailed to prevent its bridging the gap between panel and structure (Figure 8.6) as this may cancel the purpose of the gap, resulting in damage to the plaster, the infill panel and the structure.

8.3.5 Windows

It is worth observing that in the San Fernando earthquake, which caused £200 million worth of damage, glass breakage cost more than any other single item.

Window sashes should be separated from frame action except where it can be shown that no glass breakage will result. If the drift is small, sufficient protection of the glass may be achieved by windows glazed in soft putty (Figure 8.8) where the minimum clearance c all round between glass and sash is such that

$$c > \frac{\Delta_w}{2[1 + (h/b)]}$$

The failure mode of hard putty glazed windows tends to be of the explosive buckling type and should be used only where sashes are fully separated from the structure, for example when glass is in a panel or frame which is mounted on rockers or rollers as described in Section 6.4. Further discussion of window behaviour in earthquakes may be found elsewhere.[7,8]

8.3.6 Doors

Doors which are vital means of egress, particularly main doors of highly populated and emergency service buildings, should be specially designed to remain functional after a strong earthquake. For doors on rollers, the problem may not be simply a geometric one dealing with the frame drift Δ, but may also involve the dynamic behaviour of the door itself.

328

8.4 MISCELLANEOUS ARCHITECTURAL DETAILS

8.4.1 Exit requirements

Every consideration should be given to keeping the exit ways clear of obstructions or debris in the event of an earthquake. As well as the requirements for wall finishes and doors outlined in Sections 8.3.4 and 8.3.6, the following points should be considered.

Floor covers for seismic joints in corridors should be designed to take three-dimensional movements, i.e. lateral, vertical and longitudinal. Special attention should be given to the lateral movement of the joints.

Free-standing showcases or glass lay-in shelves should not be placed in public areas, especially near exit doors. Displays in wall-mounted or recessed showcases should be tied down so that they cannot come loose and break the glass front during an earthquake. Where this is impracticable tempered or laminated safety glass should be used for greater strength.

Pendant-mounted light fixtures should not be used in exit ways. Recessed or surface-mounted independently supported lights are preferred.

8.4.2 Suspended ceilings

In seismic conditions ceilings become potentially lethal. Individual tiles or plaster may jar loose from the supports and fall. Ceiling-supported light fixtures may loosen and drop out, endangering persons below. Thus alternatives to the standard ceiling construction procedures should be considered.

The horizontal components of seismic forces to which a ceiling may be subjected can be allowed for in several ways. A dimensional allowance should be made at the ceiling perimeter for this motion so as to minimize damage to the ceiling where it abuts the walls: one way of doing this is to provide a gap and a sliding cover (Figure 8.9). Some ceiling suspension systems need additional horizontal restraints at columns and other structural elements in

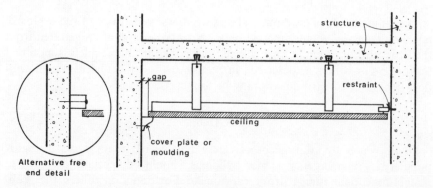

Figure 8.9 Details at periphery of suspended ceilings to prevent hammering and excessive movement

order to minimize ceiling motion in relation to the structural frame. This will reduce hammering damage to the ceiling and tiles will be less likely to fall out. The suspension system for the ceiling should also minimize vertical motion in relation to the structure.

Lighting fixtures which are dependent upon the ceiling system for support should be securely tied to the ceiling grid members. If such support is likely to be inadequate in earthquakes, the light fixtures should be supported independently from the building structure above. Diffuser grilles, if required for the air supply system, should also be hung independently.

In seismic areas, a lay-in T-bar system for ceiling construction should be avoided if at all possible, as its tiles and lighting fixtures drop out in earthquakes. In both the 1964 Alaska and the 1971 San Fernando earthquakes, the economical (and therefore popular) exposed tee grid suspended ceilings suffered the greatest damage. Evidently the differential movement between the partitions and the suspended ceilings damaged the suspension systems, and as the earthquake progressed the ceilings started to sway and were battered against the surrounding walls. This damage was aggravated when the ceilings supported lighting fixtures, and in many instances the suspension systems were so badly damaged that the lighting fixtures fell.

The need for independent support and lateral bracing of lighting fixtures mounted in suspended ceilings requires further study. The City of Los Angeles has a regulation which stipulates minimum requirements for ceiling suspension systems supporting acoustic tile ceilings and lighting fixtures. It requires that ceiling suspension systems be designed to support a minimum load of 2·5 pounds per square foot of ceiling area, except that if the suspension system also supports lighting fixtures, this requirement is increased to 4 pounds per square foot. It also stipulates that the lighting fixtures shall not exceed 50 percent of the ceiling area and that they be fastened to the web of the load-carrying member. It does not, however, require independent support of the lighting fixtures or any lateral bracing.

Damage to ceilings can also occur where sprinkler heads project below the ceiling tiles. One way of minimizing this problem is to mount the heads with a swivel joint connection so that the pipe may move with the ceiling. Figures 8.9 and 8.10 give suggestions for seismic detailing of suspended ceilings.

8.4.3 Landscape elements

An interesting feature of the 1971 San Fernando earthquake was that many of the free-standing items of landscape furniture were found upside down after the earthquake. In order for this to have occurred, the items of furniture must have been subjected to horizontal forces equal to their own weights, throwing them about dangerously. This suggests that heavy items of movable landscape furniture should be secured to the ground in strong motion areas in order to prevent personal injury.

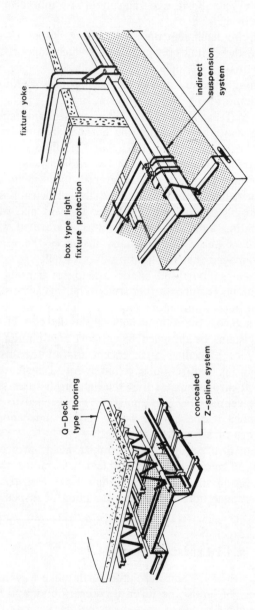

Figure 8.10 Two details of suspended ceiling construction providing movement restraint and secure tile fixing (after Berry[4])

8.4.4 Window washing rigs

Window washing rigs should be restrained close to the building against earthquake forces as well as wind. Curtain wall mullions may have to accommodate this additional load, or instead of structural mullion guides attached to the building structure, a spring-loaded roller arrangement may be included on the window washing rig provided the building structure has projecting fins between which the window washing rig can ride. The roof-mounted carriage of the window washing rig should also be secured to the building structure against seismic forces.

REFERENCES

1. Brandenburger, J. 'Internal details that permit movement', *Symposium on Design for Movement in Buildings*, The Concrete Society, London, 14th October 1969.
2. New Zealand Ministry of Works. 'Design of public buildings', *Code of Practice*, P.W. 81/10/1, 1970.
3. Housner, G. W., and Jennings, P. C. 'The San Fernando California earthquake', *Earthquake Engineering and Structural Dynamics*, **1,** No. 1, 5–32 (1972).
4. Berry, O. R. 'Architectural seismic detailing', *State of the Art Report* No. 3. Technical Committee No. 12, Architectural–Structural Interaction. IABSE–ASCE International Conference on Planning and Design of Tall Buildings, Lehigh University, August, 1972, (Conference Preprints, Reports Vol. 1a–12).
5. U.S. Department of Commerce, National Bureau of Standards, 'The San Fernando, California, earthquake of February 9, 1971', *NBS Report* 10556, March 1971.
6. Toomath, S. W. 'Architectural details for earthquake movement', *Bulln. New Zealand Society for Earthquake Engineering*, **1,** No. 1, 1968, 7 pages.
7. Bouwkamp, J. G. 'Behaviour of window panels under in-plane forces', *Bulln. Seismological Society of America*, **51,** No. 1, 85–103 (Jan. 1961).
8. Osawa, Y., Morishita, T., and Murakami, M. 'On the damage to window glass in reinforced concrete buildings during the earthquake of April 20, 1965', *Bulln. Earthquake Research Institute, University of Tokyo*, **43,** 819–827 (Dec. 1965).

Appendix A

Earthquake resistance of specific structures

The previous eight chapters describe the different phases of the design process illustrated in the figure in the introduction to this book and provide the design basis for a wide variety of structures and elements thereof. However some types of structure have problems of earthquake resistance peculiar to themselves, and are conveniently dealt with in chapters specific to the type of structure concerned. Hence this appendix provides design guidance for several types of structure, and should be used in conjunction with the previous more general chapters of this document.

A.1 EARTHQUAKE RESISTANCE OF BRIDGES

A.1.1 Introduction

Until the late 1960s reports of serious earthquake damage to bridges were relatively few compared to those of other structures. Since then earthquakes in California, Papua–New Guinea and New Zealand have added considerably to our knowledge of the seismic response of bridges.[1,2,3,4,5] These earthquakes have demonstrated that bridges are vulnerable to differential longitudinal, lateral and vertical movements at piers and abutments. Dealing with these relative movements gives major design problems at junctions between horizontal and vertical members and within the supports themselves; the deck members are generally only modestly affected by earthquake stresses.

A.1.2 Choice of structural form for bridges

As with buildings, the choice of structural form can have considerable bearing on seismic performance and costs. Unfortunately important non-seismic factors may conflict with purely seismic considerations when selecting the form of the superstructure, and compromises must be made.[6] For example it is desirable for earthquake resistance purposes to make the superstructure as continuous and redundant as possible, but deck shortening effects can cause larger pier moments in monolithic construction. On the other hand if the deck is arranged to slide over the piers, difficulties arise in providing a satisfactory anchorage structure, as the horizontal inertia forces are then

332

concentrated at fewer supports. The optimum solution for a given structure will depend on achieving a balance between pier heights, span lengths, and foundation problems.

The advantages and disadvantages of various solutions for average road bridge and elevated motorway construction are summarized in Table A.1, and have been further discussed by Chapman.[6]

A.1.3 Seismic analysis of bridges

For bridges both dynamic and equivalent-static analyses are used, the respective advantages and limitations of which are discussed in Chapter 5. For important bridges, dynamic analysis is most desirable, and where foundations are constructed in, or driven through, softer soils a dynamic response analysis of the site may be considered essential. The prediction of the magnitude and effects of differential soil movements is most important for aseismic bridge design. While such predictions are still very much a matter for the judgement of experienced earthquake engineers, they can be greatly facilitated by a site response analysis.

Although carried out more as a research project than as a design exercise, a study reported by Penzien[7] illustrates the type of information which may be derived from a response analysis of soft soils. A similar but simpler study carried out for the design of a bridge at Tamaki, Auckland, has been described by Parton et al.[8] The piles for this bridge were assumed to deform as much as the soil, as predicted by the site response analysis, and were reinforced to resist the corresponding curvatures. This involved the possibility of inelastic behaviour near the top of the pile. Two opposing views on building foundations in similar ground conditions (soft clay) near San Francisco have been described elsewhere.[9,10]

For the dynamic analysis of special bridges such as suspension and tied cantilever bridges, and complex elevated roadways, special computer programs are required.[2,11,12]

In any bridge analysis a realistic degree of damping must be allowed for. The overall damping of most ordinary bridges in earthquakes may be taken as about 5 percent of critical.[13,14] Allowance for foundation damping in softer soils is problematical, and may need careful consideration (Section 5.5). In the design of long span steel bridges the damping may be nearer 2 percent of critical, depending on the influence of concrete road surface elements.

The reduction factor on the actual earthquake loading, and the corresponding ductility demanded in the structure, as with other structures (Sections 5.1, 5.2) is a matter of economy. It is too expensive to attempt to keep most bridges elastic in strong earthquakes, and a deflection ductility factor μ in the range 4–6 is commonly taken. The New Zealand Ministry of Works,[6] for example, takes $\mu = 6$. Although the assumption of moderate ductility

Table A.1 Advantages and disadvantages of various configurations of bridge structure (after Chapman[6])

	Advantages	Disadvantages
1. *Multiple Simply Supported Spans*		
(a) *All spans separate but restrained by:*		
Longitudinally:		
(i) fixing to piers via shear keys and/or linkage bolts at each end of each span.	Good integrity in earthquake. Tolerant to differential settlement. Longitudinal forces can be shared between piers.	Provisions required for allowing superstructure to shorten.
(ii) fixing to piers via shear keys only at one end of each span with freedom to slide at the other.	Superstructure shortening effects create no problems.	Precautions necessary to prevent free span end from leaving pier—hydraulic shock absorbers may serve this purpose.
(iii) linkage bolts to adjacent spans—all spans interconnected and sliding over intermediate piers. Restraint at one abutment.	Superstructure shortening effects create no problems.	Unpredictably large horizontal inertia forces at abutment, plus small vertical reaction, likely to require heavy anchorage system.
Transversely:		
(i) fixing to piers and abutments via concrete or steel shearkeys		
(b) *In-situ concrete deck cast continuously on top of simple spans for full bridge length.*		
Horizontal restraint by:		
Longitudinally:		
(i) fixing to some or all piers via shear keys; sliding over remainder.	All advantages of 1a(i) above plus: Deck joints eliminated. Unequal transverse seismic response of piers can be redistributed.	As 1a(i) above plus: Detail required to avoid slab damage being caused by differential vertical movement of beam ends.

(ii) restraint at one abutment—sliding over intermediate piers.	As 1a(iii) above.	As 1a(iii) above.
Transversely:		
(i) fixing to some or all piers via concrete or steel shear keys.	Superstructure may be free at abutment.	
(c) *In-situ concrete deck, beam ends and diaphragms cast to create live load moment continuity between spans. Horizontal restraint by:*		
Longitudinally:		
(i) fixing to some or all piers via shear keys (non-moment connection); sliding over remaining piers.	As 1b(i) above except for differential settlement.	As 1a(i) above plus: Lacks much tolerance to differential settlement.
(ii) fixing to some or all piers via monolithic (moment resisting) connection; sliding over remaining piers.	As 1c(i) above plus: added redundancy should give more security against collapse.	As 1c(i) above plus: Increased pier stiffness leads to increased effects of shortening.
(iii) restraint at one abutment—sliding over intermediate piers.	As 1a(iii) above.	As 1a(iii) above.
Transversely:		
(i) fixing to some or all piers via shear keys—two or more bearings; or—single bearing.	(Use of alternative depends on stabilising from other piers (and torsional strength of superstructure	
(ii) fixing to some or all piers via monolithic connection		Transverse deflection of pier leads to rotation of top of single stem pier—induces torsion in s/structure and uplift at ends of curved bridge

2 *Continuous spans of various types (slabs, box girders etc.)* These, for seismic purposes, are similar to 1(c) above.

involves plastic hinging (usually in the piers), axial stresses are usually relatively small and the piers are therefore able to deform safely in this way. Any such hinges should be designed to occur in visible portions of the piers. If ductility is not available in the superstructure, correspondingly higher elastic forces will be transferred to the foundations, and a balance between these two factors will have to be found.

A.1.4 Strength design of bridges

The main overall design criterion is to prevent partial or total collapse in strong earthquakes; bridges on roads which are strategic for relief and/or economic reasons should remain open at least to light traffic at all times. Various categories of safety desirability can readily be worked out in a given area, and some local authorities stipulate them.

Local regulations governing earthquake loading for bridges exist in various places, examples of which are those for Japan[15,16] and New Zealand.[17] Since the 1971 San Fernando earthquake Californian seismic loadings for bridges[18] have been increased by a factor of 2·5.

One of the main preoccupations is to ensure that seismic hinges will form in the chosen places, i.e. generally in visible portions of the piers in order to facilitate post-earthquake repairs. Problems in doing this in reinforced concrete have been discussed in Section 6.2, such as those relating to over-strength and strain-hardening of the reinforcement. Also the design of piers to provide adequate restraint in one direction may result in its actual strength in the perpendicular direction being more than the required hinge value.[6]

A.1.5 Ductility of bridges

The ductility demand and capacity should be as suggested in Sections 6.2, 6.5 and 6.7, depending on whether reinforced concrete, prestressed concrete or steel is being used. Applying the techniques described therein, the ductility demand may be conveniently determined in terms of the horizontal deflection ductility factor μ, as relatively simple plastic mechanisms are likely in most bridges.

A.1.6 Superstructure forces on abutments

The stiff non-ductile nature of abutments compared with the piers means that they would generally carry most of the lateral seismic loads if permitted to do so. Even allowing for the higher damping of the abutments this may not be feasible, and the actual response of the abutments is in any case difficult to predict because of the difficulty of assessing how much soil acts with the abutment.[6] It is therefore good practice, on structures with two or more piers and with continuous desk diaphragm action, to separate super-structure from abutments transversely and to carry all transverse loads on the piers.

Care should be exercised in minimizing horizontal rotations about a vertical axis, and such torque should be resisted by bending rather than torsion in the piers. Where the pier and deck geometry is controlled by other considerations and torsion is high, the abutments may have to be used to resist the lateral forces.

A.1.7 Movement joints

Large amounts of seismic movement should be allowed for at movement joints in bridges as discussed in Section A.1.1 above; in strong ground motion adjacent structural components may easily undergo large inelastic oscillations out of phase. In New Zealand,[6] the clearance provided is taken as nine times the deflection under the equivalent-static loading of the code, which assumes a ductility factor of $\mu = 6$ for a single structure.

Although it is usually straightforward to provide bearings even for this order of movement, it is expensive to provide the corresponding movement joints in the deck surface. It may be reasonable on economy grounds to provide for a small fraction of this total movement in a deck joint (say twice the code deflection), and allow for the remaining movement by accepting a concentrated zone of secondary damage. This may be in the form of a 'knockout' wedge of concrete as suggested by Chapman,[6] but satisfactory detailing of such a component is not easy.

A.1.8 Holding-down bolts

Problems arise in allowing for seismic movements, and at the same time avoiding damage to holding-down bolts. Some protection may be provided by using long bolts which will yield in flexure rather than shear, and rubber packing is sometimes provided under washers. Access to nuts and bolts for repair purposes may also be desirable.

A.1.9 Stiff non-ductile structures

Situations where seismically unsatisfactory non-ductile arrangements cannot be avoided include;

(i)   simply-supported spans restrained transversely at the abutments;
(ii) bridges of one or two spans requiring lateral restraint at the abutments;
(iii) bridges with short piers where minimum pier proportions prevent ductile flexural behaviour.

If brittle failure of such bridges is likely to be catastrophic socially or economically, their design should be based on the predicted maximum elastic response. Generally such dire consequences would be unlikely, and it would
be reasonable to be more economical and accept some damage in strong
ground motion, but try to ensure that the structure would not slide off its
supports. Horizontal restraint using concrete shear keys, tension bolts or
force-limiting devices should be used.[6] A shock-absorbing device such as
described by Kitta et al.[19] may help.

A.1.10 Aseismic detailing of bridges

Satisfactory bridge detailing for earthquake resistance should be obtained
if the ductility requirements appropriate to the materials used are applied
(as in Sections 6.2, 6.5 and 6.7) in conjunction with the recommendations
on structural form and articulation set out above. Further guidance on detailing may be found explicitly or implicitly elsewhere.[16,17,20,21]

A.1.11 Shock-absorbing or force-limiting devices

Over the years the notion of shock-absorbing or force-limiting devices
has been suggested occasionally for structures of various types. Mainly
because of detailing problems or inability to deal with associated displacements or insufficient energy capacity of the devices themselves, these ideas
have not progressed very far. Recently however a major railway bridge at
Mangaweka, New Zealand, has been constructed with force-limiting devices
at the base of its high piers; this greatly reduces the pier stresses to within
elastic limits. These devices have been described by Skinner et al.[22] and
Kelly et al.[23] and their application to a bridge pier has been discussed by
Beck and Skinner.[24] Another type of shock-absorbing device has been mentioned in Section A.1.9 above.[19] Further developments of such devices should
be of considerable benefit to bridges as well as other structures in the future.

A.2 CHIMNEYS AND TOWERS

A.2.1 Introduction

Towers and industrial chimneys pose a series of specialist design and construction problems generally related to their height and slenderness. They
are vulnerable to earthquakes because they usually have only one line of
defence, the failure of any one part of the structure resulting in spectacular
failure. The earthquake resistance of these structures is discussed briefly in
this section.

A.2.2 Structural forms

A variety of forms of chimney and tower construction has been used in earthquake areas, including simple cantilevers, guyed structures, chimneys with supporting towers, and structurally combined multiple chimneys. The adoption of the latter two forms has the advantage of increasing the redundancy of the structures and hence decreasing their seismic vulnerability. The structures may be prismatic, or taper or step with height, while the inverted pendulum form is implicit for elevated water tanks (Figures A.1, A.2). Whereas stack-like structures are basically shells or tubular members, towers or supporting structures may also involve braced and unbraced frameworks.

Heights of well over 100 m (300′) are common for chimneys and towers in seismic areas, and steel, reinforced concrete or prestressed concrete are used in their construction.

A.2.3 Seismic analysis of chimneys and towers

For chimneys and towers of moderate size a dynamic earthquake analysis is highly desirable. Equivalent static loadings of codes of practice are not well suited to modelling higher mode effects which can be significant in slender structures. The controlling design criterion may be deformation rather than stress in the case of chimney linings, and wind loading may govern the design in shear or moment or both in some structures.

Unlike in building structures, it is seldom feasible to use the concept of ductility to make chimneys and towers more economical, as one plastic hinge will usually be sufficient to induce partial or total collapse. The need for elastic rather than inelastic behaviour is reflected in the relatively high loadings required by most codes for this type of structure. Only in multi-redundant supporting frames or chimney groups is ductility likely to be safely usable.

A.2.4 Framed chimney and tower structures

A few examples of the many types of framed chimney and tower structures are shown in Figure A.1. Seismic analysis may be carried out as for normal

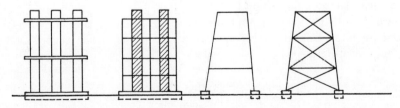

Figure A.1 Typical examples of chimneys and towers utilizing structural frame action (not to scale)

building frames, and plane frame or space frame dynamic analysis may be appropriate.

Diagonally braced towers usually have slender bracing members which are assumed to carry zero load in compression. The seismic response of lightly braced 10-storey single bay frames has been studied by Goel and Hanson,[25] who found that elastic analysis with or without viscous damping did not represent the dynamic behaviour of their frames when considerable yielding occurred in most of the members. They reported maximum ductility ratios of about seven in the bracing and five in the columns. The inference may be drawn that where only modest inelasticity is permitted in braced towers during the design earthquake, elastic analysis may provide sufficiently reliable design criteria (cf. Section 5.2.4.1). In any case a conservative design is warranted because of the vulnerability of these structures to accidental torsions arising from asymmetrical yielding of the diagonal bracing. The post-buckling behaviour of diagonally braced offshore structures has been discussed by Kallaby and Millman.[26]

Inverted pendulum structures are discussed specifically in Section A.2.6.

A.2.5 Free-standing chimneys and stack-like towers

A variety of chimney and tower structures fall into this category, as illustrated in Figure A.2. They range from simple prismatic cantilevers to tapered inverted pendulums. In this section free-standing chimneys and towers with relatively uniform distributions of mass with height will be considered, inverted pendulums being discussed in Section A.2.6. For the type of structure under consideration, Newmark and Rosenblueth[27] considered it important to take into account the very high harmonics, particularly when the fundamental period is so long that the design acceleration spectrum is hyperbolic over several natural periods of vibration. However, experienced analysts consider that no significant errors arise through considering only the first three or four modes of vibration of chimneys.[28,29] Response spectrum analyses, in which the total response is taken as the square root of the sum of the squared modal responses, appear suitable for chimneys.[27,30] In any case the desirability of dynamic analysis for chimneys is not disputed. Any such analysis should incorporate the effects of bending and shear deformations, soil-structure interaction, rotational inertia and gravitational loading, as all of these effects may be important.

Since inelastic behaviour is undesirable, these structures should be designed to remain largely elastic in strong earthquakes, and hence elastic analysis will be appropriate.

A.2.5.1 Reinforced concrete chimneys

A description of the response spectrum analysis of eight reinforced concrete chimneys up to 250 m in height has been given by Rumman,[30] who used

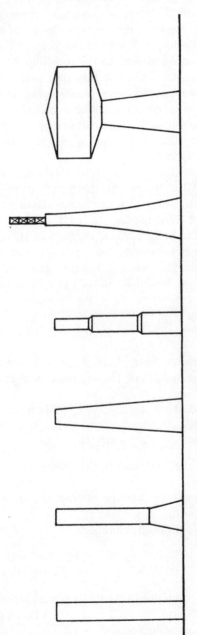

Figure A.2 Some typical free-standing chimneys and stack-like towers (not to scale)

seven earthquake inputs and a structural damping of 5 percent of critical. Rumman found that three or four modes should be taken in order to achieve satisfactory accuracy with the response spectrum technique (Section A.2.5).

Reporting in a more general paper on the design of reinforced concrete chimneys, Maugh and Rumman[28] pointed out that as the seismic moments and shears are inversely proportional to the damping, particular care should be exercised in choosing the value of this parameter. Unfortunately they gave no specific guidance on this point. A damping value of 5 percent of critical seems commonly taken for reinforced concrete chimneys, and this value is reasonably appropriate for their behaviour in the elastic range (Section 5.1.5). However the effect of the lining on the overall damping of the chimney should be considered.

An equivalent-static analysis for reinforced concrete chimneys has been recommended by the ACI,[31] which had the benefit of considering the above mentioned papers.[28,30] In order to make some allowance for the whip-lash effect of the second and third modes, the ACI suggests that 15 percent of the total horizontal shear be applied at the top of the chimney. The New Zealand code[32] has a similar device, taking 20 percent of the shear at the top. In each of these codes the remainder of the shear is distributed vertically so that the force F_x at any height h_x above the base is given by

$$F_x = \frac{(V - V_H)v_x h_x}{\Sigma v_x h_x} \tag{A.1}$$

where V is the total seismic base shear, V_H is that fraction of V applied at the top and v_x is the weight of the segment of the chimney at height h_x.

For deriving the base shear V, the ACI[31] follows the SEAOC code,[33] by taking

$$V = ZKCW \tag{A.2}$$

where Z corresponds to the American seismic zones ($Z = 1\cdot0$ for California),

  $K = 2\cdot0$, a constant for the type of structure ($K = 0\cdot67$ for a ductile moment-resisting space frame),
  W = the total weight of the chimney,
$$C = \frac{0\cdot05}{\sqrt[3]{T}} \tag{A.3}$$

Hence the base shear V is a function of the fundamental period T of the chimney. For chimneys on a rigid base, the ACI[31] gives a dimensionally inconsistent empirical formula for T equivalent to

$$T = \frac{0\cdot49H^2}{(3D_b - D_t)\sqrt{E}} \sqrt{\left(\frac{m_1}{m}\right)} \tag{A.4}$$

where

H = height (m),
D_b = external diameter at the base (m),
D_t = external diameter at the top (m),
E = modulus of elasticity (N/mm^2),
m_1 = total mass of the chimney including linings etc.,
m = total mass of the chimney structure only.

Other formulae for calculating the fundamental periods of chimneys have been proposed, such as that by Housner and Keighley[34] for tapered cantilevers. This work also presents formulae for the second and third mode periods. In an unpublished work[35] Mitchell developed a method suitable for computing the fundamental periods of cylindrical, tapered, and step-tapered chimney structures.

A useful comparative analysis of the above three methods of computing T has been made by Rinne.[36]

It should be pointed out that in the above discussion no account has been taken of the effect of subsoil flexibility or gravity effects. The significance of soil-structure interaction is discussed in Section 5.5.3, where an example of a stack-like tower was cited for which the fundamental period was $T = 1.2$ s for a rigid base and $T = 3.0$ s for a soft soil base. In slender chimneys gravity effects during seismic deformation may also be significant.

A.2.5.2 Steel chimneys

The analytical considerations for steel chimneys follow similar lines to those for reinforced concrete discussed above. Blume[29] provides an interesting discussion of the dynamic analysis of a number of steel chimney and tower structures which were damaged in the Chilean earthquakes of May 1960. He came to a similar conclusion to that regarding concrete chimneys quoted above (Section A.2.5.1) that the first three modes were significant. This does not conform to the contention of Newmark and Rosenblueth[27] as discussed in Section A.2.5.

According to Blume[29] the damping of steel chimneys is of the order of 1–2 percent of critical, including the effect of the lining. In some instances it may be feasible to reduce the seismic (and wind) response which would otherwise occur, by introducing special structural damping devices such as studied by Johns et al.[37] An increase in damping from 2–4 percent, for example, would be of considerable benefit.

Where a fully computerized dynamic analysis is not envisaged, the fundamental period of a cantilever chimney of uniform section (or any similar structure) may be derived from

$$T = 1.79 H^2 \bigg/ \sqrt{\left(\frac{v_g}{EIg}\right)} \tag{A.5}$$

where

v_g = weight per unit height (KN/m)
g = acceleration of gravity (m/sec^2)
H = height (m)
E = modulus of elasticity (KN/m^2),
I = moment of inertia of cross section (m^4).

· For chimneys with a flared base, the fundamental period may be found from

$$T = 2\pi \sqrt{\left(\frac{0 \cdot 08 \Delta}{g}\right)} \tag{A.6}$$

where Δ = the calculated deflection in metres at the top of the chimney due to 100 percent of its weight applied as a lateral load.

The structural contribution of the lining to the stiffness of the steel shell should be considered; the effect of gunited linings, for example, on the period of vibration may be considerable.[29,36] Lining should not be considered effective in buckling or yield resistance unless specially designed for composite action with the shell. A chimney shell with a critical buckling stress below the yield point is undesirable as there is little energy absorption capacity after buckling.

As mentioned above little reliance should be placed on ductile behaviour in chimneys and towers. This is particularly true for steel chimneys where yield can rapidly develop into a secondary failure. Blume[29] however makes the single concession that holding-down bolts and connections should have yield capacity above the concrete foundation surface, as this affords protection to the chimney itself through the energy absorbed in bolt stretching and controlled rocking.

Finally the remarks in Section A.2.5.1 regarding soil-structure interaction should be noted.

A.2.6 Inverted pendulum structures

Inverted pendulums consist of tower or column structures with a large concentrated mass at the top, and occur commonly in forms such as canopies, observation platforms, elevated restaurants and water towers. They may have one or more vertical supports which in some cases form frameworks (Figures A.1, A.2). The large mass at the top makes such structures especially vulnerable to earthquakes because of the accompanying horizontal inertia forces and the so-called $P \times \Delta$ effect. For this reason most codes of practice are even more conservative for inverted pendulums than other chimneys and towers.[32,33]

The seismic bending moments at the tops of columns may govern the design of the columns and of parts of the structure above.[38] Asymmetry

345

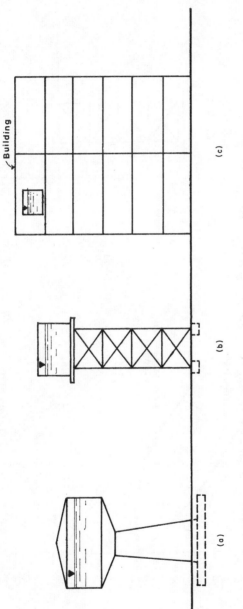

Figure A.3 Typical elevated water tank structures

346

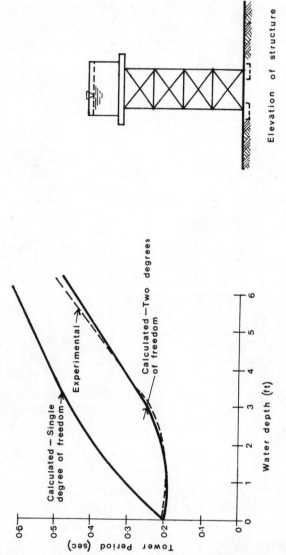

Figure A.4 Comparison of measured and calculated periods of vibration for an elevated water tank (after Boyce[40])

of live load and unintended asymmetry of structural mass distribution may induce significant moments about horizontal and vertical axes.

In these structures the previously mentioned unsuitability of inelastic behaviour is emphasized by the large mass at the top. Considering a simple inverted pendulum Newmark and Rosenblueth[27] found that the column design moment including gravity effects was almost double the moment determined without taking gravity into account.

A.2.6.1 Elevated liquid containers

Because of hydrodynamic effects it is convenient to consider elevated water tanks and other liquid containers as a special case of the inverted pendulum. These structures may be either supported on a single vertical member or a framework (Figure A.3). In either case the conclusion of both Blume et al.[39] and Boyce[40] may be applied, namely that elevated tanks should be modelled as two-degree-of-freedom structures. Boyce demonstrated this with observations of a real water tower, and also showed that large errors are involved in using a single-degree-of-freedom model (Figure A.4).

If the water is completely contained to prevent vertical motion of the water surface (sloshing), the water tower may be treated as a normal inverted pendulum (Section A.2.6). Sloshing will usually act as damping, and may result in a useful reduction in seismic response of the structure compared with the contained liquid case. However, sloshing may damage the roof of the tank or cause spillage of toxic or other liquids (Section 7.3).

The hydrodynamics of sloshing is mathematically complex,[41] but a simplified dynamic analysis has been suggested by Housner[42] as a result of a study of the great damage to elevated water tanks which occurred in the Chilean earthquakes of May 1960. However, for design office purposes the most convenient dynamic analysis for elevated tanks which has so far been developed is that of Blume et al.[39] This work presents graphs enabling the rapid determination of the complex constants used in the hydrodynamic equations.

It should also be pointed out that some computer programs have also been written for the dynamic analysis of elevated water tanks, such as that by Shepherd[43,44] for a three-storeyed cross-braced supporting tower. Computer programs written for offshore oil platforms could also be used, such as that described by Kallaby and Millman.[26]

A.3 LOW-RISE COMMERCIAL–INDUSTRIAL BUILDINGS

A.3.1 Introduction

This large rather ill-defined class of buildings has several earthquake resistance problems peculiar to itself which are not readily treated in the more general earlier chapters of this document. The buildings concerned are of one or two storeys and are used for a wide variety of purposes such as warehouses, light manufacturing, shops, supermarkets and entertainments. They represent a considerable proportion of the annual investment in new buildings in many countries; in the U.S.A., for example, about half the total value of new building construction in recent years has been spent on commercial–industrial buildings.

However, because these buildings are not easy to classify as structural types and because they are only low-rise, they are often inadequately dealt with by earthquake resistance regulations and by design and construction practice. Hence they tend to suffer in earthquakes disproportionately to their monetary value and occupancy, as occurred for example in the San Fernando, California earthquake of 1971.[45,46,47]

A.3.2 Structural form

From a design point of view the planning of the structural form is of great importance. In commercial–industrial buildings one particular aspect of structural form, namely symmetry in plan of the horizontal shear resistance (Figure 4.1) has often been neglected, or non-structure has been allowed to create serious asymmetry in the effective horizontal resistance. These matters have been discussed in Sections 4.2 and 4.4.

A.3.3 Seismic analysis

The unsatisfactory earthquake performance of commercial–industrial buildings may often be largely attributed to unsatisfactory seismic analysis. This may result in serious underestimations of the strength required in connections or the significance of asymmetries. To overcome this more use should be made of dynamic analysis. In buildings with stiff walls and diaphragm roof construction (Section 6.8.6), three-dimensional finite-element dynamic analysis will readily demonstrate the vulnerable features. Even coarse meshed elastic analysis will help provide the ground rules for this type of aseismic construction.

A.3.4 Foundations for low-rise commercial–industrial buildings

The provisions of suitable foundations for earthquake resistance of low-rise construction *on soft ground* is a basic engineering problem for commercial–industrial buildings. Because the foundation requirements for gravity and wind loading are minimal in such buildings, the extra cost for providing protection at source against differential ground movements is large compared with that for taller structures.

It should be recognized that considerable amounts of differential horizontal and vertical movement may be imposed on a low-rise structure provided with economical foundations, even when following the recommendations for shallow foundations suggested in Section 5.5.4.1. Such movements should be allowed for in the super-structure by providing suitable continuity or articulation especially at roof level (Section A.3.5).

A.3.5 Connections in low-rise commercial–industrial buildings

Commercial–industrial buildings structured entirely in one material, or in compatible materials such as masonry and *in situ* concrete, are generally more effective in earthquakes than buildings comprised of a mixture of materials, and should be dealt with as described in the appropriate sections of Chapter 6.

Many buildings of the type under consideration are built with different members in different materials. For example the walls may be of precast concrete, reinforced bricks or concrete blocks, while roofs or floors may consist of steel trusses, laminated timber beams and plywood elements. Such heterogeneous construction has proved to be particularly vulnerable at the connections between dissimilar materials.[45]

In the 1971 San Fernando earthquake the commonest failures according to Bockemohle[45] were as follows;

(i)   separation of plywood roof diaphragms from the supporting timber ledgers on the walls;

(ii)   separation of roof girders from the tops of walls, columns or pilasters;

(iii)   inadequate continuity of roof chords at dowels and laps;

(iv)   inadequate continuity around corners at roof level;

(v)   inadequate shear transmission through roof diaphragms into shear walls in buildings of irregular plan form.

A connection detail between timber roofs and concrete or masonry walls which performed well in the San Fernando earthquake[45] and in subsequent tests[48] is indicated in Figure A.5(a). An alternative detail using a steel angle ledger is shown in Figure A.5(b).

The Los Angeles City Building Department requires joist anchors that are spaced at not more than 1·2 m, and that transmit a minimum horizontal force of 4·4 kN/m.

A number of details for connection of masonry walls to different roof and floor constructions, some of which are appropriate for earthquake resistance, have been given by Amrhein.[49]

As mentioned above, care is also necessary to ensure the integrity of connections between girders and wall columns or pilasters. Holding down bolts must be adequately embedded, and the shear resistance of the column section should be provided by adequate transverse reinforcing links particularly in the region immediately below the girder support.

350

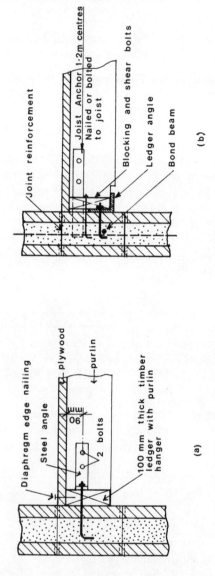

Figure A.5 Joist anchor connections between roof and wall. (Part (b) reproduced by permission of the Masonry Institute of America)

A.4 LOW-RISE HOUSING

A.4.1 Introduction

It is difficult to say exactly what low-rise housing comprises, because there is no precise definition of the term 'low-rise'. However for the purposes of this document it may be said to include one- and two-storey houses generally, and sometimes may also refer to three- or four-storey construction. We are considering housing for the design of which a structural engineer's direct involvement is marginal or non-existent.

The provision of adequate earthquake-resistant housing poses a considerable world problem. Every year sees much damage, homelessness, and loss of life due to the effect of earthquakes on housing, particularly in developing countries. Low-rise buildings are especially vulnerable because of the consequent lack of engineering design and use of lower grades of construction technology. These drawbacks are worsened in developing countries, which also suffer from having to use less suitable materials in the masonry range.

Because of the lack of engineering design, the responsibility lies on the architect, and the governmental building supervisor (if they are involved) and on the builders themselves, to produce earthquake-resistant construction. The problem is to use the available materials to the best advantage by choosing a sound structural form and by using building details which provide maximum structural continuity.

In seismic regions with reasonably advanced technologies, architects and builders are usually assisted by having to comply with building regulations which have an earthquake engineering basis. For example in New Zealand there are codes of practice for timber and masonry buildings *not* requiring specific engineering design.[50,51] Such regulations are naturally written for standard house forms, and buildings on soft or sloping ground or those using a mixture of building materials are likely to warrant specific engineering design consideration.

In less advanced countries, little engineering-based guidance may be available for the builder of low-rise dwellings. The problem of selecting construction standards appropriate to the local technology is difficult.[52,53] More should be done to discourage dangerous practices and to foster the use of seismically successful vernacular construction details. For example, in Latin America a form of construction called quincha, consisting of a timber and cane lattice plastered with mud, has proved remarkably effective in earthquakes.[53]

As with any other type of structure, the earthquake resistance of housing is a function of the strength and ductility of the materials concerned (Chapter 6). Some features of the aseismic design of low-rise housing worthy of specific mention are discussed below.

A.4.2 Symmetry in plan of low-rise housing

The building's resistance against horizontal forces should be derived from walls providing reasonably symmetrical resistance in two orthogonal directions in plan (Figure A.6(a)). If one facade only consists mainly of window and door apertures, horizontal diaphragm action at eaves level should be capable of transferring the resulting earthquake torque to the end walls at right-angles to that facade (Figure A.6(b)). It should be noted that because of the inherent high torsional flexibility of buildings with essentially only three resisting walls, this type of construction is forbidden by the Uniform Building Code[54] when using masonry walls and timber roof diaphragms;

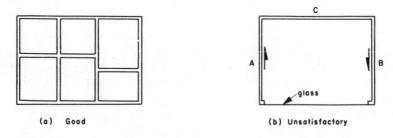

       (a)   Good   (b) Unsatisfactory

Figure A.6 Schematic plans showing layout of shear walls in low-rise housing

some short elements resisting horizontal shear must be introduced into the window facade. The resulting reduced torsions will nevertheless need to be distributed through a horizontal diaphragm. Damage arising from excessive asymmetry occurred in the San Fernando earthquake, as shown in Figures 27 and 28 of the paper by the California Institute of Technology.[47]

A.4.3 Apertures in walls

Apertures for doors or windows require care in positioning and detailing. In masonry the positioning of apertures is particularly important as discussed in Section 6.6.3. Lintels in heavy materials need careful detailing against falling during earthquakes. If it is structurally necessary for a wall to act as a whole, the effect of apertures on the integrity of the wall should be considered.

A.4.4 The strength and stiffness of timber walls

The strength and stiffness of timber walls required to act in shear in their own plane are greatly enhanced by use of panelled linings of timber such as plywood or particle board (Section 6.8) or metal cladding.[55]

A.4.5 Horizontal continuity

Horizontal continuity at floor and roof levels should be provided by special connections or lapping reinforcement, and such continuity should go around facade corners.

A.4.6 Foundations for low-rise housing

Foundation problems in low-rise housing are similar to those expressed in Section A.3.4 for commercial–industrial buildings. Also, in timber housing the substructure between the footings and the first occupied floor tends to have inadequate horizontal shear resistance, and sidesway damage (Figure A.7) occurs in earthquakes.[46,47,56] Pole frame construction as illustrated in Figure 6.35 readily overcomes this problem.

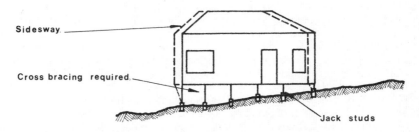

Figure A.7 Substructure in timber stud construction requiring extra horizontal shear strength

Another common failing has been that the timber structure is inadequately connected to the concrete foundation blocks or strips. The detail shown in Figure 6.41(b) for example, should be provided with adequate bolts.

A.4.7 Roofs of heavy construction

Roofs of heavy construction are a great menace causing large loss of life in earthquakes. In some underdeveloped areas massive earth and masonry roofs are the norm, but less heavy tiles can also be dangerous. Where this type of construction is unavoidable, appropriate measures should be taken to ensure the integrity of the roof during earthquakes. Apart from proper vertical support, horizontal diaphragm action at eaves level to prevent spreading and collapse of the roof is particularly valuable.

A.4.8 Chimneys and decorative masonry panels

Elements which are stiffer and heavier than the rest of the building cause a great deal of damage in earthquakes. Concrete and masonry chimneys in basically timber houses are particularly vulnerable.[46,56] In many cases the ideal solution would be to make the stiff elements structurally independent of the rest of the building, but difficulties arise in detailing the movement gaps. Otherwise the stiff and the flexible elements should be much more strongly tied together than has been common practice in the past. When a timber structure is tied to a stiff element, the latter becomes a major horizontal shear resisting element for the whole building, and the building should be designed accordingly.

REFERENCES

1. Wood, J. H., and Jennings, P. C. 'Damage to freeway structures in the San Fernando earthquake', *Bulln. New Zealand Society for Earthquake Engineering*, **4**, No. 3, 347–376 (Dec. 1971).
2. Tseng, W. S., and Penzien, J. 'Seismic response of highway overcrossings', *Proc. 5th World Conference on Earthquake Engineering, Rome*, **1**, 942–951 (1973).
3. Ellison, B. K. 'Earthquake damage to roads and bridges, Madang, R. P. N. G. Nov. 1970', *Bulln. New Zealand Society for Earthquake Engineering*, **4**, No. 2, 243–257 (April, 1971).
4. Hollings, J. P., and Fraser, I. A. N. 'Earthquake damage to three railway bridges 1968 Inangahua earthquake', *Bulln. New Zealand Society for Earthquake Engineering*, **1**, No. 2, 22–48 (Dec. 1968).
5. Wilson, J. B. 'Notes on bridges and earthquakes', *Bulln. New Zealand Society for Earthquake Engineering*, **1**, No. 2, 92–97 (Dec. 1968).
6. Chapman, H. E. 'Earthquake resistant design of bridges and the New Zealand Ministry of Works bridge design manual', *Proc. 5th World Conference on Earthquake Engineering, Rome*, **2**, 2242–2251 (1973).
7. Penzien, J. 'Soil-pile foundation interaction', in *Earthquake Engineering* (Ed. R. L. Wiegel), Prentice-Hall, Englewood Cliffs, New Jersey, 1970, Chap. 14, pp. 349–381.
8. Parton, I. M., and Melville Smith, R. W. 'Effect of soil properties on earthquake response', *Bulln. New Zealand Society for Earthquake Engineering*, **4**, No. 1, 73–93 (March 1971).
9. A.S.C.E. 'Building foundation for soft clay, earthquake area—Mat foundation', *Civil Engineering, ASCE*, **44**, No. 2, 56–57 (Feb. 1974).
10. A.S.C.E. 'Building foundation for soft clay, earthquake area—Pile foundation', *Civil Engineering, ASCE*, **44**, No. 2, 58–59 (Feb. 1974).
11. Tezcan, S. S., and Cherry, S. 'Earthquake analysis of suspension bridges', *Proc. 4th World Conference on Earthquake Engineering, Chile*, **II**, A3, 125–140 (1969).
12. Arya, A. S., and Thakkar, S. K. 'Earthquake response of a tied cantilever bridge', in 'Earthquake engineering', *Proc. 3rd European Symposium on Earthquake Engineering, Sofia*, 343–353 (1970).
13. Katayama, 'Dynamic characteristics of bridge structures', *Lecture notes*, presented at the International Institute of Seismology and Earthquake Engineering, Tokyo, 1972.
14. Charleson, 'The dynamic behaviour of bridge substructures', *Report*, to Road Research Unit of New Zealand National Roads Board, 1970.
15. Standards of Aseismic Civil Engineering Constructions in Japan published in *Earthquake regulations—a world list, 1973.* International Association for Earthquake Engineering, Tokyo, 1973.
16. J.S.C.E. *Earthquake resistant design for civil engineering structures, earth structures and foundations in Japan*, compiled by the Japan Society of Civil Engineers, 1973.

17. N.Z.M.O.W. *Highway bridge design brief*, Ministry of Works, New Zealand, Issue B, Nov, 1972, plus amendments July, 1973.
18. Pond, W. F. 'Performance of bridges during San Fernando earthquake', *Jnl. Prestressed Concrete Institute*, **17**, No. 4, 65–75 (July/Aug. 1972).
19. Kitta, T., Kodera, J., Ujiie, K., and Tada, H. 'A new type shock absorber and its effect on the response of the bridge to the earthquake', *Proc. 5th World Conference on Earthquake Engineering, Rome*, **1**, 1397–1400 (1973).
20. Inomata, S. 'Japanese practice in seismic design of prestressed bridges', *Jnl. Prestressed Concrete Institute*, **17**, 76–85 (July/Aug. 1972).
21. Japanese participants at Seminar under the Japan–U.S. Cooperative Science Program. 'Construction and Behaviour of precast concrete structures', *reports*, Parts I and II, Seattle, U.S.A., Aug. 1971.
22. Skinner, R. I., Kelly, J. M., and Heine, A. J. 'Energy absorption devices for earthquake resistant structures', *Proc. 5th World Conference on Earthquake Engineering, Rome*, **2**, 2924–2933 (1973).
23. Kelly, J. M., Skinner, R. I., and Heine, A. J. 'Mechanisms of energy absorption in special devices for use in earthquake resistant structures', *Bulln. New Zealand Society for Earthquake Engineering*, **5**, No. 3, 63–88 (Sept. 1972).
24. Beck, J. L., and Skinner, R. I. 'The seismic response of a reinforced concrete bridge pier designed to step', *Earthquake Engineering and Structural Dynamics*, **2**, No. 4, 343–358 (April/June 1974).
25. Goel, S. C., and Hanson, R. D. 'Seismic behaviour of multistorey braced steel frames', *Proc. 5th World Conference on Earthquake Engineering, Rome*, 1973, **2**, 2934–2943 (1973).
26. Kallaby, J., and Millman, D. N. 'Inelastic analysis of fixed offshore platforms for earthquake loading', *Proc. Offshore Technology Conference, Houston, Texas*, **III**, 215–227 (1975).
27. Newmark, N. M., and Rosenblueth, E. *Fundamentals of earthquake engineering*, Prentice-Hall, Englewood Cliffs, New Jersey, 1971.
28. Maugh, L. C. and Rumman, W. S. 'Dynamic design of reinforced concrete chimneys', *Jnl. of the ACI*, **64**, No. 9, 558–567 (Sept. 1967).
29. Blume, J. A. 'A structural-dynamic analysis of steel plant structures subjected to the May 1960 Chilean earthquakes', *Bulln. Seismological Society of America*, **53**, No. 2, 439–480 (Feb. 1963).
30. Rumman, W. S. 'Earthquake forces in reinforced concrete chimneys', *Jnl. Structural Division, ASCE*, **93**, No. ST6, 55–70 (Dec. 1967).
31. American Concrete Institute, *Specification for the design and construction of reinforced concrete chimneys*, (ACI307-69), American Concrete Institute, 1969.
32. N.Z. Standards Institute, 'Basic design loads; earthquake provisions', *New Zealand Standard Model Building Bylaw* (NZSS 1900), 1965, Chap. 8.
33. Seismology Committee, S.E.A.O.C. *Recommended lateral force requirements and commentary*, Structural Engineers Association of California, 1973.
34. Housner, G. W., and Keightley, W. O. 'Vibrations of linearly tapered cantilever beams', *Trans. ASCE*, **128**, Part 1, 1020–1048 (1963).
35. Mitchell, W. W. 'Determination of the period of vibration of multi-diameter columns by the method based on Rayleigh's principle', *unpublished report*, for the Engineering Department of the Standard Oil Company of California, San Francisco, 1962.
36. Rinne, J. E. 'Design of earthquake resistant structures: towers and chimneys' in *Earthquake Engineering* (Ed. R. L. Wiegel), Prentice-Hall, Englewood Cliffs, New Jersey, 495–505 (1970).
37. Johns, D. J., Britton, J., and Stoppard, G. 'On increasing the structural damping of a steel chimney', *Earthquake Engineering and Structural Dynamics*, **1**, No. 1, 93–100 (July-Sept. 1972).
38. Rascón, O. A. 'Effectos sísmicos en estructuras en forma de péndulo invertido', *Ref. Soc. Mex. Ing. Sísm.*, **3**, No. 1, 8–16 (1965).

39. Blume, J. A., and Associates. *Earthquake engineering for nuclear reactor facilities*, J. A. Blume and Associates, San Francisco, 1971, (particularly pp. 111–123).
40. Boyce, W. H. 'Vibration tests on a simple water tower', *Proc. 5th World Conference on Earthquake Engineering, Rome*, **1**, 220–225 (1973).
41. Housner, G. W. 'Dynamic analysis of fluids in containers subjected to accelerations', Appendix F in *Nuclear reactors and earthquakes*, U.S. Atomic Energy Commission, T1D—7024, 1963.
42. Housner, G. W. 'The dynamic behaviour of water tanks', *Bulln. Seismological Society of America*, **53**, No. 2, 381–387 (Feb. 1963).
43. Shepherd, R. *Seismic analyses by digital computer modelling*, pre-print for the 3rd National Computer Conference, Massey University, New Zealand, 1972.
44. Shepherd, R. 'The seismic response of elevated water tanks supported on cross braced towers', *Proc. 5th World Conference on Earthquake Engineering, Rome*, **1**, 640–649 (1973).
45. Bockemohle, L. W. 'Earthquake behaviour of commercial–industrial buildings in the San Fernando valley', *Proc. 5th World Conference on Earthquake Engineering, Rome*, **1**, 76–81 (1973).
46. National Bureau of Standards. 'The San Fernando California earthquake of February 9, 1971', *NBS Report 10556*, U.S. Department of Commerce, March, 1971.
47. Earthquake Engineering Research Laboratory, California Institute of Technology. 'Engineering features of the San Fernando earthquake', *Bulln. New Zealand Society for Earthquake Engineering*, **6**, No. 1, 22–45 (March 1973).
48. Briasco, E. *Joist anchors vs. wood ledgers*, Los Angeles Department of Building and Safety, 1971.
49. Amrhein, J. E. *Reinforced masonry engineering handbook*, Masonry Institute of America, Los Angeles, 1972, Section 8.
50. New Zealand Standards Institute. 'Construction requirements for buildings not requiring specific design—Timber', *New Zealand Model Building Bylaw*, NZSS:1900, 1965, Chap. 6.1.
51. New Zealand Standards Institute. 'Construction requirements for buildings not requiring specific design—Masonry', *New Zealand Model Building Bylaw*, NZSS:1900, 1964, Chap. 6.2.
52. Flores, R. 'An outline of earthquake protection criteria for a developing country', *Proc. 4th World Conference on Earthquake Engineering, Chile*, **III**, J4, 1–14 (1969).
53. Evans, F. W. 'Earthquake engineering for the smaller dwelling', *Proc. 5th World Conference on Earthquake Engineering, Rome*, **2**, 3010–3013 (1973).
54. International Conference of Building Officials. *Uniform Building Code*, I.C.B.O., Pasadena, California, 1971.
55. Tracey, W. J. 'A simulated earthquake test of a timber house' *Bulln. New Zealand Society for Earthquake Engineering*, **2**, No. 3, 289–294 (Sept. 1969).
56. Falconer, B. H. 'Preliminary comment on damage to buildings in the Inangahua earthquake', *Bulln. New Zealand Society for Earthquake Engineering*, **1**, No. 2, 61–71 (Dec. 1968).

Appendix B

Miscellaneous information

B.1 MODIFIED MERCALLI INTENSITY SCALE

I.   Not felt except by a very few under exceptionally favourable circumstances.

II.   Felt by persons at rest, on upper floors, or favourably placed.

III.   Felt indoors; hanging objects swing; vibration similar to passing of light trucks; duration may be estimated; may not be recognized as an earthquake.

IV.   Hanging objects swing; vibration similar to passing of heavy trucks, or sensation of a jolt similar to a heavy ball striking the walls; standing motor cars rock; windows, dishes, and doors rattle; glasses clink and crockery clashes; in the upper range of IV wooden walls and frames creak.

V.   Felt outdoors; direction may be estimated; sleepers wakened. liquids disturbed, some spilled; small unstable objects displaced or upset; doors swing, close, or open; shutters and pictures move; pendulum clocks stop, start, or change rate.

VI.   Felt by all; many frightened and run outdoors; walking unsteady; windows, dishes and glassware broken; knick-knacks, books, etc., fall from shelves and pictures from walls; Furniture moved or overturned; weak plaster and masonry D* cracked; small bells ring (church or school); trees and bushes shaken (visibly, or heard to rustle).

VII.   Difficult to stand; noticed by drivers of motor cars; hanging objects quiver; furniture broken; damage to masonry D, including cracks; weak chimneys broken at roof line; fall of plaster, loose bricks, stones, tiles, cornices (also unbraced parapets and architectural ornaments); some cracks in masonry C*; waves on ponds; water turbid with mud; small slides and caving in along sand or gravel banks; large bells ring; concrete irrigation ditches damaged.

VIII.   Steering of motor cars affected; damage to masonry C or partial collapse; some damage to masonry B*; none to masonry A*; fall of stucco and some masonry walls; twisting and fall of chimneys, factory stacks, monuments, towers and elevated tanks; frame

357

houses moved on foundations if not bolted down; loose panel walls thrown out; decayed piling broken off; branches broken from trees; changes in flow or temperature of springs and wells; cracks in wet ground and on steep slopes.

IX.   General panic; masonry D destroyed; masonry C heavily damaged, sometimes with complete collapse; masonry B seriously damaged; general damage to foundations; frame structures if not bolted shifted off foundations; frames racked; serious damage to reservoirs; underground pipes broken; conspicuous cracks in ground; in alluviated areas sand and mud ejected, earthquake fountains and sand craters appear.

X.   Most masonry and frame structures destroyed with their foundations; some well-built wooden structures and bridges destroyed; serious damage to dams, dikes and embankments; large landslides; water thrown on banks of canals, rivers, lakes, etc.; sand and mud shifted horizontally on beaches and flat land; rails bend slightly

XI.   Rails bent greatly; underground pipelines completely out of service.

XII.   Damage nearly total; large rock masses displaced; lines of sight and level distorted; objects thrown into the air.

* Masonry A, B, C and D as used in MM scale above.

Masonry A.   Good workmanship, mortar, and design; reinforced, especially laterally, and bound together by using steel, concrete, etc.; designed to resist lateral forces.

Masonry B.   Good workmanship and mortar; reinforced, but not designed in detail to resist lateral forces.

Masonry C.   Ordinary workmanship and mortar; no extreme weaknesses like failing to tie in at corners, but neither reinforced nor designed against horizontal forces.

Masonry D.   Weak materials, such as adobe; poor mortar; low standards of workmanship; weak horizontally.

B.2 QUALITY OF REINFORCEMENT FOR CONCRETE

The following notes provide some amplification of the points on reinforcement made in Section 6.2.6.

For adequate earthquake resistance, suitable quality of reinforcement must be ensured by both specification and testing. As the properties of reinforcement vary greatly between countries and manufacturers, much depends on knowing the source of the bars, and on applying the appropriate tests. Particularly in developing countries the role of the resident engineer is crucial.

Even in California there is concern amongst designers[1] at 'the lack of quality control provided by the present ASTM Standards and the lack of uniformity in the reinforcement presently available'.

In order to obtain satisfactory ductility and control of plastic hinge mechanisms the following points require consideration.

(a) Minimum yield stress

An adequate minimum yield stress (or 0·2 percent proof stress) may be ensured by specifying steel to an appropriate standard, such as BS 4449,[2] or BS 4461,[3] or ASTM A615.[4]

(b) Variability of yield stress

For control of structural collapse mechanisms a small variability in the yield stress is necessary. In California the degree of control is of little value; SEAOC[1] requires that for a given grade of steel the actual yield stress should not exceed the minimum specified yield stress (characteristic strength) by more than 124 N/mm^2 (18·000 psi). This nominal control in the scatter of yield values is essentially a compromise with manufacturing economy, the design preference being for much less variability, say about half the above value.

(c) Higher strength steels

Grades of steel with characteristic strength in excess of 410 N/mm^2 (60,000 psi) are not recommended in some earthquake areas, e.g. California and New Zealand. This is because higher strengths generally imply decreased ductility, but where adequate ductility is proven by tests, somewhat higher strengths may be used where regulations permit. For example, steels to BS 4461[3] with characteristic strengths of 425 and 460 N/mm^2 appear satisfactory (Item (d) below). Hot rolled bars of similar strength are also available but problems have been encountered with their ductility.

(d) Cold worked steels

Cold worked steels are effectively excluded from use in a number of earthquake countries. For example, in California only steel to ASTM A615 (hot rolled) is recommended.[1] In California it is also recommended that the ultimate tensile stress should not be less than 1·33 times the actual yield stress of the bar. This requirement is ostensibly to ensure adequate post-elastic energy absorption capacity in the relatively brittle American steels, but this capacity is as well provided by many other steels with better elongation characteristics such as the British steels,[2,3] but which would have less difference between the ultimate and yield points. In fact for analytical purposes elasto-plastic behaviour is more convenient. Hence the 1·33 ratio criterion given above should not be applied to adequately ductile steels such as the British ones referred to above.

Park[5] has stated the usual reasons for avoiding cold worked steels.

(1)   It is commonly held that cold worked bars are too brittle for seismic loading conditions. This may not be true for all cold worked bars; for example steel to BS 4461[3] seems to be at least as ductile as most hot rolled steels. The tests mentioned below in items (f), (g), (i) and (j) would have to be passed to ensure adequate ductility.

(2)   The lack of a yield plateau in cold worked steel is considered a disadvantage. This is certainly analytically inconvenient in that it adds further complications to the determination of plastic hinge positions and to post-elastic behaviour generally. This objection to cold worked steel is obviously rather idealistic considering the many other simplifications in seismic design, and for members nominally without plastic hinges in the seismic collapse condition (such as columns, or floor slabs) it is invalid.

(e) Substitution of higher grades of steel

The contractor should not be permitted to use other than the specified grade of steel in the members of moment resisting frames, as this is likely to cause dangerous changes to the collapse mechanism of the structure. Substitution of higher strength steel in beams is particularly undesirable.

Table B.1

	Bar Size		Elongation on $5 \cdot 65 \sqrt{S_0}$ Gauge Length (%)				
	U.S.	British	Mild Steel		High Yield Steel		
Bar No.	Diam. (mm)	Diam. (mm)	ASTM A615	BS 4449	ASTM A615	BS 4449	BS 4461
3	9·5		20		16		
4	12·7		19		14		12
5	15·9		18		13		
		16					
6	19·1		15		12		
7	22·2		14		10		
8	25·4		12	22	10	14	
9	28·7		10		8		
10	32·3		9		8		14
11	35·8		7		7		
		40					
14	43·0		7		7		
		50					
18	57·3		6		6		

(f) Elongation tests—general ductility

The most basic measure of ductility of reinforcement is its elongation at failure. Steels with good elongation behaviour are more likely to perform well in other tests of ductility discussed below than steel with low elongation. Reinforcements complying with BS 4449, BS 4461 and ASTM A615 have moderate to good elongation values, as compared in Table B.1. The American results have been converted to a gauge length of $5.65\sqrt{S_0}$ using the Oliver formula[6] in order to conform to modern international practice; this ensures geometric similarity and allows direct comparison of elongations for different specimen diameters.

It can be seen from Table B.1 that the ASTM elongation requirements are generally less stringent than the British ones, particularly for the larger diameter bars. The effects of this are discussed in various places below.

(g) Bend tests

Bend tests are most important for ensuring sufficient ductility of reinforcement in the bent condition. It is important to use test conditions appropriate to the minimum diameters permitted on site. Because ASTM A615 requires large minimum bends in construction (Item (h) below), the mandrel diameters used for bend tests on American steels are greater than for British steels.

Table B.2. Mandrel diameters for bend tests (around 180° unless otherwise stated)

Bar Number	Mild Steel		High Yield Steel		
	ASTM A615	BS 4449	ASTM A615	BS 4449	BS 4461
3, 4, 5	4ϕ	2ϕ	4ϕ	3ϕ	3ϕ
6	5ϕ	2ϕ	5ϕ	3ϕ	3ϕ
7, 8	4ϕ	2ϕ	5ϕ	3ϕ	3ϕ
9, 10, 11	5ϕ	2ϕ	8ϕ	3ϕ	3ϕ
14			10ϕ*	3ϕ	3ϕ

Notes: ϕ is the bar diameter; * around 90° bend only.

(h) Minimum bend radius

The minimum bend radius should be chosen to suit the basic ductility of the steel. As discussed in (f) above steels to ASTM A615 are not required to be as ductile as British steels; this is taken into account in that larger minimum bend diameters are required in American practice, as shown in the table below.

Bar number	Minimum mandrel diameter for bends	
	USA	Britain
3–8	6ϕ	6ϕ
9,10,11	8ϕ	6ϕ
14,18	10ϕ	6ϕ

362

(i) Resistance to brittle fracture

The brittle fracture problem arises because all carbon steels (and most other types) undergo a transition from ductile behaviour to brittle behaviour as the temperature is reduced. This property is strain rate sensitive. At slow rates of loading a steel can behave in a ductile manner, while at the same temperature but a higher loading rate it could fail with nil deformation. Stress concentrations also increase the risk of brittle fracture.

Although reinforcing steels are not normally assessed for resistance to brittle fracture, this is likely to be important when service conditions include shock loading (earthquakes) and low temperatures (say below 3–5°C). The desirability of testing will be a matter of judgement, depending on the type of structure, the climate and the seismic risk. For example important structures such as North Sea oil platforms should always be checked for brittle fracture despite the low seismic risks involved. Heated buildings in Iran, where the winters may be cold, may not merit such tests, unless important structural elements are exposed to the weather.

The brittle fracture characteristics of reinforcing bars may be assessed from their ductile/brittle transition curves, as obtained from a series of impact tests (such as the Charpy test) carried out on standard notched specimens at various temperatures. Although this simple test provides only a fairly crude check on brittle fracture, a reasonable minimum requirement would be that the steel should have a notch toughness of 27 Joules (20 ft lb) measured in the Charpy test at the minimum service temperature.[7]

(j) Strain–age embrittlement

This form of embrittlement results from cold working steel bars and subsequent ageing which tends to raise the transition temperature from ductile to brittle behaviour. The cold working may either be part of the manufacturing process, for example, steel to BS 4461[3] or will occur in subsequent bending, strain-age embrittlement may be minimized by suitable controls in the steel making processes, particularly of the free nitrogen content of the steel.

Strain-age embrittlement shows itself as otherwise unexpected fractures which generally occur following impact load at the bends. Its existence is detected by the *rebend test*, a generally suitable procedure for which may be found in BS 4449.[2] This should be followed if bending practice on site is in accordance with UK practice. If the bending practice of other countries is to be followed, the mandrel size for the rebend test should be the same as that for the appropriate bend test. Whereas this test is normally optional in non-seismic areas, it should be applied to all batches of steel for earthquake-resistant structures.

(k) Weldability

Few high tensile reinforcing bars are readily weldable. This is particularly true of hot rolled high tensile bars which generally get their high tensile strength from increased carbon content. This renders the heat affected zone of the weld liable to serious embrittlement and cracking. General information on this subject is given elsewhere.[8]

Some reinforcement suppliers are capable of given detailed technical advice on the welding of their product, but in general a competent welding engineer or an organization such as the Welding Institute should be consulted. Testing will probably be necessary.

(l) Galvanizing

A reinforcing bar can be embrittled by other processes. For example, galvanizing of pre-bent reinforcement has been shown to embrittle it by accelerated strain-ageing. Hence coatings such as galvanizing should not be applied to reinforcing bars without giving special consideration to possible embrittling effects.

(m) Welded steel fabric

Welded steel fabric (mesh) is unsuitable for earthquake resistance because of its potential brittleness. However, mesh to BS 4483[9] may be used for the control of shrinkage in non-structural elements such as ground slabs.

REFERENCES

1. Seismology Committee, SEAOC. *Recommended lateral force requirements and commentary*, Structural Engineers Association of California, 1973.
2. British Standards Institution. 'Hot rolled steel bars for the reinforcement of concrete', *British Standard* 4449, 1969.
3. British Standards Institution. 'Cold worked steel bars for the reinforcement of concrete', *British Standard* 4461, 1969.
4. American Society for Testing and Materials. 'Deformed billet steel bars for concrete reinforcement', *ASTM* A615, 1974.
5. Park, R. 'Steel in earthquake resistant structures', Metallurgy in Australasia, *Proc. 27th Annual Conference of the Australian Institute of Metals*, 1974.
6. British Standards Institution. 'Method for converting elongation measurements for steel: carbon and low alloy steels', *British Standard* 3894, Part 1, 1965.
7. Boyd, G. M. *Brittle fracture in steel structures*, Butterworths, London, 1970.
8. American Welding Society. *Recommended practice for welding reinforcing steel, metal inserts and connections in reinforced concrete construction*, AWSD12.1.
9. British Standards Institution. 'Steel fabric for the reinforcement of concrete', *British Standard* 4483, 1969.

Index